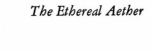

The Ethereal Aether

THE ETHEREAL AETHER

A History of the Michelson-Morley-Miller
Aether-Drift Experiments, 1880–1930

By LOYD S. SWENSON, JR.

With a Foreword by Gerald Holton

UNIVERSITY OF TEXAS PRESS
AUSTIN & LONDON

c

International Standard Book Number 0–292–72000–9
Library of Congress Catalog Card Number 74–37253
Copyright © 1972 by Loyd M. Swenson, Jr.

Composition and printing by The University of Texas Printing Division, Austin
Binding by Universal Bookbindery, Inc., San Antonio

TO JEAN,
My Partner

CONTENTS

ILLUSTRATIONS

(following page 106)

PREFACE

One of the most famous experiments in the history of science is the subject and the object of this book. It is the subject because this work is a biography, a case history in experimental science, of the half-century life from 1880 to 1930 of the celebrated Michelson-Morley experiment. It is the object because my purpose is to describe and explain historically how an experiment in physical optics conducted in 1887 in Cleveland, Ohio, was conceived, performed, and repeated until it became generally considered the primary cause and justification for Albert Einstein's first work on the theory of relativity.

Science, particularly physics, and history, especially of social and intellectual developments, in the mid–twentieth century have become inextricably intertwined. One can hardly hope to achieve an adequate understanding of the one without the other. The history of science and technology offers several avenues of approach for satisfying our quest for intelligibility in the contemporary world. When, however, the history of science is taken as the cumulative development of particular disciplines, it tends toward the celebration of progress and away from the criticism of science and its applications as human endeavor. On the other hand, when the history of science is studied as the interrelated contextual advance of various ways of knowing and doing things, the central problems in the minds of scientists who lived and loved their disciplines may be ignored. Professional historians and humanists of our times are prone to overemphasize human failure and the pessimistic view that man's pride in his science and technology is preceding his fall into urban-industrial maelstroms, if not into nuclear annihilation. This book, being primarily a study in the social history of physics, offers a combination of perspectives, showing how success emerged from failure through a specific series of episodes in physical optics.

Physicists, mathematicians, and philosophers beyond number have described, "explained," and analyzed the Michelson-Morley aether-drift experiment until one may wonder what justification there can be for another such attempt, especially when the author admits, as I do, to being a mere historian. My answer to this is simply that an extended history devoted expressly to the aether-drift experiments themselves has never been published. When interest focuses on the advent of Einstein and the relativity theory, history inevitably suffers simplification and often actual falsification. It is not the purpose of this study to correct the mistakes of commentators on Michelson-Morley but rather to penetrate beneath the myths that have overgrown this classic experiment. Since it has become so widely celebrated, a simple narrative using historical methods of synthesis and a prospective rather than a retrospective viewpoint should illuminate the Michelson-Morley experiment in several novel ways, not least of which is the story of the completion of the experiment in the 1920's.

The title of this study is meant to be puzzling and evocative. The concept of a luminiferous aether became fundamental to the wave theory of light in the nineteenth century, and both concept and theory were necessary assumptions for Albert A. Michelson's experiment and the instrument, called an interferometer, that evolved from his efforts, especially with Edward W. Morley, to measure the velocity of the earth through space. Here, however, I have juxtaposed the noun-without-a-referent and the incongruent modifier "ethereal," in order to emphasize that the failure of aether-drift experiments to detect a relative wind has led, ironically, to the adjective being considered a more precise and useful word than the noun.

Spelling "aether" with the initial diphthong may seem an archaic reversion, especially since Michelson and even Maxwell preferred the more modern form. But a short title is desirable, and the initial letter *a* will remove any confusion with the chemical fluid. Furthermore, by its etymological purity, "aether" may remind us of the Latin translation for the Greek root αἰθηρ, which means upper air or sky and relates to the refined fire of the empyrean. Old British usage also emphasizes that Victorian physics had a vested interest in speculations concerning the aether of space, attributing to nature's dislike that which many Europeans most abhorred, the vacuum.

From Empedocles and Aristotle through Descartes and Newton to Maxwell and Lodge, the age-old concept of a plenum or an aether has given one solution to the hoary philosophical problem of action-at-a-distance, across seemingly empty space. Modern physicists may find it inexcusable to use the pregnant symbol "aether," instead of their signal word "field," for spatially extended continua, but the historian wishes to recapture the very imprecision which physicists have slowly learned to avoid. The mysterious aether of space, a concept almost as penetrating and ubiquitous at the beginning of the twentieth century as the aether itself was supposed to be, provided the metaphysical metaphor that made meaningful the whole series of Michelson-Morley-Miller experiments. Just as today's physics cannot be imagined without the concept of "energy," so also our grandfathers' physics had its conceptual basis in the "aether" until Einstein's message was accepted.

Anyone who wishes to learn Einstein's ideas regarding relativity, simultaneity, and space-time geometry will be better advised to go elsewhere, preferably to Einstein's own writings, or to the libraries of literature of clarification written by physicists, mathematicians, and philosophers. This book assumes a modest familiarity with relativity on the part of the reader, yet it presumes that descriptive history as literature can be written honestly and without mathematics if time enough is taken to digest the sources and to explain the context. To read history is not an economical way to learn a discipline, if you are already equipped with the skills and techniques for advancing upon it. But if you wish to learn about men and the symbols to which they often devote their lives, then there is no substitute for history-at-large.

This study, however, is neither history-at-large nor an attempt to explain the origins of special relativity theory. While it does aim at a fairly full account of the fall from physical grace of the quintessential aether, its primary goal, as expressed in the subtitle, is to be *a history* of the Michelson-Morley aether-drift experiments. Historians seldom dare to put the definitive article *the* before their works anymore, not only because other experts and critics are so abundant, but also because historiography too has grown more sophisticated, *and* because the intellectual influence of relativity theory in physics has spread far beyond the confines of the exact sciences, even into historical seminars. To specify one's point of view, one's frame of reference, and what one postulates as his invariants is now expected everywhere.

I came into the history of science through the door of intellectual history after having discovered George Sarton's works in a library at Claremont Graduate School one day more than a decade ago. Some years previously, as an undergraduate, I had wanted to learn something about everything and so took courses to sample a university (Rice) and majored in history for the widest latitude of choice in surveys and electives. Having discovered Sarton and history of science for myself, I then learned about what James Bryant Conant called the "tactics and strategy" of science through the *Harvard Case Histories in Experimental Science.* These studies led to a dissertation in 1962 on the Michelson-Morley experiment. That was the first draft of this book.

Having moved on to work in the history of contemporary technology, I have recently returned to studies in the history of physics through a fortunate association with Project Physics, a curriculum development program that grew out of Conant's general education courses at Harvard and out of Gerald Holton's several efforts to show that within physical science there still exists natural philosophy. Thus my frame of reference, if specified in four dimensions, could be said to be anchored in history with three coordinates and fastened to physics with a fourth.

Physics is not the whole of science, of course, but it is near the center; nor is optics the whole of physics, but it too is near the center. Modern physics seems to me, a generalist and a spectator, to rest primarily on four conceptual pillars: (1) the conservation laws, (2) relativity theory, (3) quantum mechanics, and (4) symmetry principles. These few grand and beautiful generalizations about the behavior of non-human nature uphold and encompass the body of organized knowledge known as physics. Among its practitioners, however, so much more is known in detail by specialists that no one can claim in good conscience to be a scientist-in-general or even a physicist-in-general, except as an academic distinction or for pedagogical reasons. I therefore make only the mildest apology for overstepping disciplinary lines and venturing to tell a tale, true but not the whole truth, about the experimental search for the ethereal aether. Because that quest was practically barren, its later history, revolving around the conscientiousness of Dayton C. Miller, is regarded most often by physicists as an impediment to progress.

Compared to most experiments in modern physics, the Michelson-Morley-Miller experiment is simple to understand in design and performance, even though its oracular null results led into labyrinths of interpretation. From these labyrinths, it is usually thought, Einstein emerged as the prophet of a new space-time theory with revolutionary implications for science and for mankind. The mass-energy equivalence, for example, symbolized grossly by the simple equation $E = mc^2$, has been superimposed on the mushroom cloud that overcasts our age. Thus, even nuclear physics and space travel are widely believed to be related directly to Michelson's interferometer by virtue of the supposed dialectic between experiment and theory which established relativity physics.

Michelson's interferometer was an optical apparatus upon a turntable with no other moving parts. It never fulfilled the original expectations for which it was designed, but serendipity prevailed and all the exact sciences profited immeasurably from the refined delicacy of measurement that it made possible. Like the microscope and the telescope, Michelson's interferometer vastly expanded man's capacity to measure things and events on both microcosmic and macrocosmic scales. Conversely, the imponderable aether of space, by apparently eluding all attempts at interferometric measurement, made a few men think harder and more radically than ever before about the conceptual bases of physical reality. Eventually, and for other reasons as well, science, philosophy, and mankind were—indeed the cosmos as we know it was —transformed by the implications of Einstein's several theories of relativity.

The nature of the doubt which Michelson's experiment cast upon nineteenth-century physics is more nearly the theme of this book than how that doubt came to be resolved. Commentators have often said that had Michelson's experiment been done in the sixteenth or seventeenth centuries it could have been used as convincing evidence against Copernicus's heliocentric theory and against Galileo's assertion that the earth moves. Such suppositions may ignore the cumulative context, but they do illustrate the significance of cosmology. Insofar as Michelson asked a question in this experiment to which Einstein supplied an answer that eventuated in a new cosmology, the historical background to this development ought to prove of utmost interest to the philosopher resident in all men.

After the first two topically organized chapters attempt to set the stage, this book is organized chronologically to narrate how Michelson's great failure and its conceptual *raison d'être*, the aether, were born, suffered, and died. The span of time covered by each chapter differs according to the activities that must be covered, but in general we proceed chronologically by decades or half decades through the half century 1880–1930. Specialists may wish to skip the introductory background, where their knowledge may be far more profound than mine, and begin with chapter 3, where my investigation starts in depth. Chapter 6 is a special interim study designed to redress the modern tendency to view the molecule, atom, and electron as having already overthrown or undermined the aether by 1895. This I maintain is a false generalization about the period, too early by at least a decade for the majority of physical scientists then active. Chapters 8 and 12 may prove disappointing if their titles are taken out of the context of this Preface, for after 1905 this work traces not the mainstream of physics but a rivulet that ran dry.

Two aspects of this history not sufficiently developed here but important enough to have been guiding themes had they been sufficiently studied in monographic works to date are (1) the astronomical history of solar motion, proper motion, radial velocities, and stellar motion, and (2) the philosophical study of the void, together with physico-chemical notions of the vacuum and the growth of vacuum technology. Relative motion in the heavens and vacuous spaces here on earth, whether occupied by an ethereal aether or not, are topics relevant to this history that I hope to amplify at a later date.

This study will end with an attempt to assess to what degree it may be said, legitimately, from a historical standpoint, that Einstein's special theory of relativity was a lineal outgrowth of Michelson's efforts to measure the earth's velocity through space. Because historians of science have so far neglected to take careful soundings of the experimental bases for relativity theory, a number of historical fictions have been perpetuated. Einstein has been coupled with Michelson-Morley in the same way that Galileo has been tied to the leaning tower of Pisa, Newton associated with his apple tree, and Darwin with the cruise of *H.M.S. Beagle* or the suggestion from Malthus. In none of these cases does the legend grown around an incident do justice to the scientific discovery.

My list of debts for inspiration and encouragement is longer than I can fully recall, but the most salient credits may be found specified in the bibliographical essay and listing at the end of this volume. From a distance, the written words of Sarton, Conant, and Holton, and of Herbert Butterfield, I. Bernard Cohen, Charles C. Gillispie, Thomas S. Kuhn, Robert S. Shankland, Bernard Jaffe, Thomas C. Chalmers, Mary Hesse, Adolf Grünbaum, and Stanley L. Jaki have been most notable. Close enough to have been personally helpful have been Holton, Gillispie, Shankland, Dorothy Michelson Livingston, L. Pearce Williams, Keith L. Nelson, the late N. R. Hanson, several anonymous referees, and my mentors at Claremont, John B. Rae, Graydon Bell, Herbert W. Schneider, Morton Bechner, and the late Douglass Adair and Duane Roller. Nathan Reingold, while at the Library of Congress, D. T. McAllister at the Michelson Museum (China Lake, California), William B. Plum of Port Hueneme, Horace Babcock and Ira Bowen of Mount Wilson Observatory, Charles Weiner and Joan Warnow of the Niels Bohr Library of the American Institute of Physics, John N. Howard of Air Force Cambridge Research Laboratories Rayleigh Archives, Melvin Kranzberg and Robert Schofield of Case Western Reserve, Robert F. Curl, Jr., and the late William V. Houston of Rice, and my colleagues on the Harvard Project Physics staff, Alfred M. Bork, Stephen G. Brush, and Albert B. Stewart. To each of these persons I am grateful for aid and inspiration, and to many more friends and colleagues, relatives and correspondents I am thankful for sympathetic encouragement. The dedication expresses my most intimate debt, but for all errors and liabilities that remain in this work, I alone am responsible.

LOYD S. SWENSON, JR.

FOREWORD

For sheer intellectual drama, nothing can surpass the encounter between a great experiment and a great theory. This book—one of the very few full-scale case studies of its kind—deals with such an encounter: that between A. A. Michelson's experiment (including its later elaborations with E. W. Morley and D. C. Miller) and the aether theory on which nineteenth-century physics was built. It is a story with surprises and insights on a range of topics, both scientific and historical: the relations between scientific theory and experiment, the way scientific advance depends on the state of technological capability and skill, the rise of American contributions to physical science at a time when Europe still seemed to have a monopoly on it, the breakthrough of modern physics during the first decades of this century.

At the center, of course, are Michelson's invention and improvement of his interferometer, and the experiments he did with it. This device is now so familiar and easy to make in model form that it is used routinely in elementary courses. But it was and still is a lovely thing. Designed by young Michelson in response to a challenge of the foremost theoretician of the century, James Clerk Maxwell himself, the interferometer was capable of revealing to better than 0.0000001 percent the difference in the effective speeds of two beams of light that are made to run a race with each other. Until Michelson showed how this could be done, it seemed impossible; once he had shown it, it became obvious. The discussion of this simple but subtle device quickly entered into research and teaching at every level.

To this day, the interferometer is one of the highest precision instruments in science and industry, and it is usually held up as the prime example of a beautiful scientific device. The latest issue of a physics journal, lying on my desk as I write these lines, contains four separate references to Michelson's interferometer in various current forms. The future, one can be quite sure, will continue to see new uses for it. In this sense, it is analogous to Galileo's telescope, whose evidence cast fatal doubt on the physics of its time, but which then went on to have a number of lively incarnations in the new era. Or perhaps a better analogy would be to liken the invention of Michelson's interferometer to the invention of such instruments as the violin or flute, and to consider how those instruments changed and enriched music.

Michelson's experiment of measuring the speed of the expected relative drift of aether past the earth, though only one of many kinds of uses for the instrument, became itself a paradigm of good experimentation, despite the puzzles and heartbreak that the results held in store for those who believed in the aether, including Michelson himself. Tribute to his experimental genius and artistic sense came from every side—even from Einstein, who freely confessed that he loved Michelson though the latter thought his relativity theory to be a "monster."

The aether had been regarded as "God's Sensorium" by Newton, and as a chief pillar of physics by almost everyone in the nineteenth century. When Michelson, to his own and everyone else's surprise, found no clear trace of the aether, it was equivalent to a major scientific scandal. How did the scientists then behave, individually and as a profession? It is one of the tasks of the modern historian of science to pay as much attention to such sociological questions as to the more conventional ones, be they historical, epistemological, or scientifico-technical. Professor Swenson does so with full command of his material, and he gives us further proof of how wrong the old stereotype is that has scientists dispassionately, quickly, and quietly accepting the evidence of "indubitable, mere fact."

He also touches on a second scientific scandal that soon followed —the widespread acceptance of the fable that this experiment led Einstein directly to the construction of his special theory of relativity. Thus, even as highly placed an authority as R. A. Millikan could write (in 1949) about the "null" result of the Michelson-Morley experiment: "That unreasonable, apparently inexplicable experimental fact was very bothersome to 19th century physics, and so for almost 20 years after this fact came to light physicists wandered in the wilderness in the disheartening effort to make it seem reasonable. Then Einstein called out to us all, 'Let us merely accept this as an established experimental fact and from there proceed to work out its inevitable consequences,' and he went at that task himself with an energy and a capacity which very few people on earth possess. Thus was born the special theory of relativity."

The truth was quite different, and Einstein himself occasionally took the opportunity to indicate that the supposed influence of Michelson's experiments on him during his own formation of relativity theory was at best indirect. One can, however, see in retrospect that the tight linking of Michelson's experiments and Einstein's theory a posteriori in popular print was almost inevitable, and primarily for a purely pedagogic reason: the puzzling result of Michelson's experiments could be straightforwardly explained by relativity theory, and, conversely, the initially obscure and theoretically based relativity theory could be made more plausible by causing it to appear as a necessary derivation from experimental results.

This supposed symbiosis is convenient, but it is unfair to the accomplishments of both Michelson and Einstein. On the one hand, though the experiment helped him in the acceptance or justification of this theory, Einstein could, and very probably did, work without conscious reliance on the Michelson experiment. As he said in one of his interviews, "I was not conscious it had influenced me directly. . . . I guess I just took it for granted that it was true"; and in any case it was only one of a whole series of puzzling aether experiments that had accumulated by 1905, for example, those of Mascart, Fizeau, Rayleigh, Brace, and Trouton and Noble. (By his own account, the experimental results that had influenced him most were the observations on stellar aberration and Fizeau's measurements on the speed of light in moving water. "They were enough," he reported to have said in 1950.) And on the other hand, Michelson's experiment would justly have been regarded as one of the few immortal achievements of experimental science, even if relativity theory (which in a sense trivialized the result of the experiment) had never been proposed.

One does not have to agree with every paragraph of this book to see that it is full of stimulating ideas. As for me, I found myself scribbling notes and queries to myself constantly as I was reading the manuscript from which the book was fashioned; and that is one of the best tests of an important contribution. We are fortunate that Professor Swenson had both the resources and the courage to tackle this great case.

GERALD HOLTON

The Ethereal Aether

CHAPTER ONE

PERSPECTIVES ON THE AETHER

And God said, Let there be light:
and there was light.

                              Genesis 1:3

What is light? This question, like its peers—What is matter? motion? energy? mind? life? truth?—is deceptively simple and yet profoundly complex. Light has puzzled thinking men throughout history. One of the reasons why is suggested by the juxtaposition of the two words *sight* and *light*. Subjective and objective phenomena are inextricably intermixed in man's vision, making this question a true quandary. Two most ancient and perennial answers to the problem of light have postulated that it consists primarily either of waves or of particles.

In the twentieth century scientists have learned to live paradoxically with both views at once, with the so-called wave-particle duality.

As elementary particles have proliferated, the photon theory (essentially corpuscularian) has gained ascendency, but, with the advent of space science and technology, the idea of radiant energy transfer through the media (essentially undulatory) of solar and stellar plasmaspheres, magnetospheres, and gravispheres has regained prominence. Thus the uneasy balance implied by the wave-particle duality has been skewed alternately toward each side of the equation, despite the methodological sanction of the principle of complementarity and the dematerialization of the physical pictures of the world.

Physical optics, as a science distinguished from geometrical, physiological, and psychological concerns about vision, is the study of phenomena exhibited to our eyes and imagined by our brains in attempts to answer the question, What is light? While trying to separate illusions from realities and basic from superficial phenomena, scientists have discovered over the centuries certain properties of light, such as reflection, refraction, diffraction, polarization, diffusion, that seem to explain as well as to describe what light is. A highly respected present-day authority says, "Most physicists today would answer the question 'What is light?' as Newton would have: 'Light is a particular kind of matter.' The differences between light and bulk matter are now thought to flow from relatively inessential differences between their constituent particles. Particles of both kinds—of all kinds—exhibit wave properties."[1]

The historical development of physical optics to its current state is, however, a different problem. So great have been the changes in techniques of investigation and in theoretical attitudes over the past three centuries that almost all generalizations about that development must be suspect. Nevertheless, it is a common practice so far as optics is concerned to call the seventeenth century an era of plenitude in nature, the eighteenth century an era of the corpuscularian theory of

---

[1] Gerald Feinberg, "Light," *Sci. Am.* 219:50 (September, 1968). In conclusion, Feinberg says, "At present the photon theory gives an accurate description of all we know about light. The notion that light is fundamentally just another kind of matter is likely to persist in any future theory. That idea is the distinctive contribution of 20th-century physicists to the understanding of light, and it is one of which we can well be proud." All eleven articles in this special issue of *Scientific American* afford an excellent summary of the state of the art and science of optics.

light, and the nineteenth century an era dominated by the undulatory or wave theory of light.[2]

Because the twentieth century is so deeply imbued with photon theory and the notion of wave-particle duality, it is wise to review the major assumptions and achievements that established physical optics on a firm foundation and led the advances of experiment and theory so deeply into nineteenth-century wave theory. Michelson's interferometer and the Michelson-Morley experiment were conceivable only in terms of wave theory and its most basic assumption, the luminiferous aether.

At the beginning of the nineteenth century, physical science in general seemed to be purging itself gradually of dependence upon a series of hypothetical fluids that had served long and well in static, structural analyses of certain phenomena. The concept of heat as a substance called "caloric" was being challenged by the idea of heat as a process of matter in motion. The notion of fire as a substance called "phlogiston" was abolished by the chemical revolution, and certain electrical and magnetic "effluvia" likewise were being challenged more frequently as impedimenta. The "aether" had been one of the first of these subtle, ubiquitous, imponderable "fluids" to be banished in the eighteenth century, but, ironically, the concept was revived along with the wave theory early in the nineteenth century, and for a hundred years thereafter the luminiferous medium grew more "real" though not more tangible as the conceptual basis for physical action through seemingly vacuous space. Only after the advent of quantum and relativity theories, superimposed on electromagnetism, did the pendulum of optical theory swing back toward corpuscularity and then become damped down in the hysteresis of wave-particle duality.[3]

[2] For insight into the difficulties with such generalizations, see the following historical studies: Thomas S. Kuhn, *The Copernican Revolution: Planetary Astronomy in the Development of Western Thought*, and *The Structure of Scientific Revolutions*, esp. pp. 72–76; Marshall Claggett, ed., *Critical Problems in the History of Science*, esp. pp. 3–220; Charles C. Gillispie, *The Edge of Objectivity: An Essay in the History of Scientific Ideas*, esp. pp. 83–201; Stanley L. Jaki, *The Relevance of Physics*; A. I. Sabra, *Theories of Light: From Descartes to Newton*; Harold I. Sharlin, *The Convergent Century: The Unification of Science in the Nineteenth Century*; Stephen Toulmin and June Goodfield, *The Architecture of Matter*.

[3] For leads into the immense literature on philosophical implications of modern

Two of the truly titanic figures in science are Sir Isaac Newton (1642–1727) and Albert Einstein (1879–1955). The differences between these two men are commonly supposed to symbolize the progress of physical science between the seventeenth and the twentieth centuries. One of the central differences between the thought of Newton and Einstein lies in their attitudes toward basic concepts of space and time as used for the measurement of matter in motion.

Both Newton and Einstein were philosophers as well as scientists, but we honor them most as mathematical cosmologists who provided new syntheses of physical knowledge. Both were fascinated by the peculiar behavior of light, but neither Newton nor Einstein was exempt from the pressures, presuppositions, and technical capabilities of his age. Hence we cannot contrast them fairly, except in full historical context; neither can we avoid a bald recognition of their respective reputations: that Newton believed in an aether as a *Sensorium Dei*, or as a background for absolute mechanical motion in the Universe, and that Einstein abolished the belief in absolute time and space with his critique of the notion of simultaneity, thereby abandoning the aether, both as a luminiferous medium and as a metaphysical basis for distinguishing between absolute and relative motion in the cosmos knowable to man.

The characters of great men may far exceed or fall short of their reputations, but we do not expect to find the same sort of discrepancies between the character and reputation of great events in science. When we do see beyond a facade, especially in experimental physics where science should appear in one of its purest forms, we may well be surprised at how human an enterprise science is. Such a surprise is in store when we examine the great work of Albert Abraham Michelson (1852–1931), first American Nobel Prize winner for science and a

physics, one may begin with the following: Paul A. Schilpp, ed., *Albert Einstein: Philosopher-Scientist*; Mary B. Hesse, *Forces and Fields: The Concept of Action at a Distance in the History of Physics*; Adolf Grünbaum, *Philosophical Problems of Space and Time*; Max Jammer, *Concepts of Space: The History of Theories of Space in Physics*; and Stephen Toulmin, ed., *Physical Reality: Philosophical Essays on Twentieth-Century Physics*.

physicist known primarily for his measurements of the speed of light.[4]

Michelson was virtually a self-made physicist, a rugged individualist of the late nineteenth century, a sensitive and artistic type of "Herr Professor" who could paint, play billiards and tennis, calculate rapidly, and manipulate readily almost any kind of physical apparatus or delicate instrument that promised to reveal some of nature's secrets about light. Michelson was small in stature, meticulous in his dress and demeanor, and careful in his choice of friends and associates, but he was not an inspirational teacher or a prolific writer or a good theoretician. His forte was his laboratory work where optics and optical inventions could be studied and tested with accuracy and precision. His character was his lifelong quest for better answers to the question, What is light? His reputation as a master of measurement continues to win him the highest honors.[5]

Michelson became a sailor before he became a physicist, and as a naval officer in the 1870's he grew acutely aware of relative-motion problems in maneuvering ships at sea. He studied celestial navigation as a midshipman, and during his limited sea duty he had to practice the paradox of relying on the so-called fixed stars of ancient Ptolemaic astronomy to guide his way to port while he preached Copernican and Newtonian astronomy to his midshipmen students. Experiments with polar coordinate maneuvering boards to solve problems of true wind and current speed and direction were not yet far advanced, but the post–Civil War steam-and-sail navy, although in the doldrums in many respects, was beginning to develop reliable pitometer logs, current meters, anemometers, and compasses.

[4] To begin to appreciate Michelson's experimental relationship with Einstein's relativity theories, see Bernard Jaffe, *Michelson and the Speed of Light*; L. Pearce Williams, ed., *Relativity Theory: Its Origins and Impact on Modern Thought*; and Gerald Holton, "Einstein, Michelson, and the 'Crucial Experiment,'" *Isis* 60:133–197 (Summer, 1969). On the differences between Newton and his disciples, see Robert E. Schofield, *Mechanism and Materialism: British Natural Philosophy in the Age of Reason*, esp. pp. 157–190; also Arnold Thackray, *Atoms and Powers: An Essay on Newtonian Matter-Theory and the Development of Chemistry*.

[5] Most recently Michelson was elected in 1970 as the seventh scientist to be permanently enshrined in the Hall of Fame for Great Americans in New York City. See New York University News Bureau, press release, October 28, 1970.

In the midst of his nautical training and development, Michelson went back to the U.S. Naval Academy to teach physical sciences and gradually to become interested in research with physical optics. After repeating, with improvements, the latest terrestrial measurements on the velocity of light, Michelson worked with other American experts on this problem before taking a leave of absence to study in Europe under some of the world-renowned authorities on physics and optics. While in Paris and Berlin in 1880 Michelson conceived the idea of a way to measure the earth's motion through cosmic space. James Clerk Maxwell (1831–1879), the codifer of the electromagnetic theory of light, had set the problem and despaired of a solution; now Michelson, by an analogy from his experience as a conning officer at sea and as a surveyor of the velocity of light, created an instrument and an experimental design to settle the question of the speed and direction of the resultant of the earth's various motions through cosmic space.[6]

He failed, and it is the story of his and others' struggles to make sense of that failure that is told in this book. I will begin with a sketch of the history of wave theory, emphasizing the early nineteenth century, and then proceed to some perspectives on the origins of relativity and the problems of the aether.

Most historical treatments of the development of modern optics begin with a juxtaposition of the theories of Newton and Christiaan Huygens (1629–1695). These two heroes of science are often used as symbolic protagonists, Newton for the emission or corpuscular theory, and Huygens for the undulatory or wave theory. That they were contemporary rivals adds to the drama of their presumed theoretical antagonism over optics. It is folly to ascribe too much to this bipolar

---

[6] For a description of twelve known components of the earth's resultant motion in space, see *The Flammarion Book of Astronomy*, pp. 11–53. First published in Paris in 1880 as *Astronomie Populaire*, by Camille Flammarion, it has been revised and reissued by G. C. Flammarion and André Danjon and translated by A. & B. Pagel. In addition to the influence of Simon Newcomb, which may be traced through his autobiography, *The Reminiscences of an Astronomer*, and Nathan Reingold's *Science in Nineteenth Century America: A Documentary History*, see also the influential talks by John Tyndall, *Lectures on Light: Delivered in the United States in 1872–73*. See also Cecil J. Schneer, *The Search for Order: The Development of the Major Ideas in the Physical Sciences from the Earliest Times to the Present*.

characterization (since Huygens died too soon, Newton was always too ambivalent, and Descartes, Kepler, Leibniz, Hooke, and others, also, were influential). But in the wake of their subsequent influence we can recognize Huygens as the hero of the wave theorists and Newton as the champion of the corpuscular school. Both, as children of the seventeenth century and heirs of the Cartesian tradition, felt some need for an aether. Huygens, it seems, used the notion primarily to escape the alternative idea of passing light through totally empty space, the abhorrent vacuum, whereas Newton finally used the aether concept, speculatively, as the stage for the universe of mechanical action.[7]

Huygens, whose famous *Treatise on Light* was written in 1678 but not published until 1690, gave a quantitative value to Ole Roemer's (1644–1710) qualitative discovery of the finite time required for the propagation of light. Huygens's strenuous exception to Descartes's view that the speed of light is instantaneous led deeper into a comparison of sound and light as analogous vibrations in a medium: "It is inconceivable to doubt that light consists in the motion of some sort of matter. . . . Matter which exists between us and the luminous body . . . but it cannot be any transport of matter. . . . That which can lead us to comprehend it is the knowledge of . . . sound."[8] For our purposes, Huygens's greatest contributions were to develop the idea of a wave front and to furnish the wave interpretation of the phenomena with which geometrical optics had been concerned. In order to derive the first good values for the velocity of light, Huygens combined Roemer's time estimates from occultations of the moons of Jupiter with new estimates of the distance between sun and earth. Also important was his wave-model explanation for double refraction in Iceland spar, an

[7] For more adequate discussions, see A. Wolf, *A History of Science, Technology, and Philosophy in the 16th & 17th Centuries*, I, 145–187, 244–274. Cf. E. A. Burtt, *The Metaphysical Foundations of Modern Physical Science*, pp. 264–282, on Newton's use of the aether concept. Cf. A. R. Hall, *The Scientific Revolution, 1500–1800: The Foundation of Modern Scientific Attitude*, pp. 98–99. See also Wm. P. D. Wightman, *The Growth of Scientific Ideas*, pp. 99–157; Florian Cajori, *A History of Physics* . . . , pp. 109–199.

[8] Christiaan Huygens, *Treatise on Light* . . . , pp. 3–4. See also I. Bernard Cohen, *Roemer and the First Determination of the Velocity of Light*. See also Arthur E. Bell, *Christian Huygens and the Development of Science in the Seventeenth Century*, pp. 162, 177, 180, 208.

explanation that adumbrated the discovery of the polarization of light. Huygens remained a good Cartesian in his belief in the necessity for action by contact and therefore in the necessity for a universal plenum to transmit light.

Besides using the aether as a luminiferous medium, Huygens also used it to give a mechanical explanation for gravity. This ancient idea was entertained by Newton, too, who had used the "aethereal medium" as early as 1675 to provide a hypothetical cause for gravity. Dynamically considered, gravitation seemed closely associated with light, since both involved action over "empty" distances and both raised the question whether time is required for their propagation. Later in life, as his anti-Cartesian bias increased, Newton claimed to feign no more hypotheses and found less of a function for the "aethereal medium" to fulfill, until finally, by 1717, he equated the Aether with the Vacuum and rested content with a corpuscular theory of light.[9]

Sir Isaac Newton produced incalculable historical results through his grand synthesis of the laws of motion. His publication of the *Philosophiae naturalis principia mathematica* in 1687 occurred within a year of another "Glorious Revolution" in England, and in both the political and scientific spheres thereafter the world became a different place.

We cannot here do justice to Newton. His genius in synthesizing celestial and terrestrial mechanics into three "universal" laws furnished the next two centuries with so firm a mechanical framework that "matter in motion" became the basis of virtually all knowledge. The year Galileo died Newton was born, and two hundred years after the first publication of his *Principia*, Michelson and Morley published the paper that led to some embarrassing questions about Newtonian assumptions.

Although we need not go into analytical detail about Newton's famous distinctions between "absolute and relative, true and apparent, mathematical and common" ideas of time, space, and motion, it is of

[9] Sir Isaac Newton, *Opticks: Or a Treatise of the Reflections, Refractions, Inflections & Colours of Light*, p. 374. See also I. Bernard Cohen, "Newton in the Light of Recent Scholarship," *Isis* 51:507 (December, 1960); and Roger H. Stuewer, "A Critical Analysis of Newton's Work on Diffraction," *Isis* 61:188–205 (Summer, 1970).

fundamental importance to recognize how central to the theme of the *Principia* was the problem of distinguishing between true and apparent motions. "For to this end," said Newton in his first scholium, "it was that I composed it."

I. Absolute, true, and mathematical time, of itself, and from its own nature, flows equably without relation to anything external, and by another name is called duration: relative, apparent, and common time, is some sensible and external . . . measure of duration by means of motion, which is commonly used instead of true time; such as an hour, a day, a month, a year.

II. Absolute space, in its own nature, without relation to anything external, remains always similar and immovable. Relative space is some movable dimension or measure of the absolute spaces; which our senses determine by its position to bodies; . . . For if the earth, for instance, moves, a space of our air, which relatively and in respect of the earth remains always the same, will at one time be one part of the absolute space into which the air passes; at another time it will be another part of the same, and so, absolutely understood, will be continually changed.[10]

These two famous definitions, so often quoted as Newton's axiomatics, were followed by two more distinctions less well known. Newton's third point was that "place is a part of space which a body takes up, and is according to the space, either absolute or relative." Here arose a vexatious problem of correspondence between mathematical and physical descriptions, later termed the distinction between kinematics and dynamics. Newton's fourth point carried his discussion directly into the problem of relative motion: "Absolute motion is the translation of a body from one absolute place into another; and relative motion, the translation from one relative place into another." After considering compounded and complex relative motions, Newton penned a beautiful sanction for the experimental question which Michelson later asked regarding the earth and aether:

It is indeed a matter of great difficulty to discover and effectually to distinguish, the true motions of particular bodies from the apparent; because the parts of that immovable space, in which those motions are performed,

---

[10] *Sir Isaac Newton's Mathematical Principles of Natural Philosophy and His System of the World*, trans. Andrew Motte (1729), ed. Florian Cajori, pp. 6, 12.

do by no means come under the observation of our senses. Yet the thing is not altogether desperate; for we have some arguments to guide us, partly from the apparent motions, which are the differences of the true motions; partly from the forces, which are the causes and effects of true motions.[11]

These remarks by Newton have furnished the text for innumerable discussions in physics and philosophy. Ever since first minted, these speculations on relative motion and on "absolute" time and space have provoked debate. Although many popular histories of physics treat relativity theory as a revolution against Newtonian mechanics, they seldom point out explicitly that what Einstein broke in 1905 was the aether, the weakest link acknowledged by Newton in the chain of reasoning between his optical and his gravitational ideas. The aether was in the seventeenth century as all-pervasive in cosmological thought as it was thought to be in space. Newton participated in this immersion and only gradually evolved above it by separating his scientific from his theological speculations. Newton was always cautious of his use of the aether concept, and yet, as I. B. Cohen has said, it "was a central pillar of his system of nature."[12]

Fr. Francesco Maria Grimaldi's (1618–1663) experiments with diffraction were either unknown to or ignored by Huygens, but Newton did indeed take Grimaldi seriously, devoting Book III of his *Opticks* to the phenomenon of the apparent bending of light around obstacles in its way. Grimaldi's pinhole experiments, however, made less of an impression on Newton than the reciprocal relations between light rays and material bodies, such as thin films of oil or lens-and-plate experiments. Roemer's suggestion that the speed of light must

---

[11] Ibid., pp. 6, 7, 12. Cf. Cajori's note in the Appendix, pp. 639–644, which analyzes the "unfortunate" Michelson-Morley experiment.

[12] I. Bernard Cohen and Robert E. Schofield, eds., *Isaac Newton's Papers & Letters on Natural Philosophy*, pp. 7–16 and passim. See also the thorough critical analysis of publication history in Cohen's Preface to Newton, *Opticks*, pp. ix–lviii. Cf. Richard S. Westfall, "Uneasily Fitful Reflections on Fits of Easy Transmission," in Robert Palter, ed., "The *Annus Mirabilis* of Sir Isaac Newton: Tricentennial Celebration," in *The Texas Quarterly* 10:86–102 (Autumn, 1967). One of the better secondary texts is the book by A. d'Abro, *The Evolution of Scientific Thought: From Newton to Einstein*; see especially chapter 11, "The Ether," pp. 116–124. See also Marie Boas, "The Establishment of the Mechanical Philosophy," *Osiris* 10:412–541 (1952), and [Bryan Robinson] *Sir Isaac Newton's Account of the Aether* (Dublin, 1745).

be finite Newton regarded as firmly established. Propagation of light in time, as opposed to the "instantaneous" effects of gravitation, was not a physical problem for him. He seems to have assigned the imponderable aether to the realm of metaphysics, following it there himself in later years. Thereby he effectively divorced optical from gravitational theory for the next two centuries.[13]

To most eighteenth-century minds, Huygens's wave theory was severely handicapped by an insistence on too close an analogy with sound, but by the nineteenth century this analogy was found to be highly prescient. Huygens denied periodicity to his waves on the grounds that such periodicity would so confuse rays of light that we could not see anything clearly. Newton, on the other hand, could never fully accept the wave theory, because he thought it must deny the basic laws of rectilinear propagation. Huygens's acoustical analogy led him to conceive of light waves as only longitudinal compressions, and his denial of periodicity ruled out the possibility that he might have conceived of interference as part of the basic nature of light.

Only one major contribution to the theory of physical optics in the eighteenth century bears directly on the problems raised by the Michelson-Morley experiment. This was the serendipitous discovery by James Bradley (1693–1762) of the phenomenon known as astronomical aberration. Bradley was professor of astronomy at Oxford and a good Newtonian both in method and in theory, holding generally to the corpuscular view of light. In 1725 he became concerned with a problem in the positioning of his telescopes which required that they always be aimed with a certain unaccountable lead-angle.

At first these systematic adjustments were thought to be direct evidence for long-sought annual parallax differences. After experiment-

---

[13] Newton, *Opticks*, pp. 277, 280. On this general problem, which touches on the Leibniz-Newton controversy, see especially Stephen Toulmin, "Criticism in the History of Science: Newton on Absolute Space, Time and Motion," *The Philosophical Review* 68:1–29, 203–227 (1959); Norwood R. Hanson, "Waves, Particles, and Newton's 'Fits,' " *J. Hist. Ideas* 21:370–391 (July–September, 1960); Philip E. B. Jourdain, "Newton's Hypothesis of Ether and Gravitation," *The Monist* 25:79–106 (January, 1915), 234–254 (April, 1915), 418–440 (July, 1915). On Grimaldi, see the translated extract from *Physico-Mathesis de Lumine, Coloribus et Iride* (1665) in W. F. Magie, ed., *A Source Book in Physics*, pp. 294–298; and Vasco Ronchi, *Storia della Luce*, pp. 71–115.

ing thoroughly to test for parallax, Bradley guessed that the cause of
this deviation might be a nutation of the earth's axis, or an alteration
of the reference plumb line, or some still unknown refraction phenomena. Careful experiments and calculations failed to support any
of these hypotheses. At last he conjectured that "all the phaenomena
hitherto mentioned" were caused by a combination of the motion of
light with the earth's annual motion in its orbit. The emphasis Bradley
put on the observer's own motion may be seen from his specific hypothesis:

> If light be propagated in time (which I presume will be readily allowed
> by most of the philosophers of this age) then it is evident from the fore-
> going considerations, that there will be always a difference between the real
> and visible place of an object, unless the eye is moving either directly to-
> wards or from the object. And in all cases, the sine of the difference between
> the real and visible place of the object, will be to the sine of the visible in-
> clination of the object to the line in which the eye is moving, as the Velocity
> of the eye to the Velocity of light.[14]

Bradley's experimental care and theoretical insight in finding an
explanation for the apparent displacement of stars was rewarded in
1742 when he succeeded Edmond Halley (1656–1742) as Astronomer
Royal. By showing indirectly that starlight is affected by the earth's
orbital motion to the extent that the observer must correct for the error
introduced by his own motion, Bradley reduced the problem of the
parallax of the "fixed stars" to a degree unobservable for another century, while simultaneously reintroducing epicyclic movements to the
"fixed stars" as a result of the movement of the observer's earth platform.

Bradley's contention made acceptable to the scientific world at large
Roemer's suggestion of the finite speed of light, partly because aberration gave a second indirect means of estimating its velocity. In ad-

[14] James Bradley, "An Account of a New Discovered Motion of the Fixed Stars,"
Trans. R.S. 25 (1729) as reprinted in W. F. Magie, ed., *A Source Book in Physics*, p.
339. See also "A Letter from the Rev. Mr. James Bradley, Savilian Professor of
Astronomy at Oxford, and F.R.S., to Dr. Edmond Halley, Astrnom. Reg. &c., giving
an Account of a new discovered Motion of the Fix'd Stars," reprinted as Appendix to
Paul Carus, *The Principle of Relativity in the Light of the Philosophy of Science*,
p. 94.

dition, Bradley's announcement was another step in the extension of the size of the universe, a great contribution to the precision and accuracy of astronomical measurements, and the first *direct* astronomical evidence for the heliocentric system. Stellar aberration led to a new awareness of the relative aspects of the motion of light, marking another turning away from the problem of anthropocentrism in science, and encouraging belief in the doctrine of absolute space by considering the sun at rest with respect to the "fixed stars."[15]

The eighteenth century was kept quite busy verifying the multitude of predictions generated from Newtonian theory. Although a few isolated geniuses like Leonhard Euler (1707–1783) in St. Petersburg and Benjamin Franklin (1706–1790) in Philadelphia did continue to develop wave theory and the idea of an electrical aether, most European savants remained loyal to corpuscularian theories. When Joseph Priestley (1733–1804) wrote *The History and Present State of Discoveries relating to Vision, Light, and Colours* in 1772, he could say with confidence that the achievements of his immediate predecessors had authorized "us to take it for granted, that light consists of very minute particles of matter, emitted from luminous bodies."[16] A generation later Priestley's sanguine confidence would be greatly upset by a resurrection of the wave theory.

On November 12, 1801, the brilliant and versatile young physician Thomas Young (1773–1829) stood before the Royal Society to relate how his study of Newton's *Opticks*, and in particular Newton's rings, "converted that prepossession which I before entertained for the undulatory system of light into a very strong conviction of its truth and sufficiency, a conviction which has been since most strikingly confirmed by an analysis of the colours of striated substances."[17] This announce-

[15] George Sarton, "Discovery of the Aberration of Light," *Isis* 16:233–239 (November, 1931). See also Albert B. Stewart, "The Discovery of Stellar Aberration," *Sci. Am.* 210:100–108 (March, 1964).

[16] Joseph Priestley, *The History and Present State of Discoveries relating to Vision, Light, and Colours*, II, 768. This characterization of the eighteenth century was later to be amplified or revised by others, especially with regard to the role of Leonhard Euler in influencing later students of continuum mechanics: see Clifford A. Truesdell, *Essays in the History of Mechanics*, esp. chapters 2 and 3.

[17] Thomas Young, "The Bakerian Lecture: On the Theory of Light and Colours," *Phil. Trans. R.S.* 92:12–48 (1802); reprinted in Henry Crew, ed., *The Wave Theory*

ment, and his subsequent use of the priciple of interference to explain
Newton's rings, has been called the first major theoretical discovery in
optics in about a hundred years. Although Young went far out of his
way to give full, even undue, credit to Newton, he still incurred the
wrath of Newtonian orthodoxy for his irreverence in asserting as his
four famous hypotheses:

I. *A Luminiferous Ether pervades the Universe, rare and elastic in a high
degree.*
II. *Undulations are excited in this Ether whenever a Body becomes
luminous.*
III. *The Sensation of different Colours depends on the different frequency of
Vibrations, excited by Light in the Retina.*
IV. *All material Bodies have an Attraction for the ethereal Medium, by
means of which it is accumulated within their substance, and for a small
Distance around them, in a State of greater Density, but not of greater
Elasticity.*[18]

The phenomena which Grimaldi had termed "diffraction" and Newton had called "inflexion" Young now proceeded to investigate in detail. He developed a much more satisfactory explanation of the colors
of thin plates by expanding on Huygens's idea of the wave front.
Young had studied acoustics and was also impressed by the analogy
between sound and light. But his experiments and calculations were
designed to establish physical optics on the basis of the wave theory,
not vice versa:

About the same time that Newton was making his earliest experiments on
refraction, Grimaldi's treatise on light . . . appeared; it contained many interesting experiments and ingenious remarks on the effects of diffraction,
which is the name that he gave to the spreading of light in every direction,
upon its admission into a dark chamber, and on the colours which usually
accompany these effects. He had even observed that in some instances the

---

of Light: Memoirs by Huygens, Young, and Fresnel, pp. 45–61. See also Alexander
Wood, *Thomas Young: Natural Philosopher, 1773–1829.*

[18] Young, "Bakerian Lecture," *Trans. R.S.* 92: 14, 16, 18, 21. See also I. Bernard
Cohen, "The First Explanation of Interference," *Am. J. Phys.* 8:99 (April, 1940).
Cf. A. Rubinowicz, "Thomas Young and the Theory of Diffraction," *Nature* 180:
160–162 (July 27, 1957).

light of one pencil tended to extinguish that of another, but he had not in-
quired in what cases and according to what laws such an interference must
be expected.[19]

The critical feature in Young's development of Huygens's principle,
that each point on a wave front is itself the source for a new wave, was
his isolation of the fact of periodicity in light. Huygens had stopped
short here because he could see no way for waves to cross each other
from all directions without obliterating one another. Newton had
recognized periodicity, but his ambivalence toward the wave theory of
light was ascribable to his imagining waves largely as surface phenomena and to his preference for corpuscularian explanations. Young
solved these difficulties by refining the pinhole experiments and showing definitively how light under properly controlled conditions can be
made to demonstrate superposition and interference, a periodic property. Corpuscles added to corpuscles could not then be visualized as
annihilating each other, but light waves added to light waves, if they
met in plane and out of phase, could easily be imagined to cancel each
other. If they met in phase, such waves would reinforce one another
and then pass on with their original profiles undisturbed. Thus was
born the superposition principle.

The hostility, even vilification, which Thomas Young met when he
first propounded the new theory of diffraction was not due to his insistence on periodicity but rather to his insistence on an all-pervading
luminiferous aether. His presentation was less attractive to theoreticians than to experimentalists, who could immediately repeat his
simple demonstration. Yet even experimenters had a solid reason for
cautious doubt, since the phenomenon of double-refraction, at that time
a major theoretical interest, still stood outside his theory. Incredulous
smiles must have crossed some of the faces of old Newtonians when in
1803 they heard Young assert on the one hand, "the advocates for the
projectile hypothesis of light must consider which link in this chain of
reasoning they may judge to be the most feeble, for hitherto I have
advanced in this paper no general hypothesis whatever," and, on the

[19] Thomas Young, *A Course of Lectures on Natural Philosophy and the Mechanical
Arts*, ed. P. Kelland, I, 377. See also *Miscellaneous Works of the late Thomas Young,
M.D., F.R.S., &c.*, ed. George Peacock, I, 64, 180.

other hand, continue, "upon considering the phenomena of the aberration of the stars, I am disposed to believe that the luminiferous ether pervades the substance of all material bodies, with little or no resistance, as freely, perhaps, as the wind passes through a grove of trees."[20]

Meanwhile, as we can see in retrospect, other discoveries at the beginning of the nineteenth century also reinforced the evidence for the wave theory of light. Sir William Herschel (1738–1822), the prolific German-English astronomer who added so much to knowledge of our solar system, galaxy, and stellar neighborhood, in 1800 discovered invisible radiations beyond the red end of the visible spectrum. The next year Johann W. Ritter (1776–1810) found similar radiations beyond the violet end of the spectrum. The year afterward, William H. Wollaston (1766–1828) combined these extensions of Newton's spectrum into the infrared and ultraviolet regions with the insight that Newton's prismatic spectrum could be refined to show dark lines, or as he thought (erroneously), "boundaries" between colors. A larger awareness of the meaning of these discontinuities in the expanded continuum of the spectrum of radiation emerged from the works of Joseph von Fraunhofer (1787–1826), Robert W. Bunsen (1811–1899), Gustav R. Kirchhoff (1824–1887), and others. They fathered the science of spectroscopy amid the confusion of a continuum of colors shown to be discontinuous.[21]

But of course neither Young's interference principle nor the advent of spectroscopy alone was enough to change old habits of thought regarding light. There were few graphical methods of analysis for fluid-flow or energy-transfer studies before Joseph (baron) Fourier

[20] Thomas Young, "Experiments and Calculations Relative to Physical Optics," *Trans. R.S.* (1804), reprinted in Crew, ed., *The Wave Theory of Light*, pp. 74–75. For more insight into these disputations, see Gillispie, *The Edge of Objectivity*, pp. 406–420.

[21] See Harry Woolf, "The Beginnings of Astronomical Spectroscopy," in *Mélanges Alexandre Koyré*, I, 619–634. See also F. Abeles, "Instrumental Optics," in René Taton, ed., *Science in the Nineteenth Century*, pp. 144–148; and William McGucken, *Nineteenth-Century Spectroscopy: Development of the Understanding of Spectra, 1802–1897*.

(1768–1830) announced in 1807 his now famous theorem for complex periodic variations. Fourier found that any periodic function is a sum of simple sine functions; therefore, any periodic oscillation that varies repeatedly and exactly, regardless of how complicated it may appear, can be broken down into a series of simple regular wave motions. This harmonic analysis made the study of sound, heat, light, and all other periodic phenomena amenable to mathematical treatment. As the use of the sine wave proved so fundamental and ubiquitous, the analogies between different forms of wave motion likewise gained plausibility. Indeed the wave theory of heat began to reinforce the wave theory of light, and the aether began to be considered "thermiferous" as well as luminiferous.[22]

After Étienne Louis Malus (1775–1812) gave an explanation for double refraction in terms of "polarized light," in 1808, Thomas Young took time out from deciphering the Rosetta stone's hieroglyphics to suggest that all light waves might vibrate at right angles to the direction of their propagation. Young seems to have seen light as a form of energy transfer through space, and he fought throughout his productive life for recognition of the ethereal medium as imponderable, that is, weightless, matter.[23]

Young's assertions in favor of the undulatory theory remained problematic until his experimental insights were put into more elegant mathematical form by another product of the Napoleonic educational revolution in France. Augustin Fresnel (1788–1827) was trained as a civil engineer, but he became interested in optics about 1814 and shortly thereafter wrote a classic treatise on the diffraction of light,

---

[22] See E. N. da D. Andrade, *An Approach to Modern Physics*: "There is no branch of physics in which we are not brought sooner or later—generally sooner—to the study of vibrations, the study of waves, the study of processes which repeat themselves over and over again as does, to take the simplest case, the motion of a pendulum" (p. 35). See also Stephen G. Brush, "The Wave Theory of Heat: A Forgotten Stage in the Transition from the Caloric Theory to Thermodynamics," *Br. J. Hist. Sci.* 5:145–167 (1970).

[23] See Peacock, *Life of Thomas Young*. See also Isaac Asimov, *Biographical Encyclopedia of Science and Technology*, for the personal interconnections of the various contributors to the development of wave theory.

which was crowned by the Académie française in 1819. This treatise was perhaps the most important single contribution to the reestablishment of the wave theory of light.[24]

Fresnel's theory, uniting Huygens's translation of geometrical optics into wave optics with Young's principle of interference, gave for the first time a satisfactory explanation of the rectilinear propagation of light (which had been Newton's prime objection to wave theory) and the existence of diffraction fringes beyond the geometrical limits for shadows. In the process of developing the implications of this theoretical combination, Young, Fresnel, and François Arago (1786–1853) introduced a capital distinction to wave theory which moved it away from visualization by analogy with sound and toward the analogy with radiant heat. This was the idea that light consists of *transverse*, not longitudinal, vibrations. Experiments with polarization phenomena lead to this conclusion, since two oppositely polarized beams of light cannot be made to interfere. Not only are sound waves longitudinal, but transverse waves were unknown in any fluid medium, either gaseous or liquid. Thus was presented one of the greatest difficulties which the aether theory was to encounter, namely, that henceforward the aether could not be visualized as a fluid; rather it would have to be imagined as a solid![25] Thanks largely to the authority of André M. Ampère (1775–1836), the wave theory of heat also came to dominate the period between 1820 and 1850.

The year 1830 may well serve as the watershed between the fluid and the solid concepts of the nature of the luminiferous aether. Young and Fresnel had established the wave theory of light in so strong a position that by the time of Young's death in 1829, as Whittaker's history relates, "henceforth the corpuscular hypothesis was unable to recruit any adherents among the younger men." Fresnel's mathematical theory codified, as it were, Young's wide-ranging experiments, and

[24] A. Fresnel, "Memoir on the Diffraction of Light," in Crew, ed., *The Wave Theory of Light*, pp. 79–144. See also *Oeuvres complètes de Augustin Fresnel*, H. de Senarmont, ed. II, passim.

[25] F. Arago and A. J. Fresnel, "On the Action of Rays of Polarized Light upon Each Other," reprinted in Crew, ed., *The Wave Theory of Light*, pp. 145–155. Cf. Ernst Mach, *The Principles of Physical Optics: An Historical and Philosophical Treatment*, pp. 203, 210.

by the time of Fresnel's death in 1827 his work in crystal optics was widely recognized and honored as preferable to Newton's explanations in this area. The trouble with conceiving of the aether as an elastic-solid persisted, however, and made a double entendre out of its imponderability.

In his later years Fresnel had been less concerned with the kinematics than with the dynamics of his necessary aether. Indeed, he had stimulated French and English mathematicians alike to a great effort in the development of mathematical theory for wave motions (longitudinal, transverse, and torsional) in solids. Although Fresnel himself is said to have worked "backwards from the known properties of light, in the hope of arriving at a mechanism to which they could be attributed," he is credited with having succeeded "in accounting for the phenomena in terms of a few principles, but [he] was not able to specify an ether which would in turn account for these principles."[26]

In 1838 the Copernican theory of a heliocentric solar system was verified a second time by visible evidence of the earth's motion through space. In that year Friedrich Wilhelm Bessel (1784–1846) finally succeeded in determining the parallax of a so-called fixed star, 61 Cygni which turned out to be some 35 trillion miles or about six light-years from the earth. With fine precision instruments now available, several astronomers were on the verge of this discovery that Bradley and others had sought in vain for several centuries. Questions that Johann Lambert (1728–1777) had raised and Sir William Herschel had answered in the eighteenth century about the relative motions of astral bodies and about the "absolute" motion of the solar system now

[26] Sir Edmund Whittaker, *A History of the Theories of Aether and Electricity*, I, 125–126. This was first published in one volume in 1910, then revised and enlarged for the 1951 edition from which the Harper edition is taken. See also Mme M. A. Tonnelat, "The Theory of Light," in René Taton, ed., *Science in the Nineteenth Century*, pp. 156–161. See also David Brewster, *A Treatise on Optics*; George Biddell Airy, *On the Undulatory Theory of Optics*, first published in 1831; and Humphrey Lloyd, *Elementary Treatise on the Wave Theory of Light*. The mathematical contributions of Augustin-Louis Cauchy (1789–1857), Siméon Denis Poisson (1781–1840), George Green (1793–1841), and James MacCullagh (1809–1847) to the elucidation of the elastic-solid theory of the aether can only be alluded to here. For a thorough exposition of their work, see Whittaker's *History*, I, 5: 128–169. See also Hunter Rouse and Simon Ince, *History of Hydraulics*, pp. 193–213.

required new and better answers. Bessel himself gave up on the effort to use his functions and techniques to find the solar apex, but progress in the development of astronomy promised a much more sophisticated understanding of stellar and solar motions.[27] The wave theory of light, assumed by Fraunhofer in designing the instruments he made for Bessel and other astronomers to determine stellar parallax, was becoming indispensable.

By the time of Christian Johann Doppler's (1803–1853) discovery in 1842 of the effect which bears his name, the analogy of light with sound had been widely recognized as seriously at fault. But now the analogy of light with radiant heat as well as the entirely new human experience with "high-speed" travel complicated the issue. As railroads began to cover the countryside in Europe and America, people heard locomotive steam whistles dramatically change their pitch as they sped past crossings, and this accustomed scientists and the public alike to a new experience with wave phenomena. When Doppler showed that the relative speed between the source and the ear accounts for the frequency change which the ear detects in the pitch of the sound heard, his explanation was bitterly opposed by those whose "common sense" told them that the speed of sound increased with the speed of the source. It was all too easy for many people to misconstrue Doppler's explanation and their own experiences alongside railroad tracks as meaning that the velocity of propagation was being changed and not merely the frequency of reception.

The implications of the Doppler effect for light were soon explored by the French school of optical experimenters, with the result that it was eventually understood that the motion of a source of light relative to an observer modifies the apparent frequency of the light that is received. Later, after the Michelson-Morley experiment and Einstein's

[27] See William W. Campbell, *Stellar Motions: With Special Reference to Motions Determined by Means of the Spectrograph*, pp. 1–40, 127–136; for Lambert and Herschel, see excerpts in Milton K. Munitz, ed., *Theories of the Universe: From Babylonian Myth to Modern Science*, pp. 252–257, 262. On the relative-motion problem, see also Arthur Berry, *A Short History of Astronomy: From Earliest Times Through the Nineteenth Century*, pp. 101–103, 232.

postulate of relativity, the question whether velocities of source, light, and observer might not be additive after all was resurrected. The "red shift" evidence for an expanding universe was based on the Doppler principle and was hard to believe at first. Thus another inescapable problem in relative motion was introduced into the science of optics.[28]

Hippolyte Fizeau (1819–1896) was one of those who developed the implications of the Doppler effect for light, thereby reinforcing the analogy between sound and light. For our purposes, however, Fizeau is more important as the experimentalist who achieved for the first time in 1849 the direct measurement of the velocity of light on earth by terrestrial methods. Using an accurately machined cogwheel, Fizeau effectively duplicated the experiment suggested by Galileo over two hundred years earlier. This measurement over an eight-kilometer course of a fundamental constant in physics was a monumental achievement, not only for itself but also because it encouraged the search for better methods. Within two years another Frenchman, Fizeau's former collaborator, Leon Foucault (1819–1868), gained more accurate results by adapting a rotating mirror for the same purpose.[29]

In the decade of the 1840's the elastic-solid conception of the aether looked highly promising. But a number of investigators were considerably perturbed over the problem of reconciling astronomical aberration, double refraction, circular polarization, thermodynamics, and the results of refraction experiments generally with the problem of the velocity with which light travels in different kinds of media. In July, 1845, George G. Stokes (1819–1903) set out to explain the cause of aberration in accordance with undulatory theory: "I shall suppose that the earth and planets carry a portion of the ether along with them so that the ether close to their surfaces is at rest relatively to their sur-

[28] Whittaker, *History*, I, 389. See also T. P. Gill, *The Doppler Effect: An Introduction to the Theory of the Effect.*

[29] H. Fizeau, "Sur une expérience relative à la vitesse de propagation de la lumière," *Comptes Rendus* 29:90–92 (July, 1849). For a translation of this paper, see William F. Magie, ed., *A Source Book in Physics*, pp. 341–342. Galileo's worries over instantaneity of transmission of light and his suggested occultation experiment may be read in Henry Crew and A. de Salvio, trans. *Dialogues Concerning Two New Sciences*, pp. 42–44.

faces while its velocity alters as we recede from the surface till, at no great distance, it is at rest in space."[30] In several follow-up papers Stokes continued to maintain that aberration could be explained "without the startling assumption that the earth in its motion round the sun offers no resistance to the ether."[31] This "aether-drag" hypothesis was only one of a number of attempts to reconcile Fresnel's theory with what was known of astronomical movements. Stokes was proposing, in effect, to add another envelope to the earth: beyond its core and its concentric lithosphere and hydrosphere and atmosphere, there might also possibly be an "aetherosphere." Michelson was to revere Stokes's hypothesis above all others.

By mid-century the stage seemed set in physical optics for a crucial experiment which, many hoped, would settle once and for all the long-standing debate over waves versus particles and convert the few remaining corpuscular advocates. Since Newton had surmised that corpuscles of light should travel faster in a denser medium than they would travel in a less dense medium, and conversely the wave theorists had predicted an antithetical situation—that waves would travel faster in a less dense than in a more dense medium—the crucial problem had been to find a method to measure the velocity of light on earth. This Fizeau had accomplished in 1849 by means of the principle of interference in optical theory and by means of physical interference (or interruption of the beam) in his experimental design. Fizeau and Foucault, formerly friends, but now rivals, raced separately through experiments to test Arago's predictions and to be first to publish a "death blow" to the corpuscular theory. They ended in a dead heat, each getting his results published in *Comptes Rendus* on May 6, 1850. Foucault emerged the winner, however, since his method was more general, more trustworthy, and less equivocal in claiming to have made the crucial decision: "These results indicate a lesser speed of light in water

[30] George G. Stokes, "On the Aberration of Light," *Phil. Mag.*, 3d ser. 27:9–10 (July, 1845); see also Stokes, *Mathematical and Physical Papers*, I, 134.

[31] George G. Stokes, "On the Constitution of the Luminiferous Aether Viewed with Reference to the Phaenomenon of the Aberration of Light," *Phil. Mag.* 3d ser. 29:6 (July, 1846); see also his "On the Constitution of the Luminiferous Ether," *Phil. Mag.* (3) 32:343–349. (May, 1848).

than in air, and fully confirm the implications, according to the views of M. Arago, of the undulatory theory."[32]

Meanwhile, in the neighboring fields of electricity, magnetism, and heat, parallel advances were being made. Michael Faraday (1791–1867), guided by his fertile intuition, had been working steadily in the search for constructive analogies since 1831. For instance, he asked himself in his diary in 1837 about the inductive properties of the imponderable aether in relation to the atmosphere, and again in 1850 he recorded a wonder: "Does the magnetic condition of Space indicate any relation to the supposed ether, or to any other condition of space equivalent to it? . . . What a strange Magnetic system our Planet presents. First the Earth itself as a magnet—and that not unchangeable. . . . Then Space as a great and good conductor of the power, and probably permeated by the lines of force to a distance from the planet."[33] Still uncertain in 1856 as to how to handle the notion of "solid-space," Faraday confided to his diary a doubt about the report of his American rival to the honor of having discovered the principle of induced currents: "Professor [Joseph] Henry . . . says I think that lightning flashes ten miles off affected his induction coil apparatus. I do not know how; but if the results were instantaneous they are against the idea of sensible time. . . . Still, it is not absolutely certain that magnetic propagation, if in an (or the) ether, must be as quick as light, though it is likely; and the experiment is worth trying; especially as the magnetic influence is *transverse* to the electric current, which rivals light; and time is certainly required in some magnetic phenomena, as in the charging of soft iron."[34]

[32] H. Fizeau et L. Breguet, "Note sur l'expérience relative à la vitesse comparative de la lumière dans l'air et dans l'eau," *Comptes Rendus* 30:562–563 (May 6, 1850); see also pp. 771–774. L. Foucault, "Méthode générale pour mesurer la vitesse de la lumière dans l'air et les milieux transparents. Vitesses relatives de la lumière dans l'air et dans l'eau. Projet d'expérience sur la vitesse de propagation du calorique rayonnant," *Comptes Rendus* 30:551–560 (May 6, 1850). [The quotation is my translation from Foucault's paper, p. 556.]

[33] Michael Faraday, *Faraday's Diary: Being the Various Philosophical Notes of Experimental Investigation . . . during the Years 1820–1862 . . .*, ed. Thomas Martin, III, 213; V, 342.

[34] Ibid., VII, 9. See also the judgment of L. Pearce Williams, *Michael Faraday:*

Faraday had suggested in 1851 that the luminiferous aether might also be the conveyor of magnetic force, but, like Newton, he was skeptical of all imponderable concepts. He would have preferred to replace the aether by his famous "lines of force" which, from the perceptual evidence of iron filings aligned around a magnet, he thought extended indefinitely throughout all space. But the synthesis Faraday sought was only found by Maxwell.[35]

One further optical experiment must be considered before we see what happened to Faraday's speculations. The Astronomer Royal of Great Britain in 1871 was Sir George Biddell Airy (1801–1892), a man noted for precision. In that year he became interested in the possibility, expressed across the channel, that astronomical aberration might vary with the thickness of the refracting lenses in transits and telescopes. Airy built a special thirty-five-inch–long telescope whose tube was filled with distilled water expressly for testing the equality of the coefficient of sidereal aberration in water and air. He solved his technical problem to his own satisfaction, but others were to take his assertion as corroborating Fresnel's convection hypothesis also. Airy found no difference due to the medium in the tube of the aberration coefficient; this was often interpreted as a blow against Stokes's notion of a stagnant sphere of luminiferous aether around the earth.[36]

James Clerk Maxwell stands in relation to Faraday as Fresnel stood to Thomas Young. As a mathematical physicist he codified the experimental work and qualitative theory of his predecessor, also helping to forge linkages between the wave theory of light and heat. Maxwell

---

A Biography, pp. 454–457. These interactions are traced from Faraday through Zeeman by J. Brookes Spencer, "On the Varieties of Nineteenth-Century Magneto-Optical Discovery," *Isis* 61:34–51 (Spring, 1970).

[35] Whittaker, *History*, I, 170–194. See also Whittaker's praise of the 1867 elastic-solid aether theory of Joseph Boussinesq (1842–1929), ibid., pp. 167–169.

[36] George B. Airy, "On a Supposed Alteration in the Amount of Astronomical Aberration of Light Produced by the Passage of the Light through a Considerable Thickness of Refracting Medium," *Proc. R.S.* 20:35–39 (November 23, 1871). See also "Additional Note," *Proc. R.S.* 21:121 (January 16, 1873). Airy is also generally credited with the determination of the *classical apex* for solar motion, determined in 1860 with the coordinates ($\alpha = 270°$; $\delta = +30°$) established by rechecking the exact positions of the so-called fixed stars and calculating the average deviation after all corrections were applied.

brought out the first edition of his epoch-making *Treatise on Electricity and Magnetism* in the year 1873. As early as 1861 he had come to believe that the magnetic and luminiferous media were identical, and by 1865 he had derived all the essential equations of his electromagnetic theory. Maxwell began his treatise by acknowledging his heavy debt to Faraday; he ended it with a discussion of the continental objections to English proclivities for explaining action-at-a-distance by use of the imponderable medium.[37]

Maxwell felt it necessary to retain the notion of a medium primarily in order to obviate the action-at-a-distance problem. Though the partial differential equations of Maxwell's synthesis did not require a mechanical model for their elucidation, and although he himself was perhaps as sophisticated in his use of scientific models for descriptive purposes as any scientist who has ever worried over epistemology, still it is clear that Maxwell felt more comfortable with the aether than without a mechanism for radiant energy transfer. The two conceptual needs often seemed mutually inclusive where electromagnetism was concerned. Maxwell's versatile and fecund contributions to the kinetic-molecular theory of gases, to statistics, planetary theory, and to the codification of the laws of thermodynamics, make him seem more modern than he was. Maxwell himself never dared to complete the substitution of the field concept for the aether. Since his equations can stand alone and do constitute a field theory, the fact is often forgotten that he remained to the end of his life unwilling to denounce the aether concept as useless. That task was performed by Albert Einstein and the next generation.[38]

---

[37] James Clerk Maxwell, *A Treatise on Electricity and Magnetism,* p. 383; see also 3d ed. unabridged, pp. 492–493. For another example of Maxwell's viewpoint, see S. Tolver Preston, *Physics of the Ether.* See also *Origins of Clerk Maxwell's Electric Ideas: As Described in Familiar Letters to William Thomson,* ed. Joseph Larmor, p. 35. See also Alfred M. Bork, "Maxwell and the Electromagnetic Wave Equations," *Am. J. Phys.* 35:844–848 (September, 1967).

[38] Note that the book of essays by J. J. Thomson et al., *James Clerk Maxwell: A Commemoration Volume 1831–1931,* often tells us more about the authors in 1931 than about Maxwell. For example, Einstein generalizes therein that before Maxwell, physical reality was conceived as particles of matter in motion, whereas after him the real external world is conceived as "continuous fields, governed by partial differential equations, and not capable of any mechanical interpretation" (p. 71). The

The magnificent achievement of Faraday and Maxwell was to incorporate into one tight synthesis the heretofore loosely-linked, if not actually separate, phenomena of electricity, magnetism, and optics. Whereas a number of earlier investigators had thought it necessary to postulate separate media for electricity, magnetism, and visible light, now with Maxwell's synthesis the luminiferous aether became the electromagnetic aether and vice versa. Visible light became henceforth only a tiny fraction of the overall electromagnetic radiation spectrum, and radiant heat became infrared radiation. As soon as Maxwell's predictions of invisible waves were experimentally verified by Heinrich R. Hertz (1857–1894) in 1887, the major problems in physics seemed to divide into two broad categories, namely, the physics of matter and the physics of the aether.[39]

Although a few such physicist-philosophers as Maxwell, Hermann von Helmholtz (1821–1894), and especially Ernst Mach (1838–1916) did appreciate how much of a residue from old metaphysical systems continued to pervade the physics of the day, the far more common attitude in physical science was positivistic, tending to distrust the criticisms of epistemologists as worthless. Peter Tait, for example, trumpeted thus in 1874: "The fundamental notions which occur to us when we commence the study of physical science are those of Time and Space. . . . But we cannot inquire into the actual nature of either

---

problems raised by William Crookes's development of the radiometer after 1874 must be accorded a central role in Maxwell's thought toward the end of his life: see S. G. Brush and C. W. F. Everitt, "Maxwell, Osborne Reynolds, and the Radiometer," *Historical Studies in the Physical Sciences*, ed. R. McCormmach, I, 105–126; A. E. Woodruff, "Action at a Distance in Nineteenth Century Electrodynamics," *Isis* 53:439–459 (December, 1962), and "The Contributions of Hermann von Helmholtz to Electrodynamics," *Isis* 59:300–311 (Fall, 1968).

[39] For some of the more important recent articles on Maxwell's development, see Alfred M. Bork, "Maxwell and the Vector Potential," *Isis* 58:210–222 (Summer, 1967), and his forthcoming "Foundations of Electromagnetic Theory—Maxwell," in the *Sources of Science* series; Joan Bromberg, "Maxwell's Displacement Current and his Theory of Light," *Archive for History of Exact Sciences* 4:218–234 (1967); Thomas K. Simpson, "Maxwell and the Direct Experimental Test of his Electromagnetic Theory," *Isis* 57:411–432 (Winter, 1966); C. W. F. Everitt, "Maxwell's Scientific Papers," *Applied Optics* 6:639–646 (April, 1967); Joseph Turner, "Maxwell and the Method of Physical Analogy," *Br. J. for Phil. Sci.* 6:226–238 (November, 1955).

space or time, except in the way of a purely metaphysical, and therefore of necessity absolutely barren, speculation."[40] Curiously, however, this was not merely an expression of mathematical faith, for almost equally dangerous to the progress of physical science, in the view
of Tait and Thomson, was mathematical abstraction. At the beginning
of their famous abstract treatise which expressed the fundamental role
of energy in nature, Thomson and Tait wrote: "Nothing can be more
fatal to progress than a too confident reliance on mathematical symbols; for the student is only too apt to take the easier course, and consider the *formula* and not the *fact* as the physical reality."[41]

The historical necessity to maintain a broad perspective with regard
to the problems of space, time, motion, and relativity is best exempli-
fied by what Maxwell himself chose to say about these problems in
1876. Three years before his death Maxwell wrote a small introduction to the physical sciences, published posthumously as *Matter and
Motion*, which illustrates nicely most of the broad problems we have
tried to encompass in this chapter. Michelson's experience at sea, the
approximative nature of experimental physics, and a latent awareness
of the ambiguous boundary between physics and metaphysics are all
illustrated with remarkable percipiency by Maxwell several years before Michelson's aether-drift experiment was conceived:

Absolute space is conceived as remaining always similar to itself and immovable. The arrangement of the parts of space can no more be altered than
the order of the portions of time. To conceive them to move from their
places is to conceive a place to move away from itself.

But as there is nothing to distinguish one portion of time from another
except the different events which occur in them, so there is nothing to distinguish one part of space from another except its relation to the place of
material bodies. We cannot describe the time of an event except by reference to some other event, or the place of a body except by reference to some
other body. *All our knowledge, both of time and place, is essentially relative.** When a man has acquired the habit of putting words together, with-

---

[40] P. G. Tait, *Lectures on some Recent Advances in Physical Science*, p. 4. Cf. Tait,
in a different mood, publishing (anonymously) with Balfour Stewart, *The Unseen
Universe, or Physical Speculations on a Future State*, 2d ed. (New York, 1875).

[41] Lord Kelvin and Peter G. Tait, *Treatise on Natural Philosophy*, p. viii. This
was first published in 1879, before William Thomson was even knighted.

out troubling himself to form the thoughts which ought to correspond to them, it is easy for him to frame an antithesis between this relative knowledge and a so-called absolute knowledge, and to point out our ignorance of the absolute position of a point as an instance of the limitation of our faculties. Anyone, however, who will try to imagine the state of mind conscious of knowing the absolute position of a point will ever after be content with our relative knowledge.[42]

Significantly, an asterisk was inserted above, in mid-paragraph by the editor, Joseph Larmor, an aether apologist, who added to Maxwell's text this "corrective" footnote: "The position seems to be that our knowledge is relative, but needs definite space and time as a frame for its coherent expression." However, another passage, farther along, contradicts the editor's contention, because Maxwell wrote:

Our whole progress up to this point may be described as a gradual development of the doctrine of relativity of all physical phenomena. Position we must evidently acknowledge to be relative, for we cannot describe the position of a body in any terms which do not express relation. The ordinary language about motion and rest does not so completely exclude the notion of their being measured absolutely, but the reason of this is, that in our ordinary language we tacitly assume that the earth is at rest.

.   .   .   .

Our primitive notion may have been that to know absolutely where we are, and in what direction we are going, are essential elements of our knowledge as conscious beings

But this notion, though undoubtedly held by many wise men in ancient times, has been gradually dispelled from the minds of students of physics.

There are no landmarks in space; one portion of space is exactly like every other portion, so that we cannot tell where we are. We are, as it were, on an unruffled sea, without stars, compass, soundings, wind, or tide, and we cannot tell in what direction we are going. We have no log which we can cast out to take a dead reckoning by; we may compute our rate of motion with respect to the neighbouring bodies, but we do not know how these bodies may be moving in space.[43]

[42] James Clerk Maxwell, *Matter and Motion*, ed. Sir Joseph Larmor, p. 12 [italics mine].

[43] Ibid., pp. 80–81. Other aids to understanding the astrophysical background of the aether-drift problem may be the following: Colin A. Ronan, *Edmond Halley:*

Is it really impossible to know how fast and in what direction we on this spaceship earth are going? This was Michelson's grand question, derived from Maxwell himself. The answer might well be found if light itself could be employed to ask it.

Undulatory optical theory, far less mechanical than the corpuscular theory, but still far more "mechanical," in a hydrodynamical sense, than the electromagnetic optical theory of the twentieth century, reigned supreme in the last three decades of the nineteenth century. This was the situation in physical optics when Albert A. Michelson, Ensign, U.S.N., became more interested in his subject matter than in his temporary billet as an instructor at the U. S. Naval Academy in 1877.

Genius in Eclipse; Katherine B. Collier, *Cosmogonies of Our Fathers*; Robert Grant, *History of Physical Astronomy.*

CHAPTER TWO

DRAMATIS PERSONAE OF AETHER DRIFT

Although Michelson was certainly the main character in the long drama of experimental optics that lay behind relativity theory, he usually shares double billing today whenever the aether-drift experiments are remembered. The "Michelson-Morley experiment," as a titular phrase, detracts something from the credit due Michelson for his design and persistence with aether-drift interferometers. On the other hand, if a third scientist, Dayton C. Miller, had not joined Morley in the effort to find optical evidence for aether drift, it is doubtful whether the "Michelson-Morley experiment" would have gained its present celebrated status.

Three main actors played the leading roles in the history of aether-drift experimentation. Similar to "three F's" of great renown in French physical optics—Fresnel, Fizeau, and Foucault—the three American scientists with accidentally alliterative last names—Michelson, Morley, and Miller—lived intertwined personal as well as professional lives. This chapter will sketch their biographies as a necessary prelude to an appreciation of their professional pursuit of an aether drift. Because the rest of this history is so closely allied with the lives of these three protagonists, the sketches here are merely preliminary vitae.

Michelson was the innovator of the experiment; Michelson and Morley collaborated to produce the classical heritage; then Morley and Miller together defined that heritage and produced variations on its theme. Finally, Miller alone carried on the tradition of the triumvirate until the long-lived Michelson was persuaded by his disciple to return to a further refinement of his classical experiment.

ALBERT ABRAHAM MICHELSON (1852–1931)

"It will probably be generally agreed," said Robert A. Millikan, "that the three American physicists whose work has been most epoch-making and whose names are most certain to be frequently heard wherever and whenever in future years the story of physics is told are Benjamin Franklin, Josiah Willard Gibbs, and Albert A. Michelson."[1] This judgment, expressed in a eulogistic memoir by a protégé of Michelson, could certainly be disputed. Partisans for Joseph Henry, John William Draper, Henry A. Rowland, Samuel P. Langley, T. C. Mendenhall, Simon Newcomb, or Thomas Edison, to name only a few of a number of possible contenders, might wish to spread such honors more widely or in different directions. But whatever one's preference for measuring personal greatness, there can be little dispute over Arthur H. Compton's contention that Michelson's life coincided with and contributed greatly to America's coming of age in science.[2]

The elements of Michelson's early life are almost as varied as mid-nineteenth-century America itself. It is curious that such a single-

---

[1] Robert A. Millikan, "Michelson Memorial Address," *Commemorating Michelson Laboratory Dedication: May Seventh and Eighth, Nineteen Forty Eight*, p. 17. Cf. Robert A. Millikan, "Albert A. Michelson," *Biographical Memoirs*, National Academy of Sciences, 19, no. 4; Robert A. Millikan, "Albert Abraham Michelson: The First American Nobel Laureate," *Scientific Monthly* 48:16–27 (January, 1939).

[2] Arthur H. Compton, "Nobel Prize Winners in Physics," *Current History* 34:699 (August, 1931). See also reprint in Marjorie Johnston, ed., *The Cosmos of Arthur H. Compton*, pp. 193–197.

minded, purposeful individual, a superspecialist among specialists, should have developed from a boyhood so adventuresome. Exact historical knowledge of this early period of his life is at present limited, although a family biography is promised in the near future. Nevertheless, the outlines of Michelson's youth have been made clear.[3]

A boy was born in the disputed village of Strelno, Prussia, to a young couple of Prussian-Polish-Jewish extraction on December 19, 1852. Samuel and Rosalie Michelson named their firstborn son Albert and began to plan a better future for their family. Their native village in the province of Posnan, Polish in language and tradition but under Prussian hegemony, offered little of either security or opportunity in the wake of the convulsive political revolutions of 1848. Therefore Samuel Michelson emigrated with his family to New York City, where he found work for a short time as a jeweler's apprentice. Later, the lure of fortune apparently stronger than the desire for security, Samuel became infected with the fever of the gold rush to California. Heeding the advice of Horace Greeley and the invitation of a sister who had preceded him to California, Samuel converted his assets into a cargo of dry goods and embarked with his wife and two children for Panama and an Isthmian crossing. Their arrival in San Francisco about 1855 might well have been Albert's first memory in later years, but he is certain to have remembered the opening in 1856 of his father's dry-goods store at Murphy's Camp in Calaveras County. Whether jumping frogs were a better business than dry goods or whether the discovery of the Comstock Lode simply beckoned irresistibly to merchant Michelson, the family followed the mining frontier. Young Albert was left with relatives in San Francisco to attend Lincoln Grammar School there as the family relocated from the gold diggings in California to the

---

[3] To date the most trustworthy account is Bernard Jaffe, *Michelson and the Speed of Light*, pp. 35–48. Another narrative, overdramatized and often erroneous, but adequate for his youth, is John H. Wilson, Jr.'s *Albert A. Michelson: America's First Nobel Prize Winner in Physics*, pp. 13–44. Michelson's daughter, Dorothy Michelson Livingston, is presently at work on a biography of her father which will describe his early life and genealogy. Some research into the family background has been performed for her at Copernicus University in Poland. See, for example, Dorothy Michelson Livingston, "Michelson in the Navy: The Navy in Michelson," *U.S. Naval Institute Proceedings* 95:72–79 (June, 1969).

silver rush in Nevada. Here the family's business and residence were reestablished in Virginia City, and here the Michelsons found a moderate prosperity and security even while the Civil War was raging less than 3,000 miles to the eastward.[4]

Albert turned thirteen during the year of Appomattox and after Lincoln's assassination he was given the middle name Abraham. While his bustling family enjoyed booming Nevada, Albert Abraham, the oldest son, showed such scholastic promise that he was encouraged to remain in San Francisco and to matriculate at Boys' High School. There he boarded with the headmaster, Theodore Bradley, who took a personal interest in guiding his education and drilling him in science and mathematics. As a student Albert made a good impression on Mr. Bradley and was given a job as laboratory assistant, which perhaps first stimulated his aptitude for scientific pursuit.

Upon graduation in 1869, the sixteen-year-old youth decided to enter the competition for admission to the U. S. Naval Academy far across the continent. This ambition was made feasible by the fact that on May 10, 1869, the golden spike had just been driven at Promontory, Utah, linking the eastward and westward thrusts of the construction gangs engaged on the transcontinental railroad. Although Albert did well on the examination, he failed to get the appointment from Nevada. Undaunted by this he promptly boarded an eastbound train, thereby becoming one of the first transcontinental rail passengers, in order to make a special plea before President U. S. Grant for one of the ten presidential appointments-at-large. At the White House young Michelson's persistence was finally rewarded by an interview with President Grant, which resulted in a special appointment, above the usual quota, for Michelson, as a midshipman.[5]

Michelson was one of eighty-six entering "plebes," or freshmen, in

[4] Charles Michelson, Albert's younger brother, who later became a journalist for Hearst and then a ghost-writer for Franklin D. Roosevelt during the New Deal, has written a chapter on their boyhood in the West: "How I Got That Way," in his autobiography, *The Ghost Talks*, pp. 67–80. A novel by a younger sister, Miriam Michelson, *The Madigans*, uses a Virginia City setting for a family comedy. For background on California and Nevada contrasts, see Rodman W. Paul, *Mining Frontiers of the Far West, 1848–1880*, pp. 87–108.

[5] A considerable legend has grown up around this "illegal" appointment. See

the class of 1873, only twenty-nine of whom were to graduate. Class-
mates remembered his prowess in boxing, fencing, violin playing, and
painting. Michelson himself often recalled a verbal lashing given him
by Commodore Worden, the hero of the battle of the *Monitor* and
the *Virginia* (*Merrimac*), for being more interested in science than
in the sea.[6] On the other hand, another naval officer, the head of the
physics department, Lt. Comdr. William T. Sampson, later to become
the butt of the Battle of Santiago by being in the wrong place at the
wrong time, calculated correctly in encouraging cadet Michelson's
irregular interests in the laboratory. Later Michelson was to serve under
Sampson as an instructor in physical sciences at Annapolis, but as a
cadet he acquired from him discipline in physical exercise and in systematic living.

In his senior year midshipman Michelson may have attended, either
in Baltimore or in Washington, one or more of a series of popular
lectures being given by John Tyndall in which Tyndall attempted to
explain light in terms of physical science and to define physical science
in terms of light. Whether Tyndall's enthusiasm for his subject and
his profession was instrumental in spurring young Michelson to graduate in 1873 at the head of his class in optics and acoustics is not
known, but certain it is that attendance at Tyndall's lectures would
have made a snap of the cadet's first question on his final examination
in optics. The question was, "Discuss undulatory theory. What fact
proves emission theory false?"[7] At that time the satisfactory answer to
this question would have centered around the physical discussion of
interference phenomena. Easy though this examination might have
been to the senior midshipman, later, as a naval officer and as a pro-
fessor-scientist, Michelson would live out his long and productive life
in the search for ever more adequate answers to this question and its

Harvey B. Lemon, "Albert Abraham Michelson: The Man and the Man of Science,"
The American Physics Teacher (now *Am. J. Phys.*) 4:2 (February, 1936).

[6] James O'Donnell Bennett, "Superlative Americans. Second Article. Albert Abraham Michelson at 70," The Chicago Tribune, rotogravure section (1923), pp. 22–
23. See also Bradley A. Fiske, *From Midshipman to Rear Admiral*, p. 15.

[7] J. R. Smithson, "Michelson at Annapolis," *Am. J. Phys.* 18:425–427 (October,
1950); John Tyndall, *Lectures on Light: Delivered in the United States in 1872–
1873.*

near relatives: What is light? What is its speed? Does the luminiferous aether exist? If so, does it prove emission theory false? If not, what then is the ultimate nature of light?

Upon graduation, Michelson ranked ninth among twenty-nine in his overall class standing and likewise ninth in demerits; although first in optics and acoustics, he ranked twenty-fifth in seamanship, and he was "anchorman," the poorest, in historical studies. The physics course at the Naval Academy, one of the best in the nation during the years in which Michelson was in attendance, consisted of five one-semester courses lasting two years; the text was Atkinson's translation of the famous *Ganot's Treatise.*

In this college textbook, used by Michelson as both student and teacher and by thousands of other aspiring scientists in the late nineteenth century, the French author and the English editor expressed an almost messianic hope for a far-reaching integrative substance, for an aether that could make mechanics complete:

As the physical sciences extend their limits the opinion tends to prevail that there is a subtile, imponderable, and eminently elastic fluid called the ether distributed through the entire universe; it pervades the mass of all bodies, the densest and most opaque, as well as the lightest or the most transparent. It is also considered that the ultimate particles of which matter is made up are capable of definite motions varying in character and velocity, and which can be communicated to the ether. A motion of a particular kind communicated to the ether can give rise to the phenomenon of heat; a motion of the same kind, but of greater frequency, produces light; and it may be that a motion different in form or in character is the cause of electricity. Not merely do the atoms of bodies communicate motion to the atoms of the ether, but this latter can impart it to the former. Thus the atoms of bodies are at once the sources and the recipients of the motion. All physical phenomena, referred thus to a single cause, are but transformations of motion.[8]

Michelson may well have been impressed by this passage far more than he ever realized.

[8] E. Atkinson, ed. and trans., *Elementary Treatise on Physics, Experimental and Applied*, p. 3, from *Ganot's Éléments de Physique*. Adolphe Ganot (1804–1887) first wrote his *Éléments de Physique* about 1860, and Atkinson began his English translation shortly thereafter, with revised and enlarged editions being issued almost biennially throughout the century.

In those days midshipmen, after graduation, had to go through a year's course of ship familiarization before being fully commissioned as line officers. Michelson was no exception, and he spent several months aboard each of four different ships before becoming an ensign in July of 1874 and being assigned his first regular billet aboard the U.S.S. *Worcester*. But his sea duty was short-lived, since he was transferred back to the Naval Academy in December of 1875, for shore duty as an instructor in science under Commander Sampson.

For four years he taught physics and chemistry and prepared himself for further research. Marriage in the spring of 1877 to Margaret McLean Heminway (Sampson's niece) and the birth of two sons in the next two years brought added responsibilities. But, unexpectedly, Michelson's marriage also brought aid to his scientific work, since his wealthy father-in-law became deeply interested in his project to measure the speed of light—interested enough to advance two thousand dollars for the precision apparatus which Albert needed. In 1878 this kind of financial support for specialized research by an unknown young man was not available through the Navy.

In the spring of 1879 Michelson gained his first public notice as a scientist when reporters learned that the young officer had set up apparatus like Foucault's to demonstrate how to measure the speed of light by the use of a rotating mirror. Privately, Michelson had written the year before to Professor Simon Newcomb, then superintendent of the navy's *Nautical Almanac* and an astronomer with rank both in the navy and in science, to request an interview to talk over techniques: "I trust I am not taking too great a liberty in laying before you a brief account of what I have done."[9] From this initial contact there grew an important cooperation between Newcomb and Michelson in their mutual quest for an ever better evaluation of the speed of light. Later in 1879, Michelson was promoted to the rank of Master and

[9] A. A. Michelson to Simon Newcomb, April 26, 1878, in Simon Newcomb Papers (alphabetized Letters Received file), Manuscript Division, Library of Congress. For the optical parts of the 1879 experiment, see the note in Deborah J. Warner, *Alvan Clark & Sons: Artists in Optics*, p. 80. For more details on Michelson's career at the Naval Academy, see D. T. McAllister, "A Tribute to Albert A. Michelson," address delivered at the dedication of Michelson Hall, U.S. Naval Academy, Annapolis, Md., May 9, 1969.

reassigned directly to Professor Newcomb on the *Nautical Almanac* staff. Through Newcomb's patronage he published his first scientific paper and thus was launched a scientific career.[10]

The story of the Newcomb-Michelson collaboration in measuring the speed of light has been told elsewhere.[11] What needs emphasis here is that in the period between the Civil War and the establishment of the first academic graduate schools in this country, the federal government offered an opportunity to young men interested in pure science to make a living while indulging their interest in science.[12] In the spring of 1880, when Michelson sought permission from the navy to go to Europe for postgraduate study in his chosen field, he was following a rather common procedure. The fact that he was granted this leave of absence to proceed in the summer of 1880 to Germany and France for his specialized education is perhaps some measure of the pressure being brought by the scientific community, especially by Simon Newcomb, on the federal government to encourage the development of pure science in the United States.[13]

Just as Einstein's famous thought experiments often made use of familiar situations aboard trains or elevators, Michelson's most famous

[10] A. A. Michelson, "On a Method of Measuring the Velocity of Light," *Am. J. Sci.*, 3d ser. 15:394–395 (May, 1878).

[11] See Nathan Reingold, ed., *Science in Nineteenth Century America: A Documentary History*, pp. 251, 275–306. A facsimile copy of Michelson's notebook, entitled "Velocity of Light" (on cover), and "Experimental Determination of Velocity of Light" (inside), was published by Honeywell in 1965, but mistakenly advertised as a record of Michelson's 1878 work; all internal evidence points to this as a polished record of the 1879 work with USN colleagues and Professor A. M. Mayer of Stevens Institute, Hoboken, N.J.; see p. 5; cf. pp. 23, 25, 27, 39.

[12] For a full study, see A. Hunter Dupree, *Science in the Federal Government: A History of Policies and Activities to 1940*, pp. 184–194, 289–301. See also George H. Daniels, "The Process of Professionalization in American Science: The Emergent Period, 1820–1860," *Isis* 58:151–166 (Summer, 1967); Geraldine Joncich, "Scientists and the Schools of the 19th Century: The Case of American Physicists," *American Quarterly* 18:667–685 (Winter, 1966).

[13] On December 4, 1880, Newcomb delivered an influential address before the Philosophical Society of Washington in which he pressed hard for social and governmental understanding of the value of pure science. See Simon Newcomb, *The Relation of Scientific Progress*. For comparative data on the profession of physics, see the fine study by Daniel J. Kevles, "The Study of Physics in America, 1865–1916," Ph.D. diss., Princeton University, 1964.

experiment for aether drift had part of its roots at least in a familiar situation aboard ship. His experience as a young naval officer in 1874 and 1875 must have provided him with considerable food for thought experiments regarding the problems of relative motion. To a physicist and a navigating sailor the concepts of velocity, acceleration, and relative motion are fundamental. To others, in the days before freeway driving, jet air travel, and vicarious space flight became common experiences, high-speed and long-range cases of relative motion were hard to imagine. Michelson was a sailor, pilot, and navigator before he became a physicist. His greatest fame as a physicist stems from his efforts to solve certain kinds of relative-motion problems which he could not have avoided as a naval officer.

In learning to become a conning officer aboard a ship at sea, Michelson had to face a number of problems of relative motion without the aid of any fixed standards of reference except the stars. Celestial navigation was, before radio, loran, and transit satellites an indispensable tool to the mariner because it offered what can be considered a solution, for practical purposes, to the problem of "absolute" motion on the earth's surface. Using geocentric or Ptolemaic cosmology, the marine navigator must ignore Copernicus's sun-centered universe to find his way to port. Equally important to the naval ship-handler are the practical approximations required to maneuver a ship in formation. The new steam-and-sail navy in which Michelson served had simplified the conning officer's task in that he was no longer so dependent upon wind conditions, but at the same time steam propulsion complicated the relative-motion problems in maneuvering, by an extension in mobility, an increase in speed, and the need to know more accurately relative wind and current conditions to compute set and drift. In short, wind and current were reduced as problems, whereas relative motion was made more critical.

Aboard a ship at sea, it was often an officer's task to perform the simple calculation of converting the relative wind across the deck into the true wind divorced of the effects of the ship's own motion. Undoubtedly, Michelson's brief experience at sea included this routine task. We may infer, although there is little historical evidence to support it, that Michelson's sea duty—an experience that inevitably invites the idea that one's ship is the world—made him better able as a physicist to appreciate the possibility of measuring aether drift in a manner analogous to the computation of true wind speed and direction. If the earth, moving in its orbit through light-filled interplanetary space, were like a ship moving across the sea and through the air, then it should be possible somehow to measure the relative motion of the aether wind past the earth and to convert that relative motion into an "absolute" motion.[14]

Although the word *absolute* in this context refers simply to a larger frame of reference, and although the comparative *larger* is not identical with the superlative word *largest*, still, scientific and nautical progress depend to some extent on deliberately, if provisionally, equating the two. As a physicist, Michelson would know another meaning for the word *absolute*, referring not to size but to Newtonian ideas of more or most fundamental *inertial* systems of reference. In either case, the hypothetical celestial sphere that envelopes the earth with a panorama of seemingly "fixed stars" provided the most fundamental reference system. By whatever means he actually conceived of his aether-drift hypothesis, the major assumption that Michelson made in first performing his most famous experiment was that the relative motion of earth and aether ought to show the "absolute" motion of the earth and perhaps of the sun as well.[15]

At this point I will suspend this biographical sketch of Michelson's background in order to proceed with sketches of the early lives of his two colleagues in the aether-drift tests. But we should not leave Michelson without taking some note of the high reputation he achieved in the history of science. When he became in 1907 the first American Nobel laureate in science, his consummate stature was assured. Two

---

[14] For some indication of these nautical problems, see the various editions of S. T. S. Lecky, *Wrinkles in Practical Navigation*. See also William Froude, "Report on Instruments for Measuring the Speed of Ships," *Br. Ass'n, Adv. Sci. Annual Report* (1879), pp. 210–218.

[15] For the astronomical problem of relative motions, see William W. Campbell, *Stellar Motions: With Special Reference to Motions Determined by Means of the Spectrograph*. See also E. Finlay-Freundlich, "Cosmology," in Otto Neurath et al., eds., *International Encyclopedia of Unified Science*, I, Part 2, pp. 506–565; and W. M. Smart, *Stellar Kinematics*, pp. 38–68.

stanzas from an epic poem written by Edwin Herbert Lewis in January, 1923, give some measure of the esteem in which he came to be held.[16] Lewis's poem was written to celebrate the scientific accomplishments that had issued from the Ryerson Laboratory at the University of Chicago. This was where Michelson spent most of his mature years. The poem is entitled "The Ballad of Ryerson," and stanzas eight and nine refer to Michelson's life and work:

> Rude is the minstrel's measure, and rudely he
>   plucks the strings,
> But in Ryerson rainbows murmur the music of
>   heavenly things.
> Is not this stranger than heaven that a man
>   should hear around
> The whole of earth and the half of heaven and see
>   the shadow of sound?
> He gathereth up the iris from the plunging of
>   planet's rim
> With bright precision of fingers that Uriel envies
>   him.
> But when from the plunging planet he spread out
>   a hand to feel
> How fast the ether drifted back through flesh or
>   stone or steel
> The fine fiducial fingers felt no ethereal breath.
> They penciled the night in a cross of light and
>   found it still as death.
> Have the stars conspired against him? Do measure-
>   ments only seem?

[16] Edwin Herbert Lewis, *University of Chicago Poems*, pp. 23–31 (© The University of Chicago Press, 1923); see also p. 3. See also T. W. Goodspeed, *A History of the University of Chicago*. For different kinds of personal assessments, see Robert A. Millikan, "Michelson's Economic Value," *Science* (new series) 69:481–485 (May 10, 1929); Arthur S. Eddington, obituary, "Professor Albert Abraham Michelson," *Nature* 127:825 (May 30, 1931); George Ellery Hale, "Some of Michelson's Researches," *Publications of the Astronomical Society of the Pacific* 43:175–185 (June, 1931). For a recent documented assessment, see D. T. McAllister, *Albert Abraham Michelson: The Man Who Taught a World to Measure*, pp. 1–30. Another helpful biographical sketch is by Jean M. Bennett, "Albert Abraham Michelson: Nobel Prize in Physics, 1907," preprint for Fratelli Fabbri Editori.

Are time and space but shadows enmeshed in a
   private dream?

But dreaming or not, he measured. He made him a
   rainbow bar,
And first he measured the measures of man, and
   then he measured a star.
Now tell us how long is the metre, lest fire
   should steal it away?
Ye shall fashion it new, immortal, of the crimson
   cadmium ray.
Now tell us how big is Antares, a spear-point in
   the night?
Four hundred million miles across a single point
   of light.
He has taught a world to measure. They read the
   furnace and gauge
By lines of the fringe of glory that knows nor
   aging nor age.
Now this is the law of Ryerson and this is the
   price of peace—
That men shall learn to measure or ever their
   strife shall cease.
They shall measure the cost of killing, and
   measure the hearts that bleed,
And measure the earth for sowing, and measure
   the sowing of seed.

EDWARD WILLIAMS MORLEY (1838–1923)

About the time of Morley's death, Michael Pupin published an auto-biographical memoir which related a story similar in many respects to the tale that Michelson might have told of himself had he been so inclined. *From Immigrant to Inventor* was the title Pupin chose to describe his progress from a Slavic boyhood to a position of eminence as an electrical engineer and scientist. Edward W. Morley was a very different sort of person, but because his fame now derives largely from his association with Michelson, we may learn from Pupin how the collaborators were viewed in the twenties: "Our famous physicists,

Michelson and Morley, are a combination of two names better known in the world of physical science today than Castor and Pollux were known when Zeus, descending from the heights of Mount Olympus, sought the companionship of mortal men. The fame of the twins, Michelson and Morley, . . . rests upon an experimental demonstration, the importance of which was not until recently fully appreciated, the demonstration, namely, that there is no ether drift.''[17]

Pupin's lyrical characterization may be taken as an apt estimate of the nominal fame of Michelson and Morley in the post–World War I era. "Castor and Pollux" indeed, but of the two certainly much more is known about Michelson than about Morley. This derives partly from the fact that Morley was primarily a chemist, and thus his work was in a field traditionally less interesting to philosophers than physics. But it is also undeniable that Michelson's impact on science as a whole was greater than Morley's influence.

Edward Morley was, however, an important American scientist and chemist in his own right. He was more than fourteen years Michelson's senior, having been born on January 29, 1838, in Newark, New Jersey. His family was of old-line New England stock, and his father was a Congregational minister. Edward's father directed his elementary and secondary education until, at the age of nineteen, he matriculated at his father's alma mater, Williams College. While an undergraduate and a student of the famous educator Mark Hopkins, Edward's interest in science was stimulated particularly by astronomical studies. Although his health was uncertain, he finished the standard three-year college course with highest honors.

Astronomy, Calvinism, and the Union cause all vied for his primary allegiance immediately upon graduation. But the wishes of his father apparently prevailed, and Edward went directly to Andover Theological Seminary, remaining there as a theological student until ordination in the spring of 1864. Qualms of conscience apparently overcame Morley at that late date, for he joined the United States Sanitary Commission as a war-relief agent and was sent to Fort Monroe, Virginia, where he served for a year until his release in mid-1865. The

---

[17] Michael Pupin, *From Immigrant to Inventor*, p. 354.

assassination of Lincoln and the incarceration of Jefferson Davis at Fort Monroe were accomplished while Morley was still on duty.[18]

For two years following the war, Morley fought with his conscience over his obligation to find a pastorate. Meanwhile he was teaching at an academy near his home and was busy planning a marriage and a household. In 1869, Morley learned that Adelbert College, established in 1826 at Hudson, Ohio, was in need of a college chaplain and a chemistry professor. When offered the double position, Morley accepted with delight this solution to his dilemma, even though it meant that he would have to teach himself chemistry from the start. Nevertheless, his talents with piano and organ, his training for the ministry, and his interest in science could all be utilized. Shortly thereafter he became so enamored of his new subject that teaching and research in chemistry crowded out his preaching duties, and by the time Adelbert College (later to become Western Reserve) moved to Cleveland in 1882, Morley's research interests had begun to make even his teaching duties seem onerous.

In 1865 Morley had published his first scientific paper, a determination of the latitude of the Williams College Observatory. Prior to 1880 Morley had published seven papers. In 1880 and 1881 he published five papers each year on his atomic weight analyses of gases, particularly the composition of air and its oxygen content.[19]

Quantitative volumetric analysis was Morley's forte, and his efforts to refine knowledge of the composition of atmospheric air and of water were directed toward more exact and comprehensive atomic weight determinations. His research in this area began with a paper in 1879, "On the Possible Cause of Variation in the Proportion of Oxygen in the Air."[20] Throughout the next two decades he continued to work on quantitative analyses of these fundamental fluids by analyzing air

---

[18] Howard R. Williams, *Edward Williams Morley: His Influence on Science in America*, pp. 75, 78, and passim. See also Mark Hopkins Centenary Commission pamphlet, *Edward Morley* (Williamstown, Mass.: Williams College, October, 1936).

[19] H. R. Williams, *Morley*, pp. 174–178. See also Frank W. Clarke, "Edward Williams Morley, 1838–1923," *Biographical Memoirs*, National Academy of Sciences, 21, no. 10 (1927), 6 pp., with 2 pp. bibliography by Olin F. Tower, Morley's successor at Western Reserve.

[20] *Am. J. Sci.* 18:168–177 (September, 1879).

samples and water samples from all over the world. His most important
personal papers were published in the decade of the 1890's just after
the Michelson-Morley collaboration was ended. In 1889 Michelson
chose to accept a position at Clark University in Worchester, Massachusetts, where he would have graduate students. Morley did his most
important chemical research after his work with Michelson. It was
made possible by his design and construction of several new instruments, including a precision eudiometer and a differential manometer,
and by the loan of a precision Ruprecht balance from the Smithsonian
Institution. The work began in 1890, as Morley sought to measure
more exactly the combination and proportions of hydrogen and oxygen
in water. His biographer describes this work as follows: "The great research achievement in Morley's life, that upon which his fame as a
chemist so securely rests, was the determination of the densities of
hydrogen and oxygen, the weights of each that combine with the other
to form water, and the volumes of the two gases that combine to form
water. From this data he calculated the relative densities of hydrogen
and oxygen. The importance of his results lies in the fact that the
weight of the oxygen atom is the basis or standard upon which the
relative weights of all the other atoms are founded."[21]

Posthumously, Morley's fame rests with his name linked to Michelson's but it has seldom been realized that Morley's work in collaboration with Miller on the aether-drift experiment was more extensive,
if not more productive, than his collaboration with Michelson. From
the viewpoint of Morley's career, the partnership with Michelson was
more of an interlude than a climax. The two neighboring colleagues
began their active collaboration in 1885 when Michelson decided to
repeat the Fizeau "water-drag" experiment as a preliminary to the repetition in more refined form of his own original aether-drift experiment. The first impetus for the two men to cooperate arose from the
fact that Morley's well-established laboratory was better equipped than
was Michelson's at the Case School of Applied Science (later the Case

---

[21] H. R. Williams, *Morley*, p. 204. Cf. Olin F. Tower, "Edward Williams Morley," *Science* 57:431–434 (April 13, 1923), and also *J. Am. Chem. Soc.* 45:93–98
(January, 1923).

Institute of Technology) for purposes of measuring the velocity of light through water in motion. But more of this later.

Morley's biographer has contended that Morley's seniority not only in age but also in scientific experience—he had produced seventeen research papers to Michelson's eleven by 1883—together with "a most agile creative imagination," "an unexcelled mathematical ability," and greater physiological stamina made Morley "in many ways, the active complement to the more phlegmatic Michelson. They made an ideal research team." However this may be, during the years of their collaboration Michelson was having health problems and domestic difficulties that eventually led to a divorce from his wife in 1897. Both men did go on to accomplish greater work in their respective fields after becoming independent of each other, but it is hardly fair to credit the older man with "greater mental and physical agility" and to censure the ambitious Michelson for "the abrupt way in which he brought the partnership to an end."[22] Both men were largely self-educated scientists, independent in mind and spirit, and each a master of perfectionistic precision in his chosen field. Theirs was a research marriage of convenience, and the professional divorce which came in 1889 after five years of intermittent collaboration was obviously not harmful to either one.

Morley continued to teach and to do research at Western Reserve until his retirement in 1906, when he moved back to New England to live with relatives in Connecticut and Massachusetts. In his declining years, Morley was honored several times for his chemical studies, but he, like Michelson, felt dissatisfied with the aether-drift experiments. They had never been finished for all four seasons of the year, for one thing, and, for another, their presumed precision was beclouded by a failure to have preformed control tests at greater altitudes above sea level.

Morley was honored with the Rumford Medal from London's Royal Society in 1907, the same year that Michelson received his Nobel Prize.

---

[22] H. R. Williams, *Morley*, pp. 200–203. Not only were good physicists in shorter supply in America than good chemists, but Michelson could not have turned down the opportunity to work with *graduate* students in pure physics.

Although other honors were also tendered to him, Morley's major work was later overshadowed by that of such eminent physical chemists as Lord Rayleigh, Sir William Ramsay, and Theodore Richards. Morley's last years in New England were spent in puttering with pewter craftsmanship and in consulting for his alma mater, Williams College. When he died on February 24, 1923, and was buried in Pittsfield, an entirely new phase of the Michelson-Morley and Morley-Miller experiments was in progress.[23]

DAYTON CLARENCE MILLER (1866–1941)

The third member of the trio of M's, whose place is central to the history of aether-drift experimentation, was Dayton C. Miller. After Michelson left Case, his place was taken by Miller, an enthusiastic young teacher and reseacher who also learned to work well with Morley on various projects. In later years, Miller alone was to persevere in the effort to prove that an optical interferometer could show the absolute motion of the earth through the aether.

Miller came upon the aether-drift scene when he accepted a teaching position at Case School in Cleveland in 1890, only three years after the performance of the "classic" experiments of 1887. Michelson's resignation in 1889 and his move immediately to the newly established graduate school at Clark University, in Worcester, Massachusetts, left a hard-to-fill vacancy in the faculty at Case. This loss had been aggravated by a serious loss in the physical plant as well, because the physics laboratory as well as most of the college was destroyed by fire in 1886. The equipment Michelson had taken pains to procure in Europe was not easily replaced. Into this double breach stepped Miller in 1890. In so doing he also stepped into intimate association with Professor Morley and with the unsolved aether-drift problem.[24]

Miller was born on a modest farm in southern Ohio on March 13, 1866. He grew up with a variety of interests and skills appropriate to

[23] Some information on Morley's later life is available in the "Williamsiana Collection" of Williams College Library, Williamstown, Massachusetts, but most primary materials are deposited in four boxes of "Morley Papers" in the Manuscript Division of the Library of Congress.

[24] Harvey Fletcher, "Dayton Clarence Miller, 1866–1941," *Biographical Memoirs*, National Academy of Sciences, 23:61–74. See esp. p. 63.

a rather cultured and prospering rural family. He had shown mechanical aptitude and musical tastes from childhood, although most portentous was his early interest in science. Astronomy particularly fascinated him; he had been immensely impressed in his early youth by a visit to the shops of the famous Pittsburgh optical technician and instrument maker, John A. Brashear.[25] A liberal arts education at Baldwin-Wallace College in Berea, Ohio, from which he graduated in 1886, allowed him to test his various talents. But it took a year in the business world to convince Miller that his true vocation was science. He enrolled for graduate study at Princeton and spent a valuable apprenticeship there under Charles A. Young, who first set him to work with an interferometer. His task was to measure the wavelength of sodium. From this experience his course veered toward astrophysics, and within two years he had obtained his Ph.D. with a dissertation on comet behavior.[26]

Almost by accident the young bachelor scientist visited Cleveland in 1890 while awaiting the completion of his newest research apparatus. There he learned of the straitened circumstances of the Case School physics department, and, when he was offered a job teaching elementary courses in mathematics, physics, and astronomy, he accepted the position as a temporary expedient. But he liked teaching, especially the physics courses, so well that he remained at Case for the next half century.

Three years after his arrival at Case, the young professor felt securely enough established to get married. He and his bride took an apartment in the same boardinghouse where lived the childless Morleys. Thereafter until Professor Morley's retirement in 1906 the two couples, in spite of their difference in age, became deeply intimate friends, sharing the same roof, the same table, and the same pleasures of companionship, primarily music and travel to scientific conventions.

Professor Morley was, in the early 1890's, at the height of his career

[25] H. W. Mountcastle, "Dayton Clarence Miller," *Science* 93:270–272 (March 21, 1941). See also J. A. Brashear, *The Autobiography of a Man Who Loved the Stars*, ed. W. L. Scaife; and Harriet A. Gaul and Ruby Eiseman, *John Alfred Brashear: Scientist and Humanitarian, 1840–1920.*

[26] Robert S. Shankland, "Dayton Clarence Miller: Physics across Fifty Years," *Am. J. Phys.* 9:273–283 (October, 1941). See also G. Walter Stewart, "Dayton Clarence Miller," *J. Acous. Soc. of Am.* 12:477–480 (April, 1941).

and productivity as a chemist. Nevertheless, Miller was a welcome addition to his circle of friends. Morley seems to have felt quite acutely the loss of his former partner, Michelson, and he needed a kindred spirit with whom he could commiserate over philistine administrative policies and share the intellectual adventures of his day-by-day research. Miller provided for these needs and benefited in return by finding a sympathetic ear for his own troubles and joys. In colleges without graduate students, both professors found stimulation among colleagues.

In 1896, when Roentgen announced his curious discovery of an even more curious form of radiation, which he christened "X-ray," Dayton C. Miller was primed and ready to take immediate experimental advantage of the announcement. The same day that he read Roentgen's paper, Miller set up his Crookes vacuum tubes, followed the published directions, and took an X-ray picture of his wife's hand. This was one of many claims to be the first X-ray picture taken in the United States, but Miller's article "Roentgen Ray Experiments," although published immediately, was lost in the avalanche of scientific excitement and publication following closely on Roentgen's paper.[27]

A year later Morley and Miller began their active professional collaboration with a paper "On the Coefficient of Expansion of Certain Gases."[28] Both chemist and physicist found full exercise of their talents in working together, but Morley was perhaps the chief beneficiary. Two recent events had had adverse effects on the elder scientist's morale, even though two other events marked the height of his fame: the American Association for the Advancement of Science had just honored him with election to its presidency and in 1896 his magnum opus had been published by the Smithsonian Institution.[29] The first blow to his pride had come with the announcement of the discovery of the first of the inert gases, argon, by William Ramsay in 1894. Morley seemed quite depressed with the thought that his own atmospheric-gas researches had missed this chance of a fundamental

[27] Shankland, "Miller," *Am. J. Phys.* 9:276.

[28] *Proc. AAAS* 46:123 (Abstract), as cited by Williams, in *Morley*, p. 278.

[29] Edward W. Morley, "On the Densities of Oxygen and Hydrogen and the Ratio of Their Atomic Weights," *Smithsonian Institution Contributions to Knowledge*, no. 980 [pamphlet], as cited by Williams, in *Morley*, p. 276.

discovery. Second, while he was on sabbatical leave in 1895–1896, Morley's personal laboratory in Adelbert Hall had been completely dismantled without his knowledge or consent. This was a serious disruption of his work in progress and a cause for special resentment toward the administration.[30]

By the turn of the century Morley was sixty-two and Miller only thirty-four, but their mutual aid and companionship were firmly established. Each had been suggested by Michelson for the honor of heading the newly authorized National Bureau of Standards, but each declined.[31] They had collaborated on three different sets of experiments and had written their joint reports while simultaneously carrying on their individual projects. Miller was by no means intellectually dependent on Morley, but he did admire and respect the older man as a mentor.

Miller himself, like Michelson, was short, but quite handsome, and he was fastidious about his appearance. In the words of one of his students, he was "nervous, quick, and kindly, a good teacher," and possessed an "amazing manual dexterity."[32] He took great pride in his lecture demonstrations and could set up a difficult exhibit of interference effects in minutes, using only his hands and a caliper to adjust his apparatus. This rare manipulative skill, combined with his perfectionistic standards, gave Miller qualifications to serve experimental physics alongside Michelson and Morley. No one of the three was talented at physical theory or in mathematics, but all were gifted at making precision measurements. Miller, at Case, had the most leisure time to indulge his interests, because he had no graduate research to supervise and only a handful of physics majors each year.

Music had always been an important avocation to Miller, and after 1906 it became in large part his vocation as well. He was an expert flutist, and collecting flutes and player pianos became his hobby. More

---

[30] Williams, *Morley*, pp. 217–218 and 247. See also William Ramsay, *The Gases of the Atmosphere: the History of their Discoveries*, 3rd ed., (1905).

[31] Shankland, "Miller," p. 276. See also Rexmond G. Cochrane, *Measures for Progress: A History of the National Bureau of Standards*, pp. 1–50.

[32] Alfred B. Focke, personal interview with the author, May 30, 1961. Edwin C. Kemble, another of Miller's students between 1907 and 1911, remembers Miller as a "cultured gentleman" in every respect (personal interview, May 17, 1968).

important for science, he became expert in acoustics and began to investigate not only the mechanisms of flutes and automatic pianos but also the science of musical sounds.[33] In so doing he invented a kind of mechanical prototype of the modern oscilloscope (which he called a *phonodeik*) to show visual representations of sound waves.[34]

Preoccupied as he was with sound waves and their uses from 1906 through 1920, Miller felt that the analogies of acoustics with the wave theory of light, so dominant in the nineteenth century, should be reinforced. He believed the acoustic analogy still viable in the third decade of the twentieth century. The ambiguous position of optical theory in 1920, as yet only halfway accommodated by quantum theory, meant that, whether he was aware of it or not, when Miller undertook the refinement of the aether-drift tests once again, he was engaging in a risk calculated to restore the purity of unadulterated wave theory—but not without reasonable hope of success. The tragedy of his professional life seems now to have been that his great hope outran and overrode the evidential support for his rational theory. A modern scientist-philosopher phrases Miller's commitment this way: "The triumph of the Michelson-Morley experiment, despite its giving the wrong result, the tragic sacrifice of D. C. Miller's professional life to the pursuit of purely empirical tests of a great theoretical vision, are sardonic comments on the supposed supremacy of experiment over theory."[35]

With these brief background biographies we are now in a better position to understand the historical development of the Michelson-Morley hypotheses and the tests thereof. In many ways the interwoven lives of these three men gave them a kind of professional camaraderie

[33] His Lowell Lectures of 1915 were later published under the title *The Science of Musical Sounds*; see also Dayton C. Miller, "The Dayton C. Miller Collection relating to the Flute: Catalogue of Books and Literary Material relating to the Flute and other Musical Instruments—with Annotations."

[34] Shankland, "Miller," p. 277. E. C. Kemble's senior thesis in 1911 was an attempt to provide a mathematical exploration for hysteresis in the *phonodeik*.

[35] Michael Polanyi, *Personal Knowledge: Towards a Post-Critical Philosophy*, p. 167. Cf. the guilt-by-association judgment of Miller in Martin Gardner's *Fads and Fallacies in the Name of Science*, pp. 84–85. For a more sympathetic judgment, see Robert B. Lindsay and Henry Margenau, *Foundations of Physics*, pp. 326, 354, 377.

often lacking to others caught in the fractionating process of super-specialization for scientific research. The burden of accumulated knowledge in science had by the mid-nineteenth century made necessary an ever more minute division of labor. New institutions for research purposes were required. As we return now to Michelson, the chief protagonist of this story, we will find him, in the year 1880, a naval officer. This profession traditionally trains a man to be a jack-of-all-trades, but Albert A. Michelson had a magnificent obsession to become the master of one, namely, the science of optics, specifically the art of interferometry.

CHAPTER THREE

MICHELSON AND INTERFEROMETRY, 1880–1883

The initial hypothesis of the Michelson aether-drift experiment was simply stated at the beginning of the first paper Michelson published on the subject:

The undulatory theory of light assumes the existence of a medium called the ether, whose vibrations produce the phenomena of heat and light, and which is supposed to fill all space. According to Fresnel, the ether, which is enclosed in optical media, partakes of the motion of these media, to an extent depending on their indices of refraction. For air, this motion would be but a small fraction of that of the air itself and will be neglected.

Assuming then that the ether is at rest, the earth moving through it, the time required for light to pass from one point to another on the earth's surface, would depend on the direction in which it travels.[1]

[1] Albert A. Michelson, "The Relative Motion of the Earth and the Luminiferous Ether," *Am. J. Sci.*, 3d ser. 22:120 (August, 1881). See my Appendix A for the complete paper.

Following some simple calculations in which he showed the mathematical feasibility of measuring the speed of two pencils of light traveling at right angles to each other, Michelson made his grand proposal: "We could find v the velocity of the earth's motion through the ether."[2] In stark simplicity, Michelson proposed to find a speedometer, or more accurately a velocity-meter, for the earth.

Curiously, there has been considerable confusion over the question as to exactly what Michelson was setting out to prove. Because his report was written, of course, after his hypothesis had been tried and found wanting, the tentative nature of the first sentence in that report has been used as the basis for the contention that Michelson sought to prove or disprove the existence of an aether.

Other questions have also been raised in relation to Michelson's purposes: Was he concerned with testing the velocity of light moving in different directions? Did he set out from the beginning to test for absolute motion or for the relative motion of the earth and the luminiferous aether? Was it legitimate to assume that the velocity of the earth's motion relative to a stationary aether is the same as the velocity of the earth with respect to the sun? These and other questions might never have been raised had Michelson's results been positive. But the null results forced a reconsideration of the fundamental assumptions on which the experiment was based. Consequently, latter-day interpreters have suggested many such alternatives as possible restatements of the original puzzle.

In later life Michelson repeatedly reasserted that his "original" purpose was to measure optically the velocity of the earth through the solar system, but this was more true for the 1887 experiment than for the experiment of 1881. Originally, he assumed an omnipervasive aether, fixed presumably in interstellar as well as interplanetary space around our sun, which, as the carrier of light waves, might reveal the long-sought distinction between absolute and relative motion of the earth. If there were a stationary all-pervading aether, it should provide a "cosmic" standard against which the measurement of a relative

[2] Michelson, "Relative Motion," p. 120. N.B. The quest here stated is for the totality of component motions with respect to the aether, and emphatically *not* simply for the earth's orbital velocity; see also pp. 124–125.

aether drift could be made to divulge the reciprocal vector of the resultant of all the earth's various motions through space.

When A. A. Michelson, Master, USN, had obtained a leave of absence from active duty and boarded ship with his wife, Margaret, in the summer of 1880, to obtain his postgraduate education in Europe, scientific prospects were bright that all optical processes could eventually be explained in terms of mechanical, or at least hydrodynamical, models. Although a few mathematicians and physicists still had their doubts,[3] the success of undulatory theory in explaining interference phenomena was so conspicuous as to override all objections. The assumption of the existence of a medium called the aether was in no wise problematic in 1880; Michelson might well have remarked upon his arrival in England that year, as Voltaire had done before him in 1727, that at the beginning of the century the universe was as empty as an exhausted receiver; now it had filled up again.[4]

In 1874 an honored American mathematician, Professor Joseph Lovering, had expressed the general consensus in these words: "At the present moment, we find the luminiferous ether in quiet and undivided possession of the field from which the grosser material of ancient systems has been banished. The plenum reigns everywhere; the vacuum is nowhere." The real question was not, Is there an aether? but, said Lovering, "How is the ether affected by the gross matter which it invests and permeates? Does it move when they move? If not, does the relative motion between the ether and other matter change the length of the undulation or the time of oscillation? These queries cannot be satisfactorily answered by analogy, for analogy is in some respects wanting between the ether and any other substance."[5]

James Clerk Maxwell had given the most authoritative sanction for

[3] For example, Joseph Lovering, "The Mathematical and Philosophical State of the Physical Sciences," *Am. J. Sci.*, 3d ser. 8:297 (October, 1874); Ernest H. Cook, "The Existence of the Luminiferous Ether," *Phil. Mag.*, 5th ser. 7:225–239 (April, 1879).

[4] Paraphrase by Lovering, in "Mathematical and Philosophical State of the Physical Sciences," p. 299. Voltaire had contrasted France and England, and Lovering was taking the same liberty with the nineteenth century.

[5] Ibid., pp. 297, 301. For an attempt to treat the aether as if it conformed to the kinetic theory of gases, see De Volson Wood, *The Luminiferous Aether* (New York, 1886), reprinted from *Phil. Mag.*, 5th ser. 20:389–417 (November, 1885).

preference to the aether over all objections to its hypothetical and para-doxical character. His article on "Ether," first published in 1878, for the ninth edition of the *Encyclopaedia Britannica*, concluded with the sentence: "Whatever difficulties we may have in forming a consistent idea of the constitution of the ether, there can be no doubt that the interplanetary and interstellar spaces are not empty, but are occupied by a material substance or body, which is certainly the largest, and probably the most uniform body of which we have any knowledge."[6] One of the major sections of this article deals with the problem of the relative motion of the aether. It indicates very clearly Maxwell's long-time concern with astronomical aberration and the effect of the motion of the medium on the velocity of light. Even more significant, perhaps, is the fact that in this article Maxwell made explicit the basic notion which Michelson now began to nurture, namely, that the relative aether wind might possibly be utilized to give us our absolute velocity through space.

If it were possible to determine the velocity of light by observing the time it takes to travel between one station and another on the earth's surface, we might, by comparing the observed velocities in opposite directions, deter-mine the velocity of the ether with respect to these terrestrial stations. All methods, however, by which it is practicable to determine the velocity of light from terrestrial experiments depend on the measurement of the time required for the double journey from one station to the other and back again, and the increase of this time on account of a relative velocity of the ether equal to that of the earth in its orbit would be only about one hundred millionth part of the whole time of transmission, and would therefore be quite insensible.[7]

[6] James Clerk Maxwell, "Ether," *Encyclopaedia Britannica*, 9th ed., VIII, 572 (1893).

[7] Ibid., p. 570. Alfred M. Bork has recently found manuscript evidence in the Maxwell papers at Cambridge University and elsewhere to show that Maxwell himself as early as 1864 had tried to perform an "Experiment to determine whether the Motion of the Earth influences the Refraction of Light." This effort, made on April 23, 1864, and submitted as an article for the *Proceedings of the Royal Society*, was never published, apparently because the editor, G. G. Stokes, convinced Max-well that his argument was unsound. See Maxwell to W. Huggins, June 10, 1867, published in Huggins memoir in the *Transactions of the Royal Society* (1896), pp. 529–564.

By this suggestion, skeptical as it was, Maxwell shifted the problem of the relative motion of earth and aether from the realm of *what* to measure to the realm of *how* to measure. Hence, it was tacitly assumed, among those who could appreciate Maxwell's reasoning, that the luminiferous aether might provide the framework within which to measure absolute motion through space, if only there could be found a way to fulfill this will.[8]

The problem of how to measure the velocity of the earth would not have been technologically feasible before the accomplishment of the terrestrial measurement of the velocity of light. In order to evaluate the novelty of Michelson's innovation, it is necessary to study in somewhat greater detail the history of optical technology, preeminently the instruments used for measuring the speed of light on earth. Only in this way can we properly appreciate the role of the precision apparatus which Michelson devised.

It is often said that the science of thermodynamics owes more to the technology of steam than the steam engine owes to the science of thermodynamics. Likewise, in optics, the case may be made that invention is as much the mother of necessity as necessity is the mother of invention. The instrumental side of the relationship between science and technology was especially important in optics during the nineteenth century, because of the development of photography, diffraction gratings, refractometers, the radiometer, and the spectroscope. Spectroscopy and astrophysics, as autonomous sciences in their own right, were not only made possible but also made inevitable by the development of these fecund instruments. Modern interferometry, also, has an indispensable instrumental basis.[9]

---

[8] Maxwell encouraged the search for some such method by concluding this section with the words: "The whole question of the state of the luminiferous medium near the earth, and its connexion with gross matter, is very far as yet from being settled by experiment," "Ether," p. 571.

[9] Spectrum analysis came to require its own theory after Bunsen and Kirchoff in 1861 interpreted Fraunhofer's dark-bands as absorption lines. This necessitated a theory cross-referenced between optics and the mechanics of the atom. See Max Born and Emil Wolf, *Principles of Optics: Electromagnetic Theory of Propagation, Interference and Diffraction of Light*, p. xxiv. Similarly, interferometry gradually required its own theory. See S. Tolansky, *An Introduction to Interferometry*, esp. pp.

Our concern here will be with the device which eventually came to be known as an "interferometer." Michelson did not christen his apparatus with this name until about 1890, when the instrument began to be produced commercially as a standard piece of optical laboratory apparatus. At first he called his innovation an "interferential refractometer," a name in some ways inappropriate but showing its debts to earlier instruments which operated on similar principles.[10] His instrument was essentially a combination of two experimental devices, one to demonstrate the principle of interference, the other to split a beam of light by refraction and reflection into two pencils of equal intensity.

It will be recalled that Grimaldi in the seventeenth century had used a single screen with a pinhole, or double screens with very small slits, to investigate the patterns formed by interfering light waves on another observation screen. Newton himself, using only lenses, prisms, and optical flat plates, had observed and explained provisionally the curious phenomena called variously the colors of "thin plates," of "thin films," or of "Newton's rings." Grimaldi's skill and Newton's ingenuity were combined in the person of Thomas Young, whose demonstration that the use of the principle of interference could admirably explain the phenomenon of Newton's rings constituted a major theoretical, but not a technological, advance for optics.

Remarkable as were the experiments of Young and his contemporary mathematical apostle, Fresnel, their accomplishments were made with only slight modifications and improvements over those of Grimaldi and Newton. The major innovation that Young introduced was simply the addition of another, complementary slit screen to Grimaldi's arrangement. By using double pinhole screens, one after the other, two "point sources" of light could more easily be approximated, thus allowing control over the effective size of the light source and the focus of

---

84–103; C. Candler *Modern Interferometers*, pp. 108–109 and 485–488; W. H. Steel, *Interferometry*, pp. 1, 8, 75–108.

[10]An acoustic "interferometer" had been built by M. Koenig in 1874. This interference apparatus, using two U-shaped tubes to study sound "beats," is described and pictured in Edward Henry Knight, *New Mechanical Dictionary: A Description of Tools, Instruments, Machines, Processes and Engineering* (Boston, 1882), p. 582.

the fringes produced thereby. Similarly, Fresnel, in the tradition of Newton, played with mirrors and prisms in various combinations. Although there were several notable exceptions, optical experimentation until the decade of the 1850's was confined by the lack of precision apparatus. Students of the physical behavior of light were still using essentially the "hand-tool" optical equipment of Newton.[11]

About mid-century, along with the terrestrial measurement of the velocity of light, a remarkable surge in the perfection of all types of instrumentation in the sciences began to appear. Precision and accuracy of measurement, as well as a new appreciation for the necessity of correcting raw observational data for known sources of error, began to assume a fundamental place in the attention of all scientists. The industrial revolution in the machine-tool industry made these new concerns necessary by making new abilities possible.[12]

One of the most significant advances in optical instrumentation came with the development of apparatus for determining the velocity of light on earth. Although astronomers had capitalized on Roemer and Bradley by calculating this fundamental value in a number of different ways, using celestial observations, no one had ever been able to carry out a terrestrial, that is, experimental, test of the speed of light until 1849. The possibility of obtaining a precisely machined gear wheel was a necessary prerequisite to Fizeau's design for an experiment of the sort Galileo had once envisioned.[13]

Fizeau had wealth enough to provide leisure for his physical research and to purchase the expensive components of the apparatus he required. His method for measuring the velocity of light required two astronomical telescope objectives, a carbon arc-light source, an accurately machined gear wheel with 720 teeth, two nearly perfect mirrors, as-

[11] Billet's split lens, Humphrey Lloyd's mirror, and Fresnel's biprism and double mirror are often cited as distinct improvements in interferometric apparatus, but these, too, are essentially "hand-tools." See Arthur C. Hardy and Fred H. Perrin, *The Principles of Optics*, pp. 568–586; Herbert J. Cooper, *Scientific Instruments*, p. 72.

[12] See Joseph W. Roe, *English and American Tool Builders*, on Henry Maudslay and Joseph Whitworth's influence, see 98, and on the introduction of such essential precision tools as the vernier and micrometer calipers, p. 201.

[13] See Robert S. Woodbury, *History of the Gear-Cutting Machine*. See also "Renaissance Roots of Yankee Ingenuity," in Derek J. de S. Price, *Science since Babylon*, pp. 64–66.

sorted lenses of near perfection, a power source, and a tachometer. These were the major parts for an optical system representing a considerable advance in experimental design over any previous apparatus.[14]

Fizeau had designed an experiment which incorporated a mechanical occulting device with a semitransparent mirror in such a way as to send an interrupted beam of light over a five-mile course and back to an observer behind the half-silvered mirror. The greatest wonder over this experiment was inspired neither by the apparatus nor by the design, but rather by the mere fact that any apparatus whatever could furnish the means for measuring the then almost incredible speed of light. Fizeau's value for V, the mid-century symbol for the velocity of light, was somewhat high (ca. 194,000 miles per second), but the present value of c, the twentieth-century symbol for the velocity of light, thanks in part to Michelson's precision measurements, is generally given as $299,773 \pm 1$ km/sec.

In addition to the importance of Fizeau's apparatus as a landmark in the complexity of optical systems, there is one simple feature of its design which is of particular significance as a precursor of one of the most novel aspects of Michelson's interferometer. This is the semitransparent mirror arrangement or amplitude divider whereby the light source and the observer's eyepiece make concurrent use of a single diagonal glass for reflected transmission of the beam on its outward journey and for refracted reception of it on its return.[15]

Everyone is familiar with the windowpane, brightly illumined from our side, which acts as a mirror as well as a transparent glass. Fizeau refined this double property of plane-parallel plate glass by lightly silvering the front surface and utilizing only the half of the beam reflected from that surface. A decade later Fizeau used the same feature in his more complex interference apparatus designed to investigate the influence of the motion of the medium on the speed of light.[16]

---

[14] For a secondary account see Ernst Mach, *The Principles of Physical Optics: An Historical and Philosophical Treatment*, trans. J. S. Anderson and A. F. A. Young (in 1926), pp. 25, 164–178.

[15] See the diagram of this apparatus, ibid., p. 25.

[16] See the diagram of this apparatus, ibid., p. 178.

Meanwhile, in 1856, Jules C. Jamin had designed the first "interferential refractometer," using two diagonal mirror-refractors opaquely silvered on the back surface. In 1872 Eleuthère Mascart used this device and improved it to investigate relative motions of source and observers. Michelson's essential innovations, as we shall see, were to use both halves of the beam split by Fizeau's semitransparent mirror and to half-silver the back surface of the refracting mirror as Jamin did.[17]

The name "interferential refractometer" was retained by Michelson, because of the similarity of his instrumental design to that of Jamin, and because of the similarity of use, as at first conceived. Jamin used his apparatus to measure the indexes of refraction of air and water at different pressures and temperatures. Michelson saw a possible way to measure, in effect, the "absolute" index of refraction of the aether. If atmospheric air could, as he proposed, be "neglected," then the ratio of light speed in a vacuum (a theoretical value) to its speed in the aether (a hypothetical medium) ought to disclose the earth's velocity.

Here we need to clarify what precisely were Michelson's fundamental expectations of what should happen when he set up and carried through his experiment. What were the major theoretical and hypothetical links between the direct measurement of a shift in interference fringes and the indirect supposition that such a "fringe shift" *should* be direct evidence for an aether drift and therefore for the velocity of the earth's motion through the aether? These links between theory and experiment form part of the problem of measurement in modern physical science and partake of the thorny question of the distinctions to be made between quantitative and qualitative analyses. A major reason for the fame of the Michelson-Morley experiment is the fact that it has

[17] J. C. Jamin, "Description d'un nouvel appareil de recherches, fondé sur les interférences," *Comptes Rendus* 42:482–485 (1856). For a secondary description and diagram, see Born and Wolf, *Principles of Optics*, pp. 308–309. J. C. Jamin, "Mémoire sur la mesure des indices de réfraction des gaz," and "Mémoire sur les variations de l'indice de réfraction de l'eau à diverses pressions," in *Ann. de Chim. et de Phys.*, 3d ser. 49:282–303 (1857) and 3d ser. 52:162–171 (1858). See also E. E. N. Mascart, "Sur les modifications qu'éprouve la Lumière par suite du mouvement de la source lumineuse et du mouvement de l'observateur," *Annales Scientifiques de l'École Normale Supérieure*, 2d ser. 1:157–214 (1872) [Paris], and 2d ser. 4:363–420 (1874).

provided a test case for innumerable discussions in the philosophy of science as to the legitimacy of theoretical implications expected and derived from experiment.[18]

Michelson's expectation of being able to measure the aether drift, by noting how many black-banded shadows or fringes passed by the bench mark in his field of view as he rotated the apparatus through ninety degrees, was based on the laboratory experiences of a host of his predecessors and co-workers in interferometry. Fresnel, Fizeau, and Foucault of the French school of optics in particular had paved the way by working out the technical details of what should be expected from diffraction or interference fringe shifts. But it should be noted that neither Fizeau's cogwheel method for measuring the speed of light, nor Foucault's rotating-mirror method, which formed the basis of Michelson's first attempt at refining the value of V (now c) in 1878–1879, used interference fringes for their measurements. Rather, in the first case, the observer had to judge the maximum intensity of the returning light beam, whereas in the second case the observer had only to note the angular displacement of an image formed on a scale from rays reflected by the high-speed rotating mirror. Since Michelson's previous optical experience was largely limited to the latter method, in 1880 he had to learn the techniques and conditions for producing interference almost from the beginning.

Throughout Michelson's lifetime the controlled interference of light was difficult to produce even in a laboratory. Several precautionary measures were required to control the extraneous factors which prevent our experiencing interference in everyday life. As one of several manifestations of the phenomenon of diffraction, which came so late to be recognized as a basic property of light, the interference of light waves, to produce visible bands or fringes of greater and lesser intensity, required that the demonstrator be knowledgeable, skillful, and experienced. He had to learn to manipulate light in order to carry on the

---

[18] An interesting discussion of the general problem of measurement in physics from a historical point of view is Thomas S. Kuhn's "The Function of Measurement in Modern Physical Science," *Isis* 52:161–193 (June, 1961); also available in Harry Woolf, ed., *Quantification: A History of the Meaning of Measurement in the Natural and Social Sciences*, pp. 31–55.

dialogue of the science of interferometry. Indeed, while the craft was young, some argued that interferometry was less a dialogue with nature than a soliloquy on nature, and that the discipline of interferometry was less a science than an art.[19]

Michelson had to relearn what François Arago had long ago specified, that four conditions were absolutely necessary to the success of all efforts to produce interference fringes. These he described in 1927 as follows:

1. The first and most important of these is that the two interfering pencils must have a constant phase relation. . . . This can only be realized in the case of light-vibrations if the two pencils originate in the same source. Thus it is altogether impossible to observe interference with two candles as sources; for the vibrations of the individual electrons, being practically independent, give a resulting illumination which amounts to an integration due to wave-trains whose phase, amplitude, and orientation vary many millions of times per second. This condition is therefore indispensable. . . .

2. If the source is not homogeneous, that is, if it is made up of a mixture of colors (as in the case of white light), and hence of different wave-lengths, . . . it will be important to make the two paths as nearly equal as possible, unless the resulting light be examined spectroscopically. . . .

3. A third condition is that the direction of the two pencils should be nearly the same; otherwise, the interference fringes will be too narrow to distinguish.[20]

Fourth, and finally, in nearly all apparatus for producing interference, prior to the introduction of Jamin's interferential refractometer, it was necessary that the source should "be of very small dimensions (pinhole or narrow slit), otherwise the different systems of fringes are not superimposed in the same location, thus masking all evidence of interference."[21]

The slightest violation of any of these conditions for producing interference could easily embarrass the demonstrator and provoke his

[19] "A scientist once said facetiously that it [Michelson's interferometer] is a wonderful instrument if operated by Michelson" (Charles G. Fraser, *Half-Hours with Great Scientists: The Story of Physics*, p. 293).

[20] A. A. Michelson, *Studies in Optics*, pp. 17–18.

[21] Ibid., pp. 18–19. For a description of the division of amplitude principle, see W. Ewart Williams, *Applications of Interferometry*, pp. 46–57.

wrath at the recalcitrance of his apparatus. Today, more sophisticated instruments and theories have obviated these requirements to a considerable degree.[22]

Popular accounts of the Michelson-Morley experiment very often begin with the analogy of two rowers, swimmers, or boats engaged in a race on a river with a strong current. Both contestants have equal speeds and equal distances to cover, but one must cross the river and come back while the other must go down- and upstream an equal distance. The effort to "explain" the original plan of Michelson's experiment by these examples familiar in most men's experience encounters several difficulties. It is usually presumed obvious that the trip across the stream and back will take less time than the trip of the same distance with and against the current. Although the mathematics necessary to prove this result is simple, the "realistic" values usually chosen for the current speed and the racers' speed more often than not obscure the problem. Then, too, the impression is often left that this analogy illustrates straightforwardly the problem of measuring a second-order effect. These are two serious faults.

In the first place, unless one sees what happens in the limiting case, namely, when the current speed equals the racers' speed, one seldom can intuit the import of the racers' speed and direction relative to the moving current. It is only necessary to postulate the speed of the racer and the speed of the current as being of equal value in order to show that the cross-stream swimmer can at least succeed in negotiating a passage, whereas the downstream swimmer can never get back to his starting point.[23] The limiting case modification to this analogy considerably reduces its clumsiness; however, it is valid only as a first approximation and as a likeness of what an aether-drift test might have

[22] For example, the fact that highly monochromatic and spatially coherent light beams were not available, until the development of the optical maser after 1958, has been a largely unacknowledged but truly serious historical limitation on optical experimentation. See Arthur L. Schawlow, "Optical Maser," *Sci. Am.* 204:52–61 (June, 1961).

[23] This is admirably pointed out by Harvey B. Lemon in *From Galileo to the Nuclear Age: An Introduction to Physics*, p. 411 n. Historians will find Lemon's treatment of Michelson's experiment highly anachronistic in some respects, however. See esp. pp. 410–411.

sought *before* Michelson ever became interested in the problem. Maxwell had worked out the implications of a first-order measurement and had despaired of solving the problem before anticipating the possibility of a second-order measurement.

As for the second objection to this analogy, it is usually assumed that the round trip itself represents a second-order effect. Apparently Michelson himself assumed this when he first performed the experiment in 1881, but he was corrected by colleagues, one of whom, H. A. Lorentz, pointed out that the truly critical consideration for the second-order effect is not simply the round trip but the overall path length.[24] First-order effects are simple ratios which may be expressed as fractions with terms in the first power. The unique individuality of the Michelson aether-drift experiment, and, indeed, the single most important reason for its first performance, lies in the fact that it was conceived as capable of detecting a second-order effect.[25]

This means mathematically that the ratio being tested, a fraction with squared terms in both numerator and denominator, was an exceedingly small quantity. The numerator represented the square of the earth's orbital velocity through space (thought to average about 18.5 miles per second); the denominator represented the square of the velocity of light (then thought to be about 185,000 miles per second).[26] Reducing these figures to their lowest proportion, we would have a first-order effect of 1/10,000; squaring each of these terms for the second-order effect, we would have 1/100,000,000.

Given figures such as these, much more difficult to imagine then than now, it is no wonder that Maxwell, in conceiving this problem, thought terrestrial experiments hopeless to solve it. Michelson's fresh triumph with his velocity-of-light determination made him a likely candidate to take up Maxwell's challenge. In addition to his youth,

---

[24] H. A. Lorentz, *The Theory of Electrons and Its Applications to the Phenomena of Light and Radiant Heat*, p. 195.

[25] Michelson, "Relative Motion," *Am. J. Sci.*, 3d ser. 22:121.

[26] E. Atkinson, ed. and trans., *Elementary Treatise on Physics, Experimental, and Applied*, p. 502, from *Ganot's Éléments de Physique*; see also Amédée Guillemin, *The Heavens: An Illustrated Handbook of Popular Astronomy*, ed. J. N. Lockyer, R. A. Proctor, p. 15.

Michelson had the mechanical aptitude, the experimental intuition, and now in 1880 he was going to Europe to study directly under the greatest minds in optics. All this contributed to Michelson's receptivity for this problem.

We have no record of how Michelson *first* learned of Maxwell's suggestion, but we do know that he could have learned about it through at least three channels. First, as stated at the beginning of this chapter, he might have read the ninth edition of the *Encyclopaedia Britannica*, which began to appear in 1875. Second, he might have been introduced to Maxwell's calculations on second-order effects more directly through one of Maxwell's last letters, written just before he died, to the director of the *American Ephemeris*, Professor D. P. Todd.[27]

In this letter Maxwell suggested that Jupiter's satellites, if very carefully observed and timed, might show whether the velocity of light varied in different directions owing to the motion of the earth through space. It is quite likely that such a letter from a most distinguished British physicist would have circulated among the professional associates at the U.S. Naval Observatory in Washington. The staff of the *Nautical Almanac*, Professor Newcomb, and his associate Michelson, were very likely on the routing slip for this letter. But, regardless of these suppositions, a third channel is indubitable, because Michelson cited the same letter (as published in *Nature*) as the seed for his experimental design of the first aether-drift apparatus.[28]

Michelson arrived in Europe in 1880 for the fall semester at the Collège de France. His exact itinerary is not known, but he probably went first to England and then on to Paris. In later years Michelson recorded that it was in Paris in 1880 that his idea for the interferometer first came to him. His purpose, he said later, was to measure the velocity of "the Earth through the Solar System," but in 1880 his conception was phrased in terms of the luminiferous aether of space.

Alfred Cornu, under whom he first studied, may have introduced

---

[27] For the background on the discovery of this letter, see [Rollo Appleyard], "Clerk Maxwell and the Michelson Experiment," *Nature* 125:566–567 (April 12, 1930).

[28] Michelson, "Relative Motion," *Am. J. Sci.*, 3d ser. 22:120.

him to Fizeau. Six years earlier Cornu had repeated and refined Fizeau's cogwheel method for measuring the velocity of light.[29] The experimental apparatuses of Fizeau and of Cornu were virtually the same, incorporating that critical feature of any interferometer, the semi-transparent mirror. With Maxwell's problem in mind, and with Fizeau's or Cornu's apparatus before him, Michelson's idea for measuring the aether drift may well have been born. He would simply have used both halves of the split beam of light, at right angles to each other, to produce interference fringe shifts as a result of their different orientations to the direction of the earth's various motions through the aether.

Michelson's interferometer, conceived in Paris in the early fall of 1880, was developed in Berlin, where Michelson moved shortly thereafter. On November 22 of that year Michelson wrote to his former superior officer, Simon Newcomb, as follows: "I had quite a long conversation with Dr. Helmholtz concerning my proposed method for finding the motion of the earth relative to the ether, and he said he could see no objection to it, except the difficulty of keeping a constant temperature."[30] This sentence is a highly significant fragment from the historical remains of the Michelson experiment. Not only is it the earliest indication of Michelson's proposal extant, but also it contains a caveat about the principal difficulty encountered with aether-drift tests throughout the next half century. Helmholtz warned Michelson before he began that temperature changes could nullify any results he might find. Long afterward Dayton C. Miller was to have his slight positive results for aether drift invalidated on these same grounds.[31]

During the academic year 1880–1881, Michelson traveled around

[29] A. Cornu, "Détermination de la vitesse de la lumière et de la parallaxe du soleil," *Comptes Rendus* 79:1361–1365 (December 14, 1874).

[30] A. A. Michelson to Simon Newcomb, November 22, 1880, from Berlin, in Miscellaneous Letters Received, vol. 3, Records of the Naval Observatory, National Archives, Record Group No. 78. Hereafter records in the National Archives are indicated by the symbol NA, followed by the record group (RG) number. For the full letter, see Nathan Reingold, ed., *Science in Nineteenth Century America: A Documentary History*, pp. 287–288.

[31] In a personal interview with the author on July 14, 1960, Robert S. Shankland, the leading author of the definitive analysis of Miller's observations in *Rev. Mod. Phys.* 27:167 (April, 1955), was interested to learn of this historical fact. Professor

Europe, visiting various optical scientists in Germany and France. He broached his idea to Professors G. H. Quincke at Heidelberg and Cornu at l'Ecole polytechnique, both of whom had been experimenting with half-silvered mirrors, as well as to Helmholtz in Berlin. Like Helmholtz, Cornu was skeptical that this idea could be made to work, but Michelson was confident that it would. He sought and found a patron to help finance a contract which he made with the German instrument-makers, Schmidt und Haensch, to build the brass framework for his apparatus.

That patron was none other than Alexander Graham Bell, who, only four years earlier, had won international fame as the "inventor" of the telephone, when he demonstrated his talking telegraph in 1876 at the Philadelphia Centennial Exposition. Bell was at this time working on a "photophone" for the transmission of visible speech, and the firm of Schmidt und Haensch had several orders from him for research instruments as well as for improvements on the microphone.[32] Michelson's apparatus was constructed by Schmidt und Haensch on the credit account of Alexander Graham Bell.

From Heidelberg, on April 17, 1881, Michelson made an informal report to his patron which tells much of the story in terms of immediate finality:

My dear Mr. Bell,

The experiments concerning the relative motion of the earth with respect to the ether have just been brought to a successful termination. The result was, however, *negative.*

. . . .

At this season of the year the supposed motion of the solar system coincides approximately with the motion of the earth around the sun, so that the effect to be oserve [*sic* (observed)] was at its maximum, and accordingly if the ether were at rest, the motion of the earth through it should produce a displacement of the interference fringes, of *at least* one tenth the distance between the fringes; a quantity easily measurable. The

---

Shankland and his colleagues arrived at these conclusions after exhaustive statistical analysis. See Chapter 12 below.

[32] Schmidt und Haensch, "Franz Schmidt und Haensch 75 Jahre," advertising brochure, Berlin, West Germany, n.d.

actual displacement was about one one hundredth, and this, assignable to the errors of experiment.

Thus the question is solved in the negative, showing that the ether in the vicinity of the earth is moving with the earth; a result in direct variance with the generally received theory of aberration.[33]

This letter report to Bell coincides nicely with the published report of the experiment in the August issue of the *American Journal of Science*. But between April and August, Michelson had cause to re-examine the whole question of first-order versus second-order effects. This problem forms the crux of the correspondence between Simon Newcomb and Michelson immediately after Michelson's report to his superior of his first trial. Supposedly, Maxwell had considered all these theoretical implications before Michelson designed his apparatus to test the question. But Newcomb, in his first letter to Michelson, dated May 2, 1881, after congratulating Michelson on his appointment to the professorship of physics at Case School, said:

> Your arrangement for getting interference through long distances seems to me very beautiful but I do not see that it can settle the question of the motion of the ether. This motion makes itself felt by a difference in the wave lengths and velocities of propagation in different directions. But when a ray returns on its own path the retardation in one direction is compensated by the acceleration in the other. So that the result is the same as if the ether were at rest, just as in the ordinary measures of velocity. I have found that, for the same reason, a wave length measured with an instrument in which the ray returns nearly on its own path will give only a mean result. But if the wave length is measured with but a small deflection at the point of displacement as with a transparent ruled plate the effect ought to show itself. This is therefore the crucial experiment I would like to see tried.[34]

Michelson's reply to this letter is lost, possibly because Newcomb, who was at this moment deeply engaged in perfecting his own measurements of the velocity of light, may have received the letter in the field

[33] A. A. Michelson to A. G. Bell, April 17, 1881, in Bell Papers, National Geographic Society, Washington, D.C. This letter was uncovered by Nathan Reingold and published in his *Science in Nineteenth Century America*, pp. 288–290.

[34] Simon Newcomb to A. A. Michelson, May 2, 1881, in Miscellaneous Letters Sent, vol. 5, p. 440, Records of the Naval Observatory, NA, RG 78. See also Reingold, *Science*, p. 290.

and neglected to have it entered in his logbooks. At any rate, Michelson's reply must have been quite convincing, because exactly one month later Newcomb wrote back to him as follows:

When I wrote that the wave lengths in opposite directions would compensate each other I meant considering only quantities of the first order. This agrees with your formula where the difference is of the order v^2/V^2 amounting to 1/50,000,000. I supposed this difference could not be made appreciable by any experimental process, but it appears by your device to be made so, theoretically at least. Still I cannot feel sure but what some little action may come in to nullify the effect of so minute a cause and it seems to me we ought to be able to devise some way of getting a result which would depend only on quantities of the first order. The only way I know to avoid this is to measure wave lengths by an ordinary ruled transparent plate and I am surprised that no one has undertaken and published a decisive experiment of this kind.[35]

Newcomb's comments, coming as they did after the completion of the experiment in the well of the big telescope at Potsdam, and long after Michelson himself had satisfied his own qualms about these questions, probably had no other effect than to assure Michelson that his experimental results were ready for publication.

His famous first paper on this experiment, published in the *American Journal of Science* in 1881, gives full details of the experimental difficulties, a description of the apparatus, and acknowledgments of his indebtedness to Maxwell, Bell, and Professor H. C. Vogel of Potsdam. Michelson was obviously quite proud of his new instrument, suggesting that it be called an "interferential refractor." It had two great advantages over all previous refractometers—its small cost and its ability to separate widely the two pencils of interfering light. Michelson was fully aware of the primary difficulty to be expected with the instrument, namely, the effect of temperature changes. The arms were brass and about 1,000 millimeters long, corresponding to approximately

[35] Simon Newcomb to A. A. Michelson, June 2, 1881, Letters Sent, vol. 6, p. 6, Records, Naval Observatory, NA, RG 78. Michelson's reply to this letter, July 2, 1881, from Heidelberg, states respectfully his conviction that the second-order test is the only feasible one. Newcomb from Michelson, Miscellaneous Letters Received, vol. 3, [n.p.], ibid.

1,700,000 wave lengths. Concerning the "principal difficulty . . . to be feared," Michelson said, "If one arm should have a temperature only one one-hundredth of a degree higher than the other, the fringes would thereby experience a displacement three times as great as that which would result from the rotation."[36] Formidable as this objection was, Michelson was undeterred: he had covered the arms with paper boxes to guard against temperature changes induced by drafts. He even surrounded the instrument with melting ice, but to no great advantage.

Michelson loved the precision, the accuracy, which this instrument made possible. He had become almost more excited about the instrument than about the experiment. One may read into the exclamation points with which he ends two sentences in the middle of this paper his whole approach to this kind of sensitivity: "So extraordinarily sensitive was the instrument that the stamping of the pavement about 100 meters from the observatory made the fringes disappear entirely! If this was the case with the instrument constructed with a view to avoid sensitiveness, what may we not expect from one made as sensitive as possible!"[37]

Although Michelson concluded this paper with the confident assertion that the hypothesis of a stationary aether is erroneous, he did not, and we should not, infer from this anything more than that one aspect of one of the aether theories was shown to be untenable. But this happened to be a very crucial aspect. For the assumption that the all-pervasive luminiferous aether pervades the whole universe, macrocosmically and microcosmically, would eventually be shown to be equivalent to the assumption of absolute space itself.

Meanwhile, the imperative task, as Michelson saw it, was to find some reconciliation for the extra contradiction which this experiment gave to the usual explanation of aberration phenomena. So far as the aether concept itself was concerned, one more contradiction more or less would not at this stage be unassimilable. After all, Fresnel, as long ago as 1825, had wrestled with similar problems as a result of dis-

---

[36] Michelson, *Am. J. Sci.*, 3d ser. 22:125. See also Appendix A, below.

[37] Michelson, *Am. J. Sci.*, 3rd ser. 22:124. I. B. Cohen has judged the high precision of the interferometer significant enough to call this "the most important experiment of the century." See remarks in A. C. Crombie, ed., *Scientific Change*, p. 467.

crepancies in crystal optics. What was needed, therefore, was, first of all, a repetition of this experiment to rule out any experimental errors, and then, if the results were still negative, to find a means of making compatible with the usual explanation of astronomical aberration the fact that at the surface of the earth there appeared to be no aether wind whatsoever.

Michelson stayed on in Europe for another year of study, having been granted a leave of absence by Case while shopping for scientific apparatus to take back with him to Cleveland. He turned out two papers during that year of study, both related to his discovery of new uses for his interferometer.[38]

Published reactions to Michelson's aether-drift announcement were slight, for the most part reflecting the confusion which Simon Newcomb saw in the experiment. It was not at all obvious just what the proper interpretation of these results ought to be. Indeed, Lorentz was convinced that the path length was too short for a definitive second-order measurement. Also, André Potier pointed out that Michelson had made a significant error in neglecting to account for the effect of the movement of the light pencil that traveled transversely to earth's motion.[39] This meant that Michelson had overestimated by a factor of two the fringe shifts originally expected.

But it was clear from the fact that Michelson had already made a name for himself through his perfection of the velocity-of-light measurement that his results were not to be dismissed lightly.

[38] A. A. Michelson, "Interference Phenomena in a New Form of Refractometer," *Am. J. Sci.*, 3d ser. 23:395–400 (May, 1882); "An Air-Thermometer Whose Indications Are Independent of Barometric Pressure," *Am. J. Sci.*, 3d ser. 24:92 (August, 1882).

[39] A. A. Michelson, "Sur le mouvement relatif de la terre et de l'éther," *Comptes Rendus* 94:522 n. (February 20, 1882). For Potier's prior concerns with the consequences of aether drift, see his "Conséquences de la formule de Fresnel relative à l'entraînement de l'éther par les milieux transparents," *Journal de physique théorique et appliquée* [Paris] 3:201–204 (1874), and "De l'entraînement des ondes lumineuses par la materie ponderable en mouvement," *J. de physique* 5:105–108 (1876). In retrospect it is curious that there is no evidence to show that Michelson sought astronomical advice specifically upon the problems of solar and stellar proper motions.

CHAPTER FOUR

THE CLASSIC EXPERIMENTS, 1884–1890

"The" Michelson-Morley aether-drift experiment of 1887 has been celebrated as an *experimentum crucis* so often that it has achieved classic status. In spite of the facts that two different experiments were involved, each of which might be termed equally "crucial," and that the second, more famous test was never finished by Michelson and Morley, the reputation of "the" Michelson-Morley aether-drift experiment grew into a *cause célèbre* as it became a debatable piece of empirical evidence in the remolding, about a decade later, of fundamental physical theory.

The two classic experiments performed by Michelson and Morley in collaboration were, first, a careful repetition of Fizeau's famous "aether-drag" test of the Fresnel convection hypothesis and, second, a repetition of Michelson's 1881 aether-drift experiment.[1] By trying

---

[1] This chapter has profited by comparison with several articles by Robert S. Shank-

to see these events in prospect rather than in retrospect, we should gain some new insights into the history of physical science, the relationships between experiment, theory, and epistemology, and the character of scientific advance.

On taking up his duties as professor of physics at Case in the fall of 1882, Michelson found that Simon Newcomb was still ready with funds and encouragement to support a more accurate determination of the velocity of light. So Michelson set up a base line accurately surveyed for the length of half a mile along the railroad tracks bounding the south side of the neighboring campuses of Western Reserve College and Case School in Cleveland. Optical techniques had for some time been used, by people like Alexander Graham Bell, for instance, to study the properties of acoustics; now Michelson was to invert that procedure and use acoustical techniques to study the nature of light. The theoretically close analogy between sound and light drawn by scientists at that time supported the use of analogous experimental techniques.[2]

Michelson then devised a method for determining the frequency of tuning forks,[3] in order to have a method for determining the rate of rotation of his Foucault mirror. Morley may have watched, but he was not yet involved in the progress and completion of his neighbor's work. Afterward Michelson turned to another suggestion from his former navy superior, Simon Newcomb, and made experiments in the summer of 1884 on the velocity of red and blue light in distilled water and in a colorless liquid, carbon disulfide.[4] His success in this study

---

land, notably, "Michelson-Morley Experiment," *Am. J. Phys.* 32:16–35 (January, 1964) and "The Michelson-Morley Experiment," *Sci. Am.* 211:107–114 (October, 1964).

[2] For a number of examples of this cross-fertilization see the lectures by G. G. Stokes et al., *Science Lectures at South Kensington*, I, 71–72 and 173–176; II, 212–226.

[3] A. A. Michelson, "A Method for Determining the Rate of Tuning-Forks," *Am. J. Sci.*, 3d ser. 25:61–64 (January, 1883).

[4] These investigations with Newcomb were published together under Simon Newcomb, "Measures of the Velocity of Light Made under the Direction of The Secretary of the Navy during the Years 1880–82," *Astronomical Papers* prepared for the use of the *American Ephemeris and Nautical Almanac*, II, part 3, pp. 1–100. Cf. II, part 4, pp. 231–258.

contributed to the distinction between phase and group velocities and greatly enhanced his confidence in the delicate art of producing interference fringes.

During 1884 Michelson also began seriously to consider how to improve his aether-drift experiment. Sir William Thomson, later to become Lord Kelvin, came to America in the fall for a lecture tour in major cities of the East. Sometime that autumn, before Thomson's guest lectureship at Johns Hopkins University, Michelson had occasion to ask Sir William's counsel on the theoretical feasibility of repeating the aether-drift tests. He had worried about his mistake in 1881 of forgetting the effect of the earth's motion on the path of the half beam traveling perpendicularly to the earth's motion, and for that reason he wanted to check the theoretical bases of his design carefully before reinvesting in the experimental refinement.

With Thomson's advice, as well as that of Lord Rayleigh shortly thereafter, he was to begin his reexamination of the problem by repeating Fizeau's experiment on the influence of the motion of media on the propagation of light.[5] Both Thomson and Rayleigh were concerned over the discrepancy between Fizeau's moving-water or "aether-drag" experiment of 1859 and Michelson's first "aether-drift" experiment of 1881. The first seemed to confirm Fresnel's theory, the latter seemed to deny it. At issue here was Fresnel's drag coefficient, a fraction of the velocity of the moving medium that accounted for the excess aether presumed necessary in transparent substances. In air, for instance, where the index of refraction is close to unity, aether drag should be negligible, according to Fresnel's coefficient, and yet Michelson's first aether-*drift* experiment seemed to indicate that the aether *drag* in air was complete! Therefore, Michelson's first step should be to verify, if possible, Fizeau's claim to have proven Fresnel's explanation of astronomical aberration by the partial aether-drag hypothesis.

We shall return to this problem shortly. But before we see how Michelson set about improving Fizeau's famous experiment in which

---

[5] A. A. Michelson to J. Willard Gibbs, December 15, 1884, Gibbs Papers, "Scientific Correspondence," Sterling Library, Yale University. This letter is published in full in Lynde P. Wheeler, *Josiah Willard Gibbs: The History of a Great Mind*, p. 140.

the speed of light with and against a flow of water is measured and compared, we need to understand the intellectual climate of those years with respect to the aether concept. Although the notion of an omnipresent luminiferous medium seemed itself to be omnipresent, it was not so monolithic or so dogmatic a concept as is often thought.

Professor George G. Stokes, for example, in delivering the Burnett Lectures at Aberdeen in 1883 was thoroughly cautious as he began his exposition of the nature of light. Although he considered the "evidence quite overwhelming that light consists of undulations,"[6] his inductive presentation showed every effort to be fair and circumspect. At the outset Stokes admitted that aberration was more easily explained by the corpuscular theory, but in all other respects the theory of waves was so far superior that the conclusion in its favor was inescapable. And, if waves, then a medium was required, but: "We must beware of applying to the mysterious ether the gross notions which we get from the study of ponderable matter. The ether is a substance, if substance it may be called, respecting the very existence of which our senses give us no direct information: it is only through the intellect, by studying the phenomena which nature presents to us, and finding with what admirable simplicity those of light are explained by the supposition of the existence of an ether, that we become convinced that there is such a thing."[7]

Sir William Thomson, as he toured the major cities of the American Atlantic states as had Tyndall a decade earlier, demonstrated his authority and articulation at several levels of discourse. Before a public audience in Philadelphia in late September, 1884, Thomson had no qualms about being too forceful: "You can imagine particles of something, the thing whose motion constitutes light. This thing we call the luminiferous ether. That is the only substance we are confident of in dynamics. One thing we are sure of, and that is the reality and substantiality of the luminiferous ether."[8] He felt such statements were justi-

[6] George G. Stokes, *On Light: First Course on the Nature of Light Delivered at Aberdeen in November, 1883*, p. 17.

[7] Ibid., p. 80; cf. p. 25. See also P. G. Tait's elementary treatise, *Light*, pp. 3, 50, 203.

[8] Sir William Thomson [Lord Kelvin], "The Wave Theory of Light," *The Harvard Classics*, ed. Charles W. Eliot, vol. 30, *Scientific Papers*, p. 268. This is a tran-

fied in order to persuade his lay audiences that the fantastic speeds of propagation and the infinitesimal lengths of light waves could actually be manipulated and measured. The elastic-solid theory of the aether was hardly a problem compared with the difficulty of convincing nonscientists that the heat of moonlight can be measured, or that man can determine the frequency of a certain band of red light to be 400 trillion cycles per second.

The luminiferous aether was easier for Thomson to explain publicly (and privately) than Maxwell's electromagnetic theory of light. Exhibiting a large bowl of clear jelly with a small red wooden ball embedded in the surface near the center, Thomson would shake the bowl of gelatin and thereby demonstrate what he meant by an elastic-solid aether:

> When we explain the nature of electricity, we explain it by a motion of the luminiferous ether. We cannot say that it *is* electricity. What can this luminiferous ether be? It is something that the planets move through with the greatest ease. It permeates our air; it is nearly in the same condition, so far as our means of judging are concerned, in our air and in inter-planetary space. The air disturbs it but little; you may reduce air by air-pumps to the hundred-thousandth of its density, and you make little effect in the transmission of light through it. The luminiferous ether is an elastic solid, for which the nearest analogy I can give you is this jelly which you see, and the nearest analogy to the waves of light is the motion which you can imagine, of this elastic jelly, with a ball of wood floating in the middle if it.[9]

Thomson's confidence on the public platform was tempered conciderably when he conducted a conference for physics teachers, which both Michelson and Morley attended, at Johns Hopkins in October. Not that he abandoned his demonstrations by analogy with elastic-solid materials like shoemaker's wax or burgundy pitch—on the contrary, he reiterated his experiments in this direction and reinforced his demonstrations with mathematical analyses. But among his fellow scientists

---

script of the lecture delivered in Philadelphia at the Academy of Music under the auspices of the Franklin Institute, September 29, 1884.

[9] Ibid., p. 276. G. F. FitzGerald attacked Thomson's elastic-jelly model in his "On the Structure of Mechanical Models Illustrating Some Properties of the Aether," *Phil. Mag.*, 5th ser. 19:438–443 (June, 1885).

he was considerably more prudent. Whereas on Chautauqua he would ask, "What is the luminiferous aether?" and answer, "It is matter prodigiously less dense than air—millions and millions and millions of times less dense than air. . . . We believe it is a real thing . . . ,"[10] when he appeared before fellow scientists he was somewhat more cautious. In the original transcript of the *Baltimore Lectures*,[11] Thomson approaches the aether concept with the positive conditional *als ob*: "I do not say it is an elastic solid but it certainly behaves as if it were!"[12]

Significantly, Thomson shows here that he had reified the noun aether, that his qualifications were misdirected toward the nature of the substance rather than toward the behavior of the processes involved in electromagnetic phenomena. Although the syntax of language must share the blame in part for this misdirection, Thomson himself in a later session furnished an often quoted confession which epitomizes the mechanical point of view of which he was the most influential and perhaps extreme advocate at that time.

I never satisfy myself until I can make a mechanical model of a thing. If I can make a mechanical model I can understand it. As long as I cannot make a mechanical model all the way through I cannot understand, and that is why I cannot get [*sic*] the electro-magnetic theory. I firmly believe in an electro-magnetic theory of light, and that when we understand electricity and magnetism and light, we shall see them together as parts of a whole. But I want to understand light as well as I can without introducing things that we understand even less of. That is why I take plain dynamics.[13]

In 1884 there were yet very few reasons why Thomson might have been cautioned by his colleagues not to be so outspoken about the

[10] Thomson, "Wave Theory of Light," p. 285.

[11] Not to be confused with these lectures as thoroughly revised and published in 1904 under this title (see next reference). Cf. A. Macfarlane, *Lectures on Ten British Physicists of the Nineteenth Century*, p. 69.

[12] Sir William Thomson [Lord Kelvin], *Notes of Lectures on Molecular Dynamics and The Wave Theory of Light* . . . stenographically reported by A. S. Hathaway, lately Fellow in Mathematics of The Johns Hopkins University, p. 6. It is significant that Thomson knows of and credits Professor E. W. Morley for his work in physical chemistry (pp. 152 and 248) to an equal if not greater degree than he knows of or credits Michelson for his work in physics (pp. 55, 90, 249).

[13] Ibid., pp. 270–271.

speculative aether. Many physicists were still well aware of the scorn with which Faraday had been greeted less than a half century before, when he put forth his "lines of force" and his dynamic analogies for electromagnetic phenomena. But by the mid-1880's the triumph of the undulatory theory and the continuing progress of the electromagnetic theory with the consequent extension of these two, namely, the aether concept as a medium for all types of radiation, meant that the impeccable authority of success blessed Thomson's words: "When we can have actually before us a thing elastic like jelly and yielding like pitch, surely we have a large and solid ground for our faith in the speculative hypothesis of an elastic luminiferous ether, which constitutes the wave theory of light."[14]

Nonetheless, one serious reason for doubt was the unsatisfactory state of undulatory explanations for aberration. Thomson recognized this explicitly in talking with his peers about the question of the earth's motion through the aether: "The subject has not been gone into very

[14] Thomson [Lord Kelvin], "Wave Theory of Light," *Harvard Classics*, vol. 30, p. 286. Basing his theory on radiant-energy studies by Thomson and others, De Volson Wood a few years later analyzed the aether in terms of a fluid-dynamics model and came to the following conclusion concerning its properties: "That a medium whose density is such that a volume of it equal to about twenty volumes of the earth would weigh one pound, and whose tension is such that the pressure on a square mile would be about one pound, and whose specific heat is such that it would require as much heat to raise the temperature of one pound of it 1° F. as it would to raise about 2,300,000,000 tons of water the same amount, will satisfy the requirements of nature in being able to transmit a wave of light or heat 186,300 miles per second, and transmit 133 foot-pounds of heat energy from the sun to the earth each second per square foot of surface normally exposed, and also be everywhere practically non-resisting and sensibly uniform in temperature, density, and elasticity. This medium we call the Luminiferous Aether" (De Volson Wood, *The Luminiferous Aether*, p. 69).

Michelson also wrote to Lord Rayleigh about this time, less for advice than for passing along his plans. See Michelson to Rayleigh, December 9, 1884, in Michelson Correspondence file, U.S. Air Force Cambridge Research Laboratories (hereafter AFCRL), Hanscom Field, Bedford, Mass. There are eighteen letters and several cards in this file covering the years 1883–1912. I am indebted to John N. Howard for showing me these documents and many others that he has rescued, catalogued, and prepared for archival retention. See J. N. Howard, ed., *The Rayleigh Archives Dedication*, AFCRL Special Report No. 63, and "The Michelson-Rayleigh Correspondence of the AFCRL Rayleigh Archives," *Isis* 58:88–89 (Spring, 1967) with a prefatory note by Robert S. Shankland, "Rayleigh and Michelson," pp. 86–88.

fully; so that we do not know at this moment," said Thomson, whether there is a drag effect or the aether acts as a frictionless fluid.[15]

Both Michelson and Morley heard and accepted this challenge. If they could obtain another quantitative corroboration for Fizeau's experimental proof of Fresnel's hypothesis that (1) the aether is stagnant, but that (2) moving media will partially drag the aether along inside it, then they could really consider it established that Fresnel's notion of a stationary or stagnant universal medium was valid. More evidence was needed if this hypothesis was to become a reliable postulate.

THE AETHER-DRAG TEST

Michelson had already ordered his apparatus and had gone through preliminary trials on the Fizeau experiment, when, shortly before Christmas, he wrote to Josiah Willard Gibbs at Yale, asking his opinion on three theoretical points regarding the projected refinement of the aether-drift interferometer:

1st—Granting that the effect of the atmosphere may be neglected, and supposing that the earth is moving relatively to the ether at about 20 miles per second would there be a difference of about one hundred-millionth in the time required for light to return to its starting point, when the direction is parallel to earth's velocity—and that, when the direction is at right angles?

2nd—Would this necessitate a movement of the interference fringes produced by the two rays?

3rd—If these are answered in the affirmative, provided the experiment is made so far above the surface of the earth that solid matter does not intervene, What would be the result if the experiment were made in a room?[16]

---

[15] Thomson, *Notes of Lectures*, p. 8. Professor Morley seems to have been the sharpest and most active participant among the so-called 21 coefficients who attended Thomson's Baltimore Lectures. See Shankland, "Michelson-Morley," *Am. J. Phys.* 32:25.

[16] Michelson to Gibbs, December 15, 1884, Gibbs Papers, Yale University. Notice that Michelson first asks Gibbs for an option on the second-order ratio of the speed of earth $[v]$ to the speed of light $[c]$: $v^2/c^2 = 10^{-8} = 1/100,000,000$.

Unfortunately, Michelson never preserved the letters he received, and Gibbs, contrary to his custom, either did not make or did not keep copies of the letters he sent in reply to Michelson. Nevertheless, we can see from this letter that the Fizeau retest was considered only a preliminary to the aether-drift retest. These queries show also the major theoretical problems which Michelson perceived to be involved in his experimental design for aether drift. Each question was an interpretative issue for which it would be wise to have authoritative backing prior to undertaking the retests. Each question also anticipated major lines of controversy over the results of Michelson-Morley in later years.

By the spring of 1885 Michelson and Morley were active collaborators. Across the fence, at Western Reserve, they made use of Morley's laboratory in order to carry through the Fizeau experiment. The chemistry laboratory was older and better equipped with the facilities necessary for glassblowing and distilling water than was Michelson's physics laboratory at Case. Gradually through the spring and summer months Morley became more and more involved in Michelson's activities, lending his personal aid in solving some of the theoretical as well as experimental design difficulties. There were four major refinements to be made to Fizeau's apparatus, each of which could well use a chemist's expertise.[17]

By midsummer physicist and chemist were deeply engaged in the effort to remeasure the speed of light in moving water. In 1851, when Fizeau had first performed this measurement, it had been hailed as another "crucial experiment" for the wave theory of light. But by 1885 a number of other first-order experiments, designed to test Fresnel's drag coefficient for the influence of the motion of a medium on the propagation of light, had given slightly differing results, thereby confusing the issue. Michelson and Morley used a modified form of Michelson's interferometer, a constant flow of water, improved conduits to reduce turbulence, and an ingenious new method of measuring the current flow, or velocity distribution. It was a complex apparatus and a complicated experiment, however, that did not readily yield its results. Michelson and Morley designed, as may be seen from the

---

[17] A. A. Michelson and E. W. Morley, "Influence of Motion of the Medium on the Velocity of Light," *Am. J. Sci.*, 3d ser. 31:380 (May, 1886).

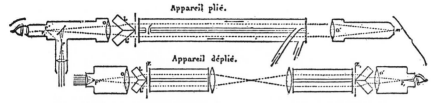

Two diagrams of Fizeau's apparatus of the 1850's for testing Fresnel's drag coefficient by comparing the velocity of light moving with and against a flow of water (courtesy Dover Publications reprint, Ernst Mach, *The Principles of Physical Optics*).

figures on page 381 of their 1886 paper (reproduced here as Appendix B), a watercourse that made Fizeau's elegant design seem crude by comparison. The reader may compare the two diagrams of Fizeau's original apparatus, reproduced here from the Dover reprint edition of Ernst Mach's *The Principles of Physical Optics* (fig. 167), with the experimental designs of Michelson and Morley. The innovations in apparatus demonstrate how meticulous improvements in design led to greater confidence in Michelson and Morley's measurements of the motion of the medium on the velocity of light.[18]

At this point an event cemented the names of the collaborators more firmly perhaps than either anticipated. Michelson suffered some sort of a "nervous breakdown" just as school was starting in September. Whether this was brought on by overwork, domestic difficulties, or both, is unknown, but the thrity-two-year-old Michelson packed his bags and went to New York, to put himself under a specialist's care, leaving Morley in full charge of the Fizeau retest. Morley told his father about it rather unsympathetically in a letter dated September 27, 1885:

Mr. Michelson of the Case School left a week ago yesterday. He shows some symptoms which point to softening of the brain; he goes for a year's rest, but it is very doubtful whether he will ever be able to do any more work. He had begun some experiments in my laboratory, which he asked

---

[18] For further concerning Fizeau's apparatus, see Ernst Mach, *The Principles of Physical Optics: An Historical and Philosophical Treatment*, p. 178. Michelson and Morley probably worked from Fizeau's paper "Sur les hypothèses relatives à l'éther lumineux," *Ann. de chim. et de phys.*, 3d ser. 57:385–404 (1859). Cf. p. 263.

me to finish, and which I consented to carry on. He had money given him
to make the experiments, and he gave me this money; the experiments are,
to see whether the motion of the medium affects the velocity of light moving
in it. I made some trials last week, and found that a good deal of modification in the apparatus would be necessary, and made drawings for some parts
of the apparatus. As soon as these are done, I shall try again. When everything is ready, it may not take more than a week to make the experiment.[19]

Morley was wrong both in supposing he could finish the retests so
soon and in thinking Michelson might be gone for good. By the end
of November, 1885, even before the final apparatus was complete,[20]
Michelson had recuperated and was back in Cleveland at work for
Case and for science. He never suffered a relapse and, if anything, his
brain was considerably hardened.

Through the winter and spring of 1886 Michelson's correspondence
with Gibbs reveals a close mutual interest in the empirical findings on
the behavior of the speed of light in moving media.[21] Since early in

[19] E. W. Morley to his father, September 27, 1885, Morley Papers, Box 2, Manuscript Division, Library of Congress. For evidence that Michelson was more unhappy
with Case than ill, see Michelson to Henry A. Rowland, November 6, 1885, from
Hotel Normandie, New York, in Rowland Papers, Johns Hopkins University Library, Baltimore.

[20] Michelson wrote Gibbs from Cleveland December 13, 1885, saying, in part: "I
have progressed so far in my repetition of Fizeau's exp. [*sic*] for testing effect of
motion of medium on the vel. [*sic*] of light, that I hope during the next week to get
at least a qualitative result—and perhaps a preliminary quantitative one. I have often
heard vague statements of objections to Fizeau's method but so far have not been able
to get any definite info. [*sic*]. If such is within your reach would you be kind enough
to tell me the essence of it" (Gibbs Papers, Sterling Library, Yale University, by permission).

[21] There are four letters in this time period, the most pertinent of which is dated
"Mch '86 [*sic*]" and reproduced here by permission of the Sterling Library, Yale
University:

My dear Prof. Gibbs,

Your welcome letter was duly received, and I have delayed answering till my
experiments were completed. My result fully confirms the work of Fizeau and the
result found for $\dfrac{a^2-1}{u^2}$ was .434 which is almost exactly the number for this ex-

pression when for a we put the index of refraction of water. I had heard that the
relation between maximum and mean velocity of liquids in tubes had been

the decade Gibbs had been concerned chiefly with optics and wave analysis, toying with the idea that light might have inertia. When Michelson initiated correspondence with him, Gibbs was involved in trying to draw some order out of the confusing plethora of elastic-solid theories of light. As one of the few men in America thoroughly conversant with the mathematical intricacies necessary to full appreciation of Maxwell's electromagnetic theory, Gibbs, the unobtrusive and retiring Yale professor, was at the forefront in the intellectual conflict over the ethereal aether.

Judging solely by Michelson's replies to his counsel and questions, it appears that Gibbs must have expressed considerable doubt whether either the Fizeau aether-drag experiment or the Michelson aether-drift experiment could be accepted at face value as proving anything at all about the luminiferous medium. At any rate, shortly after the announcements of the Michelson-Morley and Hertz experiments, Gibbs published two abstruse papers in which he analyzed their respective significance. Appearing a year apart, these two papers compared the "elastic" and "electric" theories of light with remarkable justice to each.[22] Gibbs paid tribute to the ingenuity of Sir William Thomson's latest quasi-labile aether theory but he left no doubt that, from the point of view of his mathematical physics, the preferable theory was still the "electrical" (i.e., electromagnetic) one. He avoided the use of the word *aether* altogether, preferring the phrase *luminiferous medium* where nonmathematical language was inescapable. Implicit in his treatment was the "field" concept; but he, too, like Maxwell, refused to

---

worked out—but have not been able to find it—so I made an experimental determination and found the ratio to be 1.165.

I think my result shows that your estimate of Thomson's work is correct. The number .434 is correct to within 2 or 3% and I can say with a good deal of confidence that it is not one half. I also repeated the experiment with air with a negative result. I expect to publish data in a few weeks.

[22] Josiah Willard Gibbs, "A Comparison of the Elastic and the Electrical Theories of Light with Respect to the Law of Double Refraction and the Dispersion of Colors" (1888), and "A Comparison of the Electric Theory of Light and Sir William Thomson's . . . Quasi-Labile Ether" (1889), *The Collected Works of J. Willard Gibbs*, II, 221–280.

make an explicit commitment. Gibbs's advice to Michelson in 1886 may have been skeptical, but it must not have been hostile, since two and three years later he was still giving Thomson the benefit of his doubts about the clumsy plenum idea.

Michelson's proprietary interest in the Fizeau rerun was so strong that he never mentioned Morley's aid in the extant communications with Gibbs, but when they had finished in April, 1886, the two men had their results published under a double by-line. Particularly noteworthy is their conclusion in the report of this experiment. After criticizing Fresnel's explanation of double refraction by his famous "drag coefficient" for a variable aether within moving transparent bodies, they describe their repetition, refinements, sixty-five trials, and precautions. Then they conclude by declaring that Fizeau's announced result, confirming Fresnel's stationary aether, was "essentially correct; and that *the luminiferous ether is entirely unaffected by the motion of the matter which it permeates.*"[23]

This conclusion is notable on at least two counts: first, the strength of its conviction is unusual; and, second, the syntax of its construction reflects the basic presuppositions of the classic test for aether drift which was to follow. That "the luminiferous aether" should be the *subject* of the sentence was not surprising in 1886. But what is significant is the inversion of subject and final clause which we must make in order to understand the purport of this preliminary test. Michelson and Morley were seeking to learn whether the motion of a gross material medium like water (which, insofar as it is transparent, was considered intramolecularly permeated by the luminiferous aether) would appreciably affect the velocity of light. Since light was conceived to be always and everywhere borne by the aether, rarest of all mediums, the question whether the luminiferous medium is in any way disturbed or carried along, retarded or made turbulent, by the motion of matter (whether in solid, liquid, or gaseous state) through it, is tantamount to the question whether the velocity of light is accelerated or retarded

[23] Michelson and Morley, "Influence of Motion," *Am. J. Sci.* 31:386. See also exchange of letters, Michelson and Morley and Thomson, March 22, 1886, and April 3, 1886, as published in Silvanus P. Thompson, *The Life of William Thomson, Baron Kelvin of Largs*, II, 857; cf. Shankland, "Michelson-Morley," *Am. J. Phys.* 32:28.

(and if so to what degree?) over a measured course where the course itself is in motion.

Michelson and Morley found Fizeau's confirmation of Fresnel's convection coefficient to be valid; a current of water does in fact, depending upon the direction it is flowing, speed up or slow down a light beam by a fraction of the current speed. This was another signal triumph for aether theory because it means that Fresnel's hypothesis was considered corroborated: there is a *partial* drag, but since it is so fractional, and since it is in accord with Fresnel's predictions, based on the assumption of stagnancy for a universal medium, the conclusion seemed established that matter in motion through the aether does not in any way disturb the aether that would be present in the absence of matter. Interpreting the results of their work in this way led to the assumption that the earth must move through space or the stagnant aether without inducing any frictional or electromagnetic stresses or strains whatsoever. *"The luminiferous ether is entirely unaffected by the motion of the matter which it permeates."* These are strong words, but without such confidence there would have been no point in pursuing their quest for a more refined method to measure the earth's absolute velocity through space.[24]

THE AETHER-DRIFT TEST

The latter half of the year 1886—almost two full centuries after the first publication of Newton's *Principia*—found Michelson mulling over the problems of absolute and relative motion, while his colleague Morley was finishing a series of tests on the percentage of oxygen in atmospheric air, from samples taken at various points around the world. Michelson's enthusiasm for the possibilities of his interferometer may have been dampened somewhat by Gibbs, but a Dutch theorist almost his own age had taken serious note of his 1881 effort at measuring

[24] In the same issue of one of the journals in which Michelson and Morley reported this result, there appeared another article by another Michelson, from Russia (no known relation), that represented an attempt to find the law of black-body radiation: see Wladimir Michelson, "Essai théorique sur la distribution de l'énergie dans les spectres des solides," *Journal de physique théorique et appliquée*, 2d ser. 6:467–479 (1887); cf. ibid., p. 442.

the aether drift and in 1886 published a "very searching analysis"[25] of Michelson's original experiment, which must have been quite encouraging.

Hendrik Antoon Lorentz (1853–1928), as a matter of fact, had explicitly called for a repetition and refinement of this experiment, since his own analysis had shown that the quantity to be expected had only half the value Michelson had originally supposed. Thus the "classic" Michelson-Morley experiment of 1887, even before it was undertaken, was related directly to the work of Lorentz. For years afterward Lorentz would play the major role in interpreting its results.

Lorentz's article was not the only encouraging word. John William Strutt, the English peer and wave theorist better known by his title, Lord Rayleigh, had written from England to reassure his fellow worker in the techniques of refractometry. Michelson's reply, dated March 6, 1887, is especially revealing: "I have never been fully satisfied with my Potsdam experiment, even taking into account the correction which H. A. Lorentz points out. . . . I have repeatedly tried to interest my scientific friends in this experiment without avail, and the reason for my never publishing the correction was (I am ashamed to confess it) that I was discouraged at the slight attention the work received, and did not think it worth while. Your letter has, however, once more fired my enthusiasm and has decided me to begin the work at once."[26] Allowing for a modicum of flattery in this letter, it is still significant that Michelson so greatly appreciated Lord Rayleigh's encouragement at this juncture. Few indeed were the persons who thought of the Michelson-Morley experiment as "critical" before it was performed. No one saw this as a "crossroads" experiment at the time.

[25] This is Michelson's own expression of appreciation for the paper by H. A. Lorentz, "De l'influence du mouvement de la terre sur les phénomènes lumineux," *Archives Néerlandaises des sciences exactes et naturelles* 21:103–176 ([Haarlem] 1887); cited in the introduction to the "classic" aether-drift report by Michelson and Morley (see Appendix C).

[26] Michelson to Lord Rayleigh, March 6, 1887, quoted in Florian Cajori, *A History of Physics: In its Elementary Branches including the Evolution of Physical Laboratories*, p. 199. This lengthy and important letter has been republished in full in Shankland, "Michelson-Morley," *Am. J. Phys.* 32:29, and the original is deposited temporarily with the AFCRL Rayleigh Archives.

It is tempting to think that, since "the great majority of nineteenth century physicists . . . were more confident of the existence of the luminiferous ether than of matter,"[27] the consensus of informed opinion regarding the ontological status of the aether must have been virtually unanimous and therefore dogmatic. This view does violence to the healthy scientific skepticism which, more often than not, prevailed. Dogmatism was more evident among the journalists than among the journeymen of science.

Even the most vocal and famous champions of the "realist" school of interpretation *nearly* always qualified their claims, if only slightly, with an appropriate doubt as to the substantive nature of the aether. Most student of optics were, it is true, quite confident, but not doctrinaire, in their claims for the aether. The definitions in Rodwell's *Dictionary of Science* in 1886 reflect this mood:

Ether, Luminiferous. . . . The medium whose vibrations are supposed to cause light. It is believed to pervade all space, and to be imponderable and infinitely elastic. (See *Undulatory Theory of Light.*)

.   .   .   .

Undulatory Theory of Light. The theory of light generally adopted at the present day. . . . The analogy which exists between the phenomena of light and sound, as well as the remarkable concordance between the observed phenomena of light and those predicted by mathematical investigation render it in the highest degree probable that the undulatory theory of light is very near the true one.[28]

There is no reason to believe that either Michelson or Morley began work on their experiment in the spring of 1887 with any intention to prove or disprove the existence of the aether per se. Theirs was an investigation into the behavior of light, and although they had every right, according to Fresnel's undulatory theory, to hope for some experimental verification of the hypothesis of a ubiquitous luminiferous aether, this was not their main concern, nor was the theory of aberra-

[27] Florian Cajori, *History of Physics*, p. 197.
[28] G. F. Rodwell, *Dictionary of Science: Comprising Astronomy, Chemistry, Dynamics, Electricity, Heat, Hydrodynamics, Hydrostatics, Light, Magnetism, Meteorology, Pneumatics, Sound, and Statics; Preceded by an Essay on the History of the Physical Sciences*, pp. 287, 663.

tion their chief interest, despite the fact that their final report begins with a discussion on aberration. In 1880 Michelson had preferred to think that he was looking for the sum-total velocity of the earth's motion through space, whereas now in 1887 his main quest was for a decisive, unequivocal test of the hypothesis that the all-pervasive aether was stagnant, at least with respect to the sun of our solar system.[29] Morley, as we shall see, was even more modest in his expectation of what this experiment ought to prove.

In setting out to perfect the marvelously sensitive instrument first constructed for him by Berlin instrument-makers, Michelson knew exactly wherein his technological difficulties lay. His earlier trials had been handicapped by apparatus at once too sensitive to environmental vibrations and to distortions caused by rotation, while, on the other hand, not sensitive enough to detect with the required margin of precision the extraordinarily small quantity to be observed. The first two difficulties were solved by mounting the optical parts on a massive stone slab floating on a mercury ring-bed. The third was obviated by increasing the optical path length, through repeated reflections, to a distance ten times longer than at first used.

The optical parts for the new interferometer were made by the master craftsman, John A. Brashear of Pittsburgh,[30] who, although self-taught as a maker of astronomical instruments, was nevertheless without a peer in the delicate arts of lens-grinding, mirror-polishing and speculum metal casting. Brashear furnished sixteen small speculum metal mirrors, about two inches in diameter, and two plane-parallel glass plates cut from the same piece, one being carefully half-silvered by Brashear's own newly developed front-face silvering method,[31] the

---

[29] A. A. Michelson and E. W. Morley, "On the Relative Motion of the Earth and the Luminiferous Ether," *Am. J. Sci.*, 3d ser. 34:333–336 (November, 1887). This paper was published in Britain in *Phil. Mag.*, 5th ser. 24:449–463 (December, 1887). See also my Appendix C for complete paper in facsimile.

[30] Dayton C. Miller, "The Ether-Drift Experiment and the Determination of the Absolute Motion of the Earth," *Rev. Mod. Phys.* 5:205 (July, 1933). For John A. Brashear, see *The Autobiography of a Man Who Loved the Stars*, ed. W. L. Scaife.

[31] H. A. Gaul and R. Eiseman, *John Alfred Brashear: Scientist and Humanitarian, 1840–1920*, p. 60. See also R. H. Tourin, "Optical History of a Machine Tool Builder [Warner and Swasey]," *Applied Optics* 3:1385–1386 (December, 1964).

other left clear to be used merely as a compensator. When a small telescope and an Argand lamp that burned salt for a yellow sodium light source were added to these parts, the critical pieces for the new instrument were virtually complete. Only the most delicate feature remained to be manufactured, namely, the micrometer screw used to adjust one of the mirrors beside the telescope so that the fringes could be focused and centered before observations and after a shift. Where or how this screw was made is unknown, but some idea of its importance to the interferometer can be gained by noting that it had to be machined and handfinished, with one thousand threads to the inch, in order that one turn of the screw might alter the overall path length by one thousand wave lengths![32]

The heart of the Michelson interferometer, namely, the beam splitter or semitransparent mirror, which divided the source beam into two evenly intense pencils of light traveling at right angles to each other, was apparently never a problem in manufacture or in maintenance. Michelson's design was simplicity itself, capitalizing on reflection and refraction at a forty-five-degree angle and on the ability of the human eye to distinguish images both beyond and before such a pane of glass.

On October 27, 1886, just as Michelson and Morley were literally laying the brick foundations for their retest of the 1881 aether-drift experiment, a vicious fire struck the Case main building where Michelson's laboratory and apparatus were located. Students were able to salvage some of the apparatus, but new foundations had to be laid for the experiment in the basement of nearby Adelbert Hall of Western Reserve University.

By the middle of April, 1887, Michelson and Morley had begun the assembly of these optical parts in their cast-iron holders on their base of sandstone. Morley reported the project to his father as follows:

Michelson and I have begun a new experiment. It is to see if light travels with the same velocity in all directions. We have not got the apparatus done yet, and shall not be likely to get it done for a month or two. Then we shall have to make observations for a few minutes every month for a year. We

[32] Michelson and Morley, "Relative Motion," *Am. J. Sci.*, 3d ser. 34:339. On the importance of screws for optics, see Albert G. Ingalls, "Ruling Engines," *Sci. Am.* 186:45–54 (June, 1952).

have a stone on which the optical parts of the apparatus are to be fixed. This stone is five feet square and about fourteen inches thick. This we shall have to support so that it can be turned around and used in different positions, and yet it must not be strained differently in the different positions. Now since a strain of half a pound would make our observations different useless [*sic*], we have to support it so that its axis of rotation is rigorously [vertical]. My way to secure this was to float the stone on mercury. This we accomplish by having an annular trough full of mercury with an annular float on it, on which the stone is placed. A pivot in the centre makes the float keep concentric with the trough. In this way I have no doubt we shall get decisive results.[33]

In the light of what followed, this letter is curious on several counts. Morley's statement of the purpose of the experiment is certainly oversimplified, but it is not necessarily misleading, for, as we have seen, Michelson's original ambition to find a speedometer for the celestial velocity of the earth had long since been tempered by failure and by second thoughts. From an operational standpoint, in setting out to test Fresnel's theory of aberration, based on the hypothesis of a stationary aether, they were indeed merely trying to see, as Morley had written his father, "if light travels with the same velocity in all directions."

Most curious, however, considering their later behavior, is Morley's explicit recognition of the necessity "to make observations for a few minutes every month for a year." Michelson and Morley failed even to attempt this part of the task. At stake here was the necessity of ensuring against the possibility that at any one season of the year when this experiment might be conducted, the orbital component of the earth's motion might by chance be canceled out by the component motion of the solar system as a whole. In their historic paper Michelson and Morley show their awareness that they have considered only the orbital and diurnal motions of the earth,[34] but their promise to repeat the experiment at intervals of three months throughout a year's cycle was apparently never fulfilled. We may well ask why.

Preparations were completed shortly after the end of the spring semester in 1887. A new brick pier had been laid in a basement room in

---

[33] E. W. Morley to his father, April 17, 1887, Morley Papers.
[34] Michelson and Morley, "Relative Motion," *Am. J. Sci.* 34:341.

the southeast corner of Adelbert Hall. Settled on this firm foundation, about three feet above the floor, was the cast-iron annular trough filled with some two hundred pounds of expensive mercury. A major portion of the appropriation supplied by the National Academy of Sciences from the Bache Fund had been invested in the purchase of this precious element to act as shock absorber and an almost frictionless bearing. Floating on an oversize wagon-wheel rim was the massive sandstone block with four small mirrors carefully affixed in a chevron-shaped arrangement at each corner. A large wooden box had been constructed to cover the whole slab in order to protect the optical parts and the pathway from disturbances caused by air currents and changes in temperature. At one corner the Argand lamp with its slit aperture was attached; from an adjacent corner protruded the observer's spyglass, almost at eye level, and the micrometer adjuster screw attached to the one movable mirror. By a gentle push of the observer's hand the apparatus could be made to rotate very slowly, going through one complete turn in about six minutes. The observer, once his fringes had been adjusted for best "seeing" and the block had started revolving on its almost frictionless liquid bearings, need not touch the apparatus at all, but could merely walk around with it and watch for any shift of the interference bands past a fiducial mark in his field of view.

Finally, in early July of 1887, all was ready, and at noon on July 8, 9, and 11, and for one hour in the evening of July 8, 9, and 12, Michelson and Morley performed the simple yet amazing observations which were eventually to immortalize them.[35] On the basis of their experience with the Fizeau retest, Michelson and Morley had great confidence in their ability to judge observed shifts in interference patterns as small as one-tenth of the distance between fringes. They had designed the aether-drift equipment, as they had the water-drag apparatus, in theoretical expectation of seeing four-tenths of a fringe shift when the apparatus moved through ninety degrees. But they were sadly disappointed at the time.

Michelson walked the circuit and called off his estimates of fringe shift at each of sixteen equidistant compass points, while, usually,

---

[35] Miller, "Ether-Drift Experiment," *Rev. Mod. Phys.* 5:206.

Morley sat by and recorded the data. Their entire series of final observations consisted of only thirty-six turns of the interferometer, covering only six hours' duration over a five-day period in the summer of
1887. If Dayton C. Miller's history can be trusted, and there is no
reason to doubt it, they "never repeated the ether-drift experiment at
any other time, not withstanding many printed statements to the contrary."[36]

Astonishing as this lack of persistence may be, it is not the most peculiar revelation of the history of the original Michelson-Morley collaborative effort to detect aether drift. The fact that what was really
intended as a preliminary effort came, by default, to be accepted as a
final report is most astonishing. This brief series of observations was
conclusive enough, the authors soon came to believe, for them to
publish the results as a falsification of the stagnant-aether hypothesis.
Their claim rested on the basis that the predicted effect did not have
the expected magnitude.

Writers on relativity have so often celebrated the "null," or "negative," results of the experiment that it needs to be emphasized once
again that "the indicated effect was not zero."[37] It might as well have
been zero, but in truth it was merely far less than what was expected
on the basis of Fresnel's theory of astronomical aberration. Instead of a
maximum fringe shift due to orbital motion of something like four-
tenths of a band, Michelson saw at most a shift of one hundredth of a
band. Thus, observing less than a twentieth of the predicted shift, they
could hardly do otherwise than conclude, in accord with experimental
error,

that if there is any displacement due to the relative motion of the earth and
the luminiferous ether, this cannot be much greater than 0.01 of the distance
between fringes. . . . The relative velocity of the earth and the ether is
probably less than one-sixth the earth's orbital velocity, and certainly less
than one-fourth. . . . if there be any relative motion between the earth and

---

[36] Ibid. An important letter in the AFCRL Rayleigh Archives fully corroborates
Miller's judgment and the other evidence: see A. A. Michelson to Lord Rayleigh,
August 17, 1887; the letter was reproduced in facsimile by R. S. Shankland in 1966
for the Michelson award ceremonies.

[37] Miller, "Ether-Drift Experiment," *Rev. Mod. Phys.* 5:206.

the luminiferous ether, it must be small; quite small enough entirely to refute Fresnel's explanation of aberration.[38]

The conclusions reached by Michelson and Morley in the summer of 1887 were a painful accommodation to seemingly contradictory experimental evidence. They were, as Michelson said later, "founded on so much pure assumption"[39] as to be regarded with suspicion, and yet the care lavished on the design and construction of this second Michelson interferometer was so great as to preclude, for the moment at least, any doubts about the precision and capability of the instrument. If there had been a stagnant aether, according to Fresnel's conception, this interferometer must surely have been able to sense a relative aether wind as the earth hurtles around the sun.

Or so it would seem. Michelson wrote to Lord Rayleigh in mid-August: "If the ether does slip past [the earth] the relative velocity is less than one sixth of the earth's velocity."[40] If we should ask why, in the face of the disappointing results from an experiment in which they had placed such confidence, they failed to carry through their promise to perform fall, winter, and spring tests, a likely answer can be found in their own reconsideration of what they had started.

Their original paper is inconspicuously but significantly divided into two parts: the first nine pages describe the experiment and offer the conclusions cited above; the last four pages appear as a "Supplement," apparently appended just before the November issue of the *American Journal of Science* went to press. The final sentence of the first section is a subjunctive-mood critique of Lorentz's theory as to why the aether seems to be at rest near the earth's surface. In first submitting their empirical findings for publication, then, Michelson and Morley seem to have been temporarily overwhelmed by the mass of untested assumptions underlying the aether and undulatory theories. Neither man had much confidence in his ability to theorize. Even so, they were not yet discouraged enough to have thought it necessary to call off their projected seasonal tests. Sufficient discouragement to call

[38] Michelson and Morley, "Relative Motion," *Am. J. Sci.* 34:340–341.
[39] A. A. Michelson, *Studies in Optics*, p. 154.
[40] Michelson to Rayleigh, August 17, 1887, in Shankland, *Am. J. Phys.* 32:32; original letter in AFCRL Rayleigh Archives.

a moratorium on further testing did come a short time later, however, because their "supplement" begins by stating: "It is obvious from what has gone before that it would be hopeless to attempt to solve the question of the motion of the solar system by observations of optical phenomena *at the surface of the earth*."[41]

In addition to their rationalization for not having found a significant aether drift, Michelson and Morley had another reason for abandoning further tests. Even before the July experiments,[42] Michelson had discovered another possible use for his instrument which promised definite and immediate results of very great concern to the whole scientific world, as well as practical applications of inestimable import. This new use for which preliminary experiments were run in June, 1887, established the feasibility of using the light wave as "the actual and practical standard of length."[43] The "interferential comparer," as Michelson now called it, was perfectly suitable to substitute a changeless natural unit, say a wavelength of sodium light, for the replacement or insurance of the arbitrary meter-bars and yardsticks in the vaults of national bureaus of standards. Fantastic precision and ready availability, plus the advantage that *counting* could by this method replace *comparison* as the basis for mensuration, made this new idea a pregnant scheme, tantalizing and exciting to Michelson and Morley, a worthy substitute project in the face of their other discouragement.

The very next month after the appearance of their report on the "classic" aether-drift experiment, Michelson and Morley had their initial paper on this metrological innovation published. The immediate reaction of the scientific community was intensely favorable to the suggestion to measure the meter in terms of light wavelengths.

Another distraction helped divert their attention throughout the spring and summer of 1888. Michelson and Morley, like other Cleveland scientists, were busily preparing for the annual meeting of the

[41] Michelson and Morley, "Relative Motion," *Am. J. Sci.* 34:341. Cf. reprints in *Phil. Mag.*, 5th ser. 24:449–463 (December, 1887); *J. Physique* (Paris) 1:444 (1888); *Sidereal Messenger* 6:306 (1887).

[42] A. A. Michelson and E. W. Morley, "On a Method of Making the Wave Length of Sodium the Actual and Practical Standard of Length," *Am. J. Sci.*, 3d ser. 34:427 (December, 1887).

[43] Ibid.

American Association for the Advancement of Science, to be held in their city in August. But they were not so busy as to give up intermediate testing of their new measuring apparatus.[44] Michelson delivered at the meeting an address entitled, "A Plea for Light Waves,"[45] in which he outlined numerous possibilities for the use of his "interferential refractometer" or "comparer" and enthusiastically touted its capabilities. He barely hinted at the circumstances of its origin and omitted any reference to the aether-drift failure of the previous year in this address.[46] He had become preoccupied with the meter instead of with the aether.

Thus, Michelson and Morley quickly and naturally shifted their interests from the abortive aether-drift problem to the much more promising, though hardly less challenging, metrological problem. Not only could they rest assured of positive results with this new project, but they would not need to worry much about their assumptions or interpretations in this endeavor. Victorian positivism could certainly be expected to hail the feat of measuring common length standards in terms of light waves as the highest kind of scientific achievement.

[44] A. A. Michelson and E. W. Morley, "On the Feasibility of Establishing a Light Wave as the Ultimate Standard of Length," *Am. J. Sci.*, 3d ser. 38:181–186 (September, 1889).

[45] A. A. Michelson, "A Plea for Light Waves," *Proc. AAAS* 37:67–68 (May, 1889). Michelson was beginning to learn from Rayleigh about this time the more subtle differences between phase velocity and group velocity of waves: see A. A. Michelson to Lord Rayleigh, December 18, 1888, January 30, 1889, and December 25, 1889, in AFCRL Rayleigh Archives. See also the retiring presidential address on undulatory theory by Samuel P. Langley, "The History of a Doctrine," *Am. J. Sci.*, 3d ser. 37:1–23 (January, 1889).

[46] Michelson's "Plea," shows in clear relief his double concern with applied optics and with optical theory. But most conspicuous is his fertile imagination in finding new uses for his instrument. On another hand, it should be noted that neither Michelson nor Morley seems to have made more than perfunctory efforts to become acquainted with astronomers' studies of the solar-apex problem, of star drift and proper motion. See, for example, Robert Grant, *History of Physical Astronomy*, pp. 552–556; Agnes M. Clerke, *A Popular History of Astronomy During the Nineteenth Century*.

CHAPTER FIVE

CONTRADICTIONS AND CONTRACTIONS, 1890–1900

Immediate reactions to the experiment performed in Cleveland in July of 1887 were, with one exception, negligible. But not so negligible were the results of Michelson's method of producing interference fringes. A growing number of specialists in optics were excited by the possibilities of applying interferometric techniques to their own experimental problems. By the early years of the decade, 1890–1900, interferometry had advanced far enough to attract excellent scientific talent, and in the following few years interferometry became one of the most highly developed techniques of physics.

The one exception, noted above, to the general lack of published comment and concern about Michelson and Morley's work of 1887 was a short letter by an Irishman to the editor of the immature American journal *Science.* Appearing as a single paragraph in the issue for May 17, 1889, the letter expressed "much interest" in the "wonderfully delicate experiment" of Michelson and Morley and suggested that "almost the only hypothesis" that could reconcile it with the theory of aberration would be "that the length of material bodies changes, according as they are moving through the ether or across it, by an amount depending on the square of the ratio of their velocity to that of light."[1] The author, as well as anyone else who saw this letter, promptly forgot it, so we shall, too, for a few more pages.

In the spring of 1889, during his last semester at Case before moving on to Clark University, Michelson was awarded the coveted Rumford Medal by the American Academy of Arts and Sciences for his various contributions to the science of optics. At the presentation ceremonies on April 10, Professor Joseph Lovering delivered an address extolling Michelson's recent researches on light,[2] concluding his remarks by saying, "We may thank Professor Michelson not only for what he has established, but also for what he has unsettled."[3] Lovering was referring to Michelson's work in the determination of the velocity of light, and then to his prescient discussion of the discrepancies in undulatory theory revealed by the unsuccessful aether-drift tests.

Michelson had acknowledged the year before, in his address before the Cleveland meeting of the AAAS, that his interferometer had a number of competitors and that much progress had already been made by co-workers in refractometry, especially by Quincke, Jamin, Ketteler, and Mascart. He also contended that optics (except for the technological advances of Rowland's diffraction gratings and Langley's bolometer) had remained almost static for two decades. In pleading for an increase in the use of his own interferometric techniques, he expressly hoped to help complete and perfect the edifice of optical theory. Little

[1] George Francis FitzGerald, "The Ether and the Earth's Atmosphere," *Science,* 1st ser. 13:390 (May 17, 1889). This letter was rediscovered recently by Stephen G. Brush as noted on page 108.

[2] Joseph Lovering, "Michelson's Recent Researches on Light," *Annual Report, . . . Smithsonian Institution, . . . 1889,* pp. 449–468.

[3] Ibid., p. 468. This paper is especially valuable as a historical review of various anticipations of the Michelson-Morley experiment.

could he foresee that the astonishing increase in accuracy of measurement which his new instrument and techniques made possible would in another twenty years help inaugurate another astounding change, a change so great as to be one of kind rather than of degree. The building of optical theory was to be overshadowed by a new high-rise wing of different architecture.

While Michelson and Morley had been sounding for an aether drift, across the Atlantic in Germany, at Karlsruhe, another young physicist interested in the implications of Maxwell's electromagnetic theory was hard at work seeking experimental proof of that theory's main predictions. Heinrich R. Hertz (1857–1894), under the tutelage of Helmholtz, had been convinced that Faraday and Maxwell were essentially correct in rejecting any and all actions-at-a-distance. Both the master, Helmholtz, and his student sought to determine whether electric charges undergoing acceleration actually emitted radiations like visible, or more likely invisible, light waves in the surrounding space. In 1887, while Michelson and Morley were being grievously disappointed in their search for some optical properties of the aether, Heinrich Hertz's experiments and explanations were demonstrating that electromagnetic signals could be sent and received (without wires) through virtually "empty" space. Was this not proof that an aether exists?[4]

Hertz's discovery of wireless waves attracted by sheer dramatic impact all the world's scientific interest. Here was positive proof by experimental demonstration that Maxwell's electromagnetic theory was essentially correct—that electricity, magnetism, and light itself are all fundamental manifestations of one and the same physical phenomenon. As Hertz extended his researches throughout 1888 and others joined in the corroboration of "Hertzian waves," there seemed to be less reason than ever for doubting the identical nature of electromagnetic waves and "aetherial oscillations." In his experimental autobiography

---

[4] G. W. de Tunzelmann, "Hertz's Researches on Electrical Oscillation," *Annual Report, . . . Smithsonian Institution, . . . 1889*, pp. 145–203. This article is a good example of the early reaction to Hertz's discovery, expressing simultaneously elation over wireless waves and caution toward the aether. See also Charles Süsskind, "Observations of Electromagnetic Wave Radiation before Hertz," *Isis* 55:32–42 (March, 1964).

of 1892, Hertz had this to say about his researches: "Casting now a glance backwards we see that by the experiments above sketched the propagation in time of a supposed action-at-a-distance is for the first time proved. . . . This fact forms the philosophic result of the experiments; and, indeed, in a certain sense the most important result. The proof includes a recognition of the fact that electric forces can disentangle themselves from material bodies and can continue to subsist as conditions or changes in the state of space."[5] It is some measure of Hertz's acumen that, although he believed his work to be the experimental capstone of the undulatory theory, he personally was very dubious about asserting the relationship between wave and medium.[6] But this did not prevent his friend and mentor, Helmholtz, from taking the extra step in interpreting the significance of Hertz's experimental work: "There can no longer be any doubt that light waves consist of electric vibrations in the all-pervading ether, and that the latter possesses the properties of an insulator and a magnetic medium."[7]

The researches of Hertz on the mysteries propounded by Maxwell, coming as they did at the same time as the Michelson-Morley effort, were more than adequate to compensate for the failure of the optical tests for aether drift. Hertz won as decisive a victory as any in the history of experimental science. By vastly expanding the known range of electromagnetic oscillations, he exhibited visible light as a small portion, a mere octave, of the grand spectrum of electromagnetic radiations.

Thus, Michelson's failure and Hertz's success, taken together in the decade following 1887, had countervailing effects on physical theory with regard to the ontology of the aether. The former, by falsifying a prediction, demanded an explanation; whereas the latter, by verifying

[5] Heinrich Hertz, *Electric Waves: Being Researches on the Propagation of Electric Action with Finite Velocity Through Space*, trans. D. E. Jones, p. 19.

[6] Ibid., pp. 20–28. See also *Gesammelte Werke von Heinrich Hertz*, ed. Philipp Lenard, I, 339, and II, 354.

[7] Hermann von Helmholtz's Preface to Heinrich Hertz, *The Principles of Mechanics Presented in a New Form*, trans. D. E. Jones and J. T. Wally, p. xxxii (not paginated). See also Th. Des Coudres, "Ueber das Verhalten des Lichtäthers bei den Bewegungen der Erde," *Ann. d. Phys.* 38:71–79 (1889), for an early German notice of Michelson-Morley in conjunction with Hertz's implications.

a prediction, clarified a wide range of experience (by identifying the laws of light with those of electromagnetism). The subsumption of the phenomena of optics under the phenomena of electromagnetism cleared the way for the advent of relativity by raising the importance of the seemingly constant velocity of light. In turn, relativity would eventually provide an explanation for Michelson-Morley. But for the moment, in spite of what Hertz said, his experimental deeds were taken as a vindication of the faith in the electromagnetic aether.

Heinrich Hertz was one of those rare individuals in science whose talents were as well adapted to experimental as to theoretical pursuits. His early death in 1894 at the age of thirty-seven cut short a career so promising that he was widely mourned by his colleagues. In the last four years of his life, Hertz devoted himself almost exclusively to a reexamination and a reinterpretation of the meaning of his own experimental findings in the light of Maxwell's, Helmholtz's, and Mach's theoretical structures.[8]

Hertz obeyed an astringent demand within himself to find and outline a new physical epistemology based on purified concepts and the closest possible shave with Occam's Razor. His critique finally led, in *The Principles of Mechanics*, to a systematic program for the elimination of the obscurities grown around the Newtonian concept of "force." By reducing the formal structure of mechanics to three fundamental notions, "space," "time," and "mass," and one fundamental law, a variant of "least action," he proved it possible to eliminate the concept of "force" entirely.[9] In so doing, Hertz contributed greatly to the growing linguistic reform movement in late nineteenth century physics. His assertion that "it is premature to attempt to base the equations of motion of the ether upon the laws of mechanics until we have obtained a perfect agreement as to what is understood by this name [i.e., *mechanics*],"[10] began to be taken quite seriously, even though his own reconstruction of those laws was not.

[8] See the Introductory Essay by Robert S. Cohen to Hertz, *The Principles of Mechanics*, pp. i–xx (not paginated).

[9] See Max Jammer, *Concepts of Force: A Study in the Foundations of Dynamics*, for a historicocritical analysis of this concept before and after Hertz.

[10] Hertz, *The Principles of Mechanics*, p. xxxvii (not paginated). Paul Drude

Representative of the general reaction to the experiments of Hertz and to those of Michelson in the scientific community in the opening years of the Gay Nineties is the treatment given these developments by Oliver Lodge in the second edition of his influential *Modern Views of Electricity*, published in 1892. In this work Hertz is lauded as the major stimulus to the enormous progress in radiation studies during the previous three years, while Michelson, although praised for his "consummate skill and refined appliances" in his velocity-of-light and moving-media experiments, is dismissed with a footnote in regard to the aether-drift test. On the one hand, Lodge affirms in his text: "We have now a real undulatory theory of light, no longer based on analogy with sound, and its inception and early development are among the most tremendous of the many achievements of the latter half of the nineteenth century."[11] On the other hand, Lodge admitted, in a footnote, but judged rather lightly, that the Michelson-Morley experiment and others "having reference to the theory of aberration and the motion of the ether near the earth are more puzzling, and seem discordant with ordinarily received notions at present." Thus, to Hertz went the major credit for establishing the physical reality of the undulatory theory of electromagnetism and, by implication, the reality of the aether. The lack of an aether drift was "puzzling" but in no way a basic threat to the new reality.

As Robert Millikan once noted,[12] the three editions of Lodge's *Modern Views of Electricity* were "most influential" in disseminating the ethereal theory of electricity. As early as 1882 Lodge had confidently *defined* the aether as one continuous substance filling all space: which can vibrate as light; which can be sheared into positive and negative electricity; which in whirls constitutes matter; and which transmits by continuity and not by impact, every action and reaction of which matter is capable.[13] Although more cautious in the 1889 pre-

---

codified Maxwell and Hertz for German students in his *Physik des Aethers auf elektromagnetischer Grundlage.*

[11] [Sir] Oliver Lodge, *Modern Views of Electricity*, 336; cf. pp. 327, 332, 403 n.

[12] Robert Andrews Millikan, *Electrons (+ and —), Protons, Photons, Neutrons, and Cosmic Rays*, p. 17.

[13] Lodge, *Modern Views*, p. 416 (refers to an earlier address by Lodge).

face to the first edition of *Modern Views of Electricity*, Lodge had remained true to his definitions, claiming that the question, What is ether? constituted "*the* question of the physical world at the present time."[14]

Lodge's faith never faltered—not even in the face of the most humiliating professional nonconformity during the 1930's. Sir Oliver lived until 1940 and his loyalty to the aether, in one form or another, like that of Michelson, remained firm to the end. But in 1889 Lodge was a highly respected physicist whose experimental investigations into optics and early radio research made him an authority of first rank on the aether of space. A versatile mind, a quick wit, and a literate and facile pen combined to enhance his popularity.

Ironically, Lodge himself, unintentionally, furnished the most important piece of research tending to corroborate the results of the Michelson-Morley experiment in the few years immediately following its announcement. Lodge designed a special machine to test for ethereal viscosity, being convinced by analogy that the electromagnetic fluid aether ought to show this property. Very impressive were these experiments, since Lodge, a leading exponent of the physics of the aether, herewith had to admit his inability to detect any manifestation of the retardation of light due to materially induced convection currents in the aether. In short, since Michelson had explained his negative results by invoking Stokes's hypothesis of "aether drag," Lodge set out to test this thesis and found it untenable—at least, on the scale and within the limits of his laboratory equipment.

As noted above, it is the problem of accounting for astronomical aberration, then thought to be exclusively an optical problem, which most concerned those who took the Michelson-Morley experiment seriously. If there is no aether drift near the surface of the earth, then how can we account for the fact that telescopes must be positioned with an extra compensation in perfect correlation with the resultant of the earth's orbital motion vector? Airy's water-filled telescope, in 1871, as we have seen, had supposedly ruled out any cause for this necessary correction other than a drifting luminiferous aether.

[14] Ibid., p. x. See also Lodge's *Advancing Science: Being Personal Reminiscences of the British Association in the Nineteenth Century*, p. 100.

Now, Lodge, in 1891 and the years following, sought to answer the question, Can matter carry the neighboring ether with it when it moves?[15] He did this by putting two large steel disks, each a yard in diameter and clamped to the other with an inch-wide space in between, into very rapid rotation about a vertical axis, around and within which he directed a split beam of light, following the techniques of Michelson. Had he seen from this arrangement any significant fringe shift, he would have obtained some evidence for the supposition that the rotating steel had caused the path length of one of the half-beams to be lengthened because of the aether drag induced by the revolving disks. Again, the interferometer yielded no such evidence. Lodge concluded for the moment that "these two experiments are at present in conflict,"[16] and that Michelson-Morley "may have to be explained away,"[17] since it seemed to be the only trustworthy experiment tending against the nonviscous vortex-ring theory of the aether. Michelson only hinted at this as one possible alternative in 1897.

The experimental work of Hertz and Lodge made up the major immediate corroborations of the Michelson-Morley experiment before 1897, but in the context of the time neither the one nor the other could be properly weighted. Hertz seemed to counterbalance, Lodge to countermand, the aether-drift tests. Clearly there was great need for some new way of interpreting these results.

In the third (1911) edition of his influential book *The Grammar of Science*, first published in 1892, Karl Pearson looked back over the span of years to the time of the book's first publication and remarked that the beginning of the last decade of the nineteenth century marked a basic transition: "It is not too much to say that at that period an epoch closed which was initiated by Copernicus."[18] Pearson used this state-

---

[15] Sir Oliver Lodge, *The Ether of Space*, p. 70. Cf. *Nature* 46:164–165 (June 16, 1892). For better illustrations, see Sir Oliver Lodge, *Der Weltäther*, trans. Hilde Barkhausen.

[16] [Sir] Oliver Lodge, quoted in "Minutes of the Physical Society Meeting, May 27, 1892," *Nature* 46:165 (June 16, 1892).

[17] [Sir] Oliver Lodge, "Aberration Problems: A Discussion concerning the Motion of the Ether Near the Earth and concerning the Connexion between Ether and Gross Matter; with Some New Experiments," *Trans. R.S.* 184:753 (July,1893).

[18] Karl Pearson, *The Grammar of Science*, p. 355.

ment to call attention to the fact that since 1892 the climate of opinion concerning the nature of science had so changed as to make his once radical treatise now appear prosaic. Pearson's contemporaneous account of this historical change reads as follows:

The end of the nineteenth century . . . marks the advent of experimental knowledge requiring an entire revision of the hypotheses and theories as to the constitution of matter. . . . Whereas through the greater part of the nineteenth century, "matter" was the concept which was looked upon as fundamental in physical science, of which there was a curious accidental property called electricity, it now appears that electricity must be more fundamental than matter, in the sense that our once elementary matter must now be conceived as a manifestation of extremely complex electrical phenomena.[19]

Pearson's own linguistic analysis of the physical concept of the aether as it appeared in 1892 must surely be accorded a prominent place in the historical change he later described. Unlike some critics of a more radical empiricist persuasion, Pearson was never reluctant to use the term *aether*, but he was always extremely cautious to qualify it as primarily a conceptual tool used to express continuity. As a specialist in statistics, Pearson was careful to point out the limiting character of the concepts of a *perfect fluid* and of a perfect jelly. In analyzing the vortex-ring theory and his own aether-squirt notion of the atom, Pearson was at once sympathetic and analytical:

Our conceptions of the ether are at present very ill defined. We are agreed that it must be conceived as a medium which resists strain, but we are not certain how to represent best the relative motions that follow on relative change in the position of the ether-elements. We are not yet satisfied with a perfect fluid conception of the ether.

Treating the ether not as a conception but as a phenomenon, we find it difficult to realize how a *continuous* and *same* medium could offer any resistence to a sliding motion of its parts. . . . It is not a metaphysical quibble

[19] Ibid., pp. 356–357; cf. p. 206 n. For a similar American analysis, see A. E. Dolbear, *Matter, Ether, and Motion*, pp. 26–43.

Albert Abraham Michelson, about 1887, when he and Edward Morley were refining the design for their aether-drift experiment. Michelson went on to achieve greater fame in optics and in 1907 became the first American to receive a Nobel Prize for science.

Courtesy Dorothy Michelson Livingston.

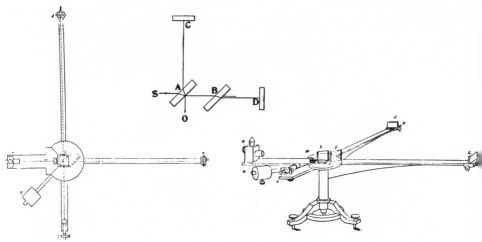

These diagrams describe the original interferometer, a brass apparatus with two arms each a meter in length, which was made for Michelson in Germany. With it he hoped to detect a relative aether wind that could be analyzed to measure the resultant velocity of the earth as it hurtles with many component motions through space.

Drawings by A. A. Michelson. University of Chicago Press (1902), and American Journal of Science, *July, 1881.*

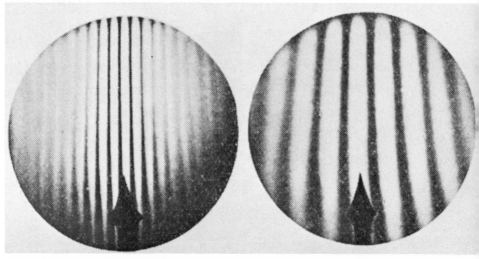

The fields of view of the interference fringes as photographed through the observer's telescope, on narrow and broad magnification, indicate the difficulty of estimating the interferometer's expected fringe shift of one-tenth of a band of light past the fiducial mark.

From D. C. Miller, "Ether-Drift Experiment," Rev. Mod Phys. *5 (1933).*

Edward Williams Morley was professor of chemistry at Western Reserve University, but his careful work on atomic weights of oxygen and hydrogen has been overshadowed .by his reputation as Michelson's collaborator on the classic aether-drift tests of 1886–1887.

Courtesy Case Western Reserve University Archives.

This recently discovered photograph of the "classic" Michelson-Morley interferometer (the famous sandstone slab floating on mercury in an annular trough) served as the model for the orthogonal drawing that graced their aether-drift paper published in the *American Journal of Science* and in the *Philosophical Magazine* (London) in 1887.

Michelson Museum, D. T. McAllister, Curator. Courtesy the Hale Observatories.

A photograph of the basic Morley-Miller interferometer, showing the structural-steel base built in Cleveland in 1903 to replace wooden cross-arm instruments first used by Morley and Miller in 1902.

Courtesy G. P. Putnam's Sons. Charles Lane Poor, Gravitation versus Relativity *(New York, 1922).*

For tests of the Fitzgerald-Lorentz contraction hypothesis, Morley and Miller in 1904 set up this trussed apparatus of tubes by which wooden or metal rods could determine the optical path length, thus perhaps showing a fringe shift due to a differential in material contractions.

From D. C. Miller, Rev. Mod. Phys. *5 (1933).*

Dayton Clarence Miller, professor of physics at Case School, was an expert in wave mechanics and an amateur musician. He had obtained his Ph.D. in astronomy from Princeton and had moved toward astrophysics and physics at Case, thence into optics and acoustics.

After obtaining small "positive effects" in April, 1921, with the steel-base interferometer last used in 1905, Miller had this nonmagnetic single piece of reinforced concrete cast as a base for his observations in December, 1921, on Mount Wilson.

From D. C. Miller, Rev. Mod. Phys. 5 (1933).

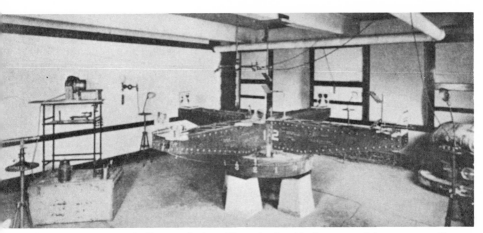

During 1922 and 1923 Miller worked assiduously in this basement laboratory at Case School, testing his experimental design. Einstein and Lorentz visited him here, encouraging him to persevere.

From D. C. Miller, Rev. Mod. Phys. 5 (1933).

Miller's "hot-dog stand," an observation hut on campus at Cleveland, which, for seasonal tests in 1923, enclosed the interferometer while maintaining a translucent optical plane. The gazebo contrasts nicely with the Gothic masonry of Amasa Stone chapel. A gift from Mr. Eckstein Case made possible the "portability" of this interferometer and its protective covering to Mount Wilson.

Courtesy Case Western Reserve University Archives.

A view of the interior of Miller's hut at the 6,000-foot Mount Wilson Observatory, showing the aether-drift apparatus essentially as used between 1924 and 1926 for the first systematically seasonal tests. With this apparatus Miller claimed to have found the earth's "absolute motion." Meteorological instruments on the turntable and insulated glassbox sheathing for the optical pathways indicate precautions taken.

Courtesy the Hale Observatories.

The massive "invar" interferometer in this quarter-scale model was Michelson's first response to the challenge laid down by Miller's "discovery" of a "meaningful" aether drift, but stress analyses showed that precision optical work would be virtually impossible on this mammoth. Note the bucket seat for the observer, opposite the cantilever weight at right.

Michelson Museum. Courtesy the Hale Observatories.

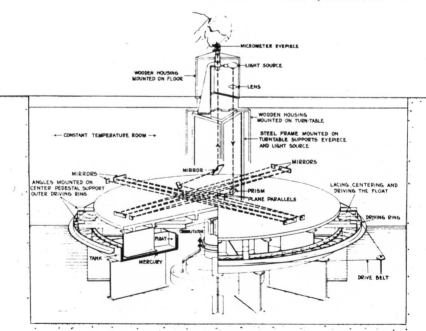

Using the bedplate built to polish the 100-inch mirror for the Hooker telescope, Francis G. Pease and Fred Pearson devised a substitute interferometer to recheck Miller's results. Under Michelson's direction, this schematic design was placed in a sealed, constant-temperature room, with both the light source and the observer above the ceiling.

From Nature 123 (1929).

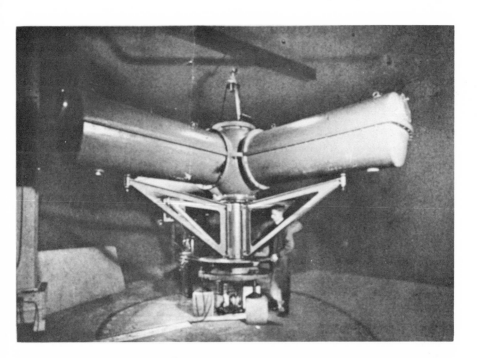

This interferometer, designed by Georg Joos and built at the Zeiss plant in Jena, was the most elaborate of all optical aether-drift instruments. It made maximum use of quartz, had helium-filled pathways, and operated automatically, with remote-controlled photography.

Michelson and Albert Einstein, shown here at their first and only meeting, at the Cal Tech Atheneum, January, 1931. Their host, Robert A. Millikan, third Nobel laureate in the picture, is on Einstein's left. In the back row, left to right, are Walter S. Adams, Director of Mount Wilson, Walter M. Mayer, Einstein's assistant, and Max Farrand, Director of the Huntington Library.

when we demand that two things shall not occupy the same space, but that when motion begins there shall be some *where* unoccupied for some *thing* to move into. The obvious fact is that while in conception we can represent the moving parts of the ether *as points* . . . yet when we project that ether into the phenomenal world it is at once recognized as a conceptual limit unparalleled in perceptual experience, and we do not feel at home with it.[20]

Perhaps the case could be made that Pearson is here attacking a straw man, because, as we have seen, even the most vocal exponents of the reality of the aether qualified their claims for it according to time and place and audience. But there is, nonetheless, the undeniable fact that such physicists as Kelvin, Stokes, Larmor, and Lodge were definitely in search of ever more perfect mechanical models with which to bolster their faith in a perceptual aether.

It is this attitudinal set, this psychological orientation, which, more than anything else, points up the differences between traditional physicists and epistemological innovationists like Ernst Mach. For instance, when Oliver Lodge read, before the Royal Society in 1892, the results of his experiments on ethereal viscosity, he professed a faith, certainly, but not a dogma, "The one thing in the way of the simple doctrine of an ether undisturbed by motion is Michelson's experiment, viz., the absence of a second-order effect due to terrestrial movement through free ether. This experiment may have to be explained away."[21] Naturally enough, then, Lodge and his fellow believers would enthusiastically receive any explanation of the failure to detect aether drift which would not be too disruptive of the remainder of the theory. But how disruptive is too disruptive? Sometime during the early spring of 1892, Lodge was entertaining in his Liverpool study Professor

[20] Pearson, *Grammar of Science*, p. 298. See also Alfred M. Bork, "The Fourth Dimension in Nineteenth Century Physics," *Isis* 55:326–338 (September, 1964).

[21] Lodge, *Trans. R.S.* 184:753. Even the chief apostle of Maxwell's mathematics, Oliver Heaviside, recognized this contradiction and maintained some reverence for the aether: see his *Electromagnetic Theory*, I, x, 308, 466. The ethereal background as a vacuous dielectric is even discernible in J. J. Thomson's *Notes on Recent Researches in Electricity and Magnetism*, pp. 2, 4. During 1891–1894 a debate over choice of notational systems also complicated the development of a consensus over the dielectric aether: see Alfred M. Bork, " 'Vectors Versus Quaternions': The Letters in *Nature*," *Am. J. Phys.* 34:202–211 (March, 1966); and Michael J. Crowe, *A History of Vector Analysis*, pp. 182–225.

George Francis FitzGerald of Trinity College, Dublin, when the
occasion arose for FitzGerald to impress Lodge with his idea that a
way out of the difficulty might be found by supposing the size of bodies
to be a function of their velocity through the aether.[22] Should this be
the case, should material bodies actually change their dimensions depending upon their absolute motion through the aether, then one could
argue that the arms of Michelson's interferometer were distorted just
enough to compensate exactly for the otherwise detectable aether drift.

Surely the idea that length might be relative to velocity was disturbing, and highly so, to most minds in 1892, but certainly not to all.
Fully ten years before, FitzGerald (1851–1901) had been engaged in
the study of electromagnetic effects due to the motion of the earth.[23]
This series of studies, along with a continuing interest in optics and
astrophysics, had led FitzGerald to an intimate appreciation of the dis-
parate problems of molecular binding forces and of aberration. He had
written up and submitted his contraction hypothesis in 1889, but he
did not know whether it had been published. His was one of the few
minds of the time capable of combining these separate ideas to suggest
a new solution: the possibility that "rigid rods" might not be absolutely rigid, but only approximately so. In the latter case it was quite
possible, though not clearly probable, that all matter might experience
electromagnetic distortion due to its motion through space.

Across the channel, at the University of Leiden in Holland, H. A.
Lorentz, some months later and independently of FitzGerald, hit upon
the same possible solution for the Michelson anomaly. As we have
seen, Lorentz had been following the Michelson experiment closely
from the beginning. In recent years his theoretical interests had tended,

---

[22] Lodge, *The Ether of Space*, p. 68. For the problem of reconstructing priority
credits for the contraction idea, see Alfred M. Bork, "The 'FitzGerald' Contraction,"
Isis 57:199–207 (Summer, 1966), and Stephen G. Brush, "Note on the History of
the FitzGerald-Lorentz Contraction," *Isis* 58:230–232 (Summer, 1967).

[23] See George F. FitzGerald's paper, "On Electromagnetic Effects Due to the
Motion of the Earth," *Trans. R.S.* (Dublin) (March 5, 1882), reprinted in *The
Scientific Writings of the Late George Francis FitzGerald*, ed. Joseph Larmor, p. 111.
See also FitzGerald's Letter to the Editor, *Nature* 40:32–34 (May 9, 1889) and his
paper, "On the Structure of Mechanical Models illustrating some Properties of the
Aether," *Phil. Mag.*, 5th ser. 19:438–443 (June, 1885).

through study of the works of Henri Poincaré and Heinrich Hertz, toward development of a mathematical structure for the growing mass of experimental data on the electrodynamics of moving electric charges. In 1891 Lorentz first made what some scholars believe was the decisive break in separating the concept of the electromagnetic field from ponderable matter and, in 1892, he published his basic paper outlining an embryonic theory of electrons as charged particles interacting with the medium (i.e., the field) in which they moved. Combining the best concepts of such continental theorists as Weber, Riemann, and Clausius with British theory, especially Maxwell's theory on electromagnetism, Lorentz participated in and stimulated further the growing tendency to see evidence for the atomic nature of electricity.[24] In working out his blend of the evidence for the propagation of influence from one electron to another through a Maxwellian medium, Lorentz, like FitzGerald, was in a perfect position to expand his long concern with Michelson's result. This example of simultaneous invention, like many others in the history of science, seems to have arisen from poor communications, despite the intensely social nature of the scientific enterprise.

Although it has often been asserted that FitzGerald's contraction hypothesis was "a pure flight of scientific fancy," it is unjust to ascribe too much imagination and too little responsibility to his contraction hypothesis. FitzGerald's approach to the problem of the aether was not naive or crudely materialistic, nor was he overly obsessed with an imperative need for explaining Michelson-Morley. He looked upon the aether as *sui generis*, substantial perhaps but not necessarily material. As early as 1878 FitzGerald is reported to have remarked that Maxwell's electromagnetic theory, if it "induced us to emancipate ourselves from the thralldom of a material ether, might possibly lead to

[24] H. A. Lorentz to Lord Rayleigh, August 18, 1892, AFCRL Rayleigh Archives. Lorentz's memoir appeared in *Archives néerlandaises* 25:363 (1892). See also Tetu Hirosige, "Lorentz's Theory of Electrons and the Development of the Concept of Electromagnetic Field," *Japanese Studies in the History of Science*, 1:101–110 (1964), a paper first brought to my attention by A. E. Woodruff. See also a forthcoming paper by Kenneth F. Schaffner, "The Lorentz-FitzGerald Contraction and the Lorentz Electron Theory." Cf. the standard account in Sir Edmund Whittaker, *A History of Theories of Aether and Electricity*, I, 393.

most important results in the theoretic interpretation of nature."[25] To
reiterate as he had to Lodge in 1892 that the length of the arm of the
Michelson-Morley interferometer in the direction of the earth's motion might be materially contracted by a one hundred-millionth part
of its former length, was indeed daring, but not at all inconceivable,
especially after the reinforcement brought by Lorentz.[26]

It has become customary, largely in the wake of Sir Edmund Whittaker's history, to regard the FitzGerald-Lorentz contraction hypothesis as a strictly ad hoc interpretation of the Michelson experiment.
There can be no question that this is what it became once the theory of
Special Relativity attracted a following. After 1905 it became clear that
the suggestion put forward by FitzGerald and Lorentz had the effect of
a special dispensation from Newton's laws of mechanics. But in 1895,
and still more so when first suggested, it was still quite possible that
material contraction might be worked into the classical mechanical
theories. For instance, Lord Rayleigh himself published, in March,
1892, a paper written in 1887 but obviously revised for publication, in
which he reviewed the whole problem of aberration and the status of
current knowledge with respect to it. Having expressed his opinion
that "stellar aberration in itself need present no particular difficulty
on wave theory,"[27] and having reviewed Michelson's 1881 experiment
and Lorentz's correction, Rayleigh permitted the following judgment
to stand, knowing full well about the 1887 results: "Under these circumstances Michelson's results can hardly be regarded as weighing
heavily in the scale. It is much to be wished that the experiments should
be repeated with such improvements as experience suggests. In observations spread over a year, the effects, if any, due to the earth's mo-

---

[25] Quoted by Sir Edmund Whittaker, in "G. F. FitzGerald," *Lives in Science: A
Scientific American Book*, p. 80.

[26] Lorentz sharply separated his conceptions of aether and matter, unlike Maxwell,
who tended to define the aether as a dielectric; but Lorentz was apparently unable to
extend his theory to cover second-order phenomena until 1904. See Gerald Holton,
"Einstein, Michelson, and the 'Crucial Experiment,'" *Isis* 60:170–175 (Summer,
1969).

[27] Lord Rayleigh, "Aberration," *Nature* 45:499 (March 24, 1892). For the importance of Rayleigh's opinions, see Robert B. Lindsay, ed., *Lord Rayleigh: The Man and
His Work*, pp. 3–25.

tion in its orbit, and to that of the solar system through space, would be separated."[28]

Over a year later, the venerable Sir George G. Stokes argued in his presidential address at the Victoria Institute that the aether-drag hypothesis he had originated a half century earlier correlated perfectly as an explanation for the precession of Encke's comet.[29] This, in addition to explaining the negative result of the Michelson-Morley experiment, added considerable confidence to the drag hypothesis.

While physics was becoming a more professional discipline and enjoying a gradual consolidation of the kinetic-molecular theory of gases, along with a codification of statistical mechanics, the early years of the last decade of the nineteenth century witnessed the advent of a new vogue in radiation research. New vacuum techniques and the experiments of Eugen Goldstein, Sir William Crookes, Arthur Schuster, and J. J. Thomson with cathode rays made possible a kind of electromagnetic "optics." The Stefan-Boltzmann law helped to forge the concept of energy and its transfer into the central link between thermodynamics and electrodynamics. The magnetic rotation of light, Hall's phenomenon, and canal rays were especially popular areas for experimentation. In 1893, William Thomson, now officially Lord Kelvin, baron of Largs, expressed a prevalent view: "If a first step towards understanding the relations between ether and ponderable matter is to be made, it seems to me that the most hopeful foundation for it is knowledge derived from experiments on electricity in high vacuum."[30]

---

[28] Rayleigh, "Aberration," p. 502. A year later Michelson wrote to ask Rayleigh's opinion of the contraction hypothesis and to suggest a repetition using different kinds of matter: A. A. Michelson to Lord Rayleigh, May 17, 1893, AFCRL Rayleigh Archives.

[29] Sir George G. Stokes, "The Luminiferous Aether," reprinted in the *Annual Report, . . . Smithsonian Institution, . . . 1893*, p. 117. Cf. R. T. Glazebrook [Review of Optical Theories], *Br. Ass'n. Adv. Sci. Annual Report* (1893), pp. 671–681. For character sketches of Stokes and many other physicists of his generation, see Arthur Schuster, *Biographical Fragments*, esp. pp. 221, 228, and 238, 243.

[30] Lord Kelvin, quoted by Whittaker, in *History . . . of Aether and Electricity*, I, 357. See also Stephen G. Brush, "Foundations of Statistical Mechanics, 1845–1915," *Archive for History of Exact Sciences* 4(3):145–183 (1967); David L. Anderson, *The*

Kelvin's confidence was not misplaced, although it may have been misdirected. Certainly the almost accidental discovery, by Professor W. K. Roentgen, in November 1895, of "X-Rays," as he was wont to distinguish them from canal or cathode rays, was a more spectacular addition to the speculations on the nature of aether than anything since Hertz's experiments in 1887. Roentgen's own immediate reaction to his discovery was to suggest that these all-penetrating x-rays, a kind of superultraviolet radiation, might be the long-sought longitudinal vibrations of the aether:

> Now it has been long known that besides the transverse light vibrations, longitudinal vibrations might take place in the ether and according to the view of different physicists *must* take place. Certainly their existence has not up till now been made evident, and their properties have not on that account been experimentally investigated.
>
> May not the new rays be due to longitudinal vibrations in the ether?
>
> I must admit that I have put more and more faith in this idea in the course of my research, and it behooves me therefore to announce my suspicion, although I know well that this explanation requires further corroboration.[31]

The announcement from Würzburg in November, 1895, that Roentgen had discovered a mysterious new penetrative radiation that would make it possible to see through opaque objects was front-page news all over the Western world. We have seen how Dayton C. Miller in Cleveland seized the opportunity to retest this surprising announcement by photographing the bones in his wife's hand. Scientific excitement was so great that within a year x-rays had captured the attention of almost every physicist.[32] Albert A. Michelson was no exception.

Now firmly and comfortably seated as the first professor of physics

---

Discovery of the Electron: The Development of the Atomic Concept of Electricity, chap. 2.

[31] Wilhelm Konrad Roentgen, "On a New Form of Radiation," *Nature* 53:274 (1896), facsimile reprint of excerpts in Mitchell Wilson, *American Science and Invention: A Pictorial History*, p. 331. For a full facsimile and discussion, see Bern Dibner, *The New Rays of Professor Roentgen*, especially pp. 52–54.

[32] See G. E. M. Jauncey, "The Birth and Early Infancy of X-Rays," *Am. J. Phys.* 13: 362–379 (December, 1945); cf. E. C. Watson, "The Discovery of X-Rays," *Am. J. Phys.* 13:281–291 (October, 1945).

at the new University of Chicago, Michelson watched with great interest the fast-growing body of knowledge concerning the behavior of
x-rays. Only a year after the original announcement, he felt that
enough evidence had accumulated to enable him to propose "A Theory
of the 'X-Rays.' "[33] As Michelson saw it, there were four principal
facts to be covered: first, the generation of rays at the cathode of a
vacuum tube; second, their propagation in straight lines but without
interference, reflection, refraction, or polarization phenomena; third,
the density of the medium; and fourth, the production of fluorescence.
In this paper Michelson observed that two theories were presently
current, first Roentgen's original suggestion that the x-rays were longitudinal wave motions in the aether, and secondly, the ballistic theory
that x-rays consisted of projected particles. Both these theories he
found objectionable, the first on the grounds that there was yet no
reason to suppose x-rays to be periodic. Michelson demurred at accepting Roentgen's original hypothesis and showed his own definitional
bias toward optical phenomena: "The absence of interference, reflection, and refraction is . . . a very formidable difficulty."[34]

Michelson expressed great astonishment at the fact that x-rays required a high-vacuum tube in order to be formed and then were able
to pass through anything. He compared this behavior to the aether-
vortex theory using the familiar analogy of a smoke ring going
through a wire mesh. Taking the early work of J. J. Thomson on the
speed of cathode and x-rays, Michelson felt confident that, since they
were apparently so slow, any connection with visible light waves must
be indirect.[35] Analogy suggested the working hypothesis, which he
proposed with some confidence: that the hydrodynamic view of the
aether would be the most fruitful approach to explaining the kinetic
properities of Roentgen rays. This paper represents one of the very
few excursions by Michelson into the speculative theoretical realm of

[33] *Am. J. Sci.*, 4th ser. 1:312–314 (April, 1896).
[34] Ibid., p. 313.
[35] Ibid., p. 314. See also Anderson, *The Discovery of the Electron*, p. 40. For another attempt to check Roentgen's suggestion of longitudinal aether waves using a
Michelson interferometer, see Richard Threlfall and James A. Pollack, "On some Experiments with Röntgen's Radiation," *Phil. Mag.*, 5th ser. 42:453–463 (December,
1896).

physics. It was premature, barren, and not in line with the mainstream of future development. These considerations show why Michelson is honored primarily as an experimentalist.

In the wake of the furor aroused by the new radiation discoveries of Roentgen, Becquerel, and the Curies, American physicists, except for B. B. Boltwood, tended to learn slowly the main lessons of radio-activity.[36] The Zeeman effect, showing how Fraunhofer lines could be split into component parts by magnetic fields, was also announced in 1896 and was more rapidly assimilated. Though less spectacular, the Zeeman effect was probably no less important a discovery than that of the x-ray, since the theory of electrons gained great prestige thereby, at the expense of the aether theories. Toward the clarification of the Zeeman effect, Michelson turned next with far greater success.[37]

The mathematical historian Whittaker has called the decade of the 1890's the "age of Lorentz." This appellation certainly seems justified, since there was no other individual so influential in structuring the course of theoretical physics as was Lorentz during these years. In 1895 he published, from Brill's presses in Leiden, the influential treatise which young Einstein was shortly to rely on heavily as the basis for his thought experiments. One biographer of Einstein reports that as a sophomore student, unaware as yet of Lorentz's concern with Michelson-Morley, Einstein planned "to construct an apparatus which would accurately measure the earth's movement against the ether."[38] Whether his reading of Lorentz, or the teaching of August Föppl, was more influential, the young Einstein turned toward theory instead of experiment, especially fascinated by the conceptual problems of rela-tive and absolute motion in space.[39]

[36] See Lawrence Badash, "The Early Developments in Radioactivity: With Em-phasis on Contributions from the United States," Ph.D. diss., Yale University, 1964.

[37] For a summary of the several papers Michelson contributed on this subject, see his book, *Light Waves and Their Uses*, pp. 107–126. A new adaptation of the inter-ferometer was required, and Michelson supplied it with his "echelon spectroscope."

[38] Anton Reiser, *Albert Einstein: A Biographical Portrait*, p. 52. The author was Rudolf Kayser, one of Einstein's stepsons-in-law, writing under a pseudonym.

[39] Toward the end of his life, Einstein honored Lorentz above all other influences, both personally and theoretically: "For me personally he meant more than all the others I have met on my life's journey," he wrote for the *festschrift*, *H. A. Lorentz: Impressions of His Life and Work*, ed. G. L. DeHaas-Lorentz, p. 8. More signifi-

Certainly it was the Lorentz treatise,[40] the *Versuch* of 1895, that did most to publicize the problems raised by the negative result of the Michelson-Morley experiment. In five prominent pages of his book, Lorentz reviewed the 1881 and 1887 experiments and his own contribution to the retest, then stated his inability to accept Michelson's interpretation of the aether-drift experiment. "Now, does this result entitle us to assume that the ether takes part in the motion of the earth, and therefore that the theory of aberration given by Stokes is the correct one?"[41] Lorentz thought not, preferring rather "to remove the contradiction between Fresnel's theory and Michelson's result" than to attempt to circumvent the aberration difficulties. The solution was simple. Assume material contraction dependent upon velocity of translation through the aether, and "the result of the Michelson experiment is explained completely." Lorentz continued:

> Thus one would have to imagine that the motion of a solid body (such as a brass rod or the stone disc employed in the later experiments) through the resting ether exerts upon the dimensions of that body an influence which varies according to the orientation of the body with respect to the direction of motion.

> .   .   .   .

> Surprising as this hypothesis may appear at first sight, yet we shall have to

---

cantly, Einstein's retrospection relates to our concerns: "The only phenomenon whose explanation could not be given completely—i.e., without additional assumptions— was the famous Michelson-Morley experiment. But it would have been unthinkable for this experiment to lead to the special relativity theory without the localization of the electromagnetic field in empty space. The really essential step forward, indeed, was precisely Lorentz' having reduced the facts to Maxwell's equations concerning empty space, or—as it was then called—the ether" (ibid., p. 7). See also Gerald Holton, "Influences on Einstein's Early Work in Relativity Theory," *The American Scholar* 37:59–79 (Winter, 1967–1968) for the role of August Föppl in Einstein's development.

[40] H. A. Lorentz, *Versuch einer Theorie der electrischen und optischen Erscheinungen in bewegten Körpern*; also in *H. A. Lorentz: Collected Papers*, eds. P. Zeeman and A. D. Fokker, V, 1–137. A section from this treatise, entitled "Michelson's Interference Experiment," was translated by W. Perrett and G. B. Jeffrey to begin *The Principle of Relativity: A Collection of Original Memoirs on the Special and General Theory of Relativity*. References in notes below are from the Dover reprint (New York, 1923).

[41] Lorentz in *The Principle of Relativity*, p. 4.

admit that it is by no means far-fetched, as soon as we assume that molecular forces are also transmitted through the ether, like the electric and magnetic forces of which we are able at the present time to make this assertion definitely. . . . Now, since the form and dimensions of a solid body are ultimately conditioned by the intensity of molecular actions, there cannot fail to be a change of dimensions as well.[42]

As these quotations show, Lorentz continued to take Michelson-Morley very seriously, but the so-called ad hoc contraction hypothesis was not a completely isolated interpretation of Michelson's null results. The transformation equations which Lorentz proceeded to derive also had their base in the "electric and magnetic forces" which supplied supporting analogies. In addition to the fact that "the lengthenings and shortenings in question are extraordinarily small,"[43] the contraction hypothesis now led Lorentz to restate his major idea. This was, as one of his students has said, "the fundamental and total separation of ether and matter."[44] Henceforward, Lorentzians would draw a dichotomy between the imponderable aether and ponderable matter.

In the years after their professional divorce, Michelson and Morley had gone their separate ways, pursuing their own research interests. In 1895, Morley had been elected president of the AAAS, after having received due recognition for his work in determining atomic weights.[45] Michelson had enhanced his reputation by his thorough work in Paris on the meter bar. After his return from France, Michelson had been one of the prize catches made by William Rainey Harper in 1894 during his raids to recruit the best possible staff for the new University of Chicago. Now, in the latter half of the decade, each man was to return partially to the experimental problems they had attacked jointly ten years earlier.

[42] Ibid., pp. 5, 6.

[43] Ibid., p. 6.

[44] A. D. Fokker, "Scientific Work," in *H. A. Lorentz (festschrift)*, ed. G. L. De-Haas-Lorentz, p. 56.

[45] Morley's address as retiring president in August, 1896, is interesting for its examination of Prout's hypothesis on hydrogen as the primordial quintessence. The aether, of course, had also been hypostasized by some, but Morley was skeptical of both claims. "A Completed Chapter in the History of the Atomic Theory," *Proc. AAAS* 45:1–22 (January, 1897).

Having retired from some of his administrative duties, Edward Morley felt free to return to a project he had initiated in 1889 with Henry T. Eddy of Minneapolis. In that year Eddy had speculated that the velocity of light in a magnetic field might vary in accordance with Hall's effect.[46] Eddy had come to Cleveland in 1890 for a preliminary trial run with Morley, but immediate frustrations and intervening delays postponed definitive tests until 1897–1898. By this time Eddy was unavailable and Dayton C. Miller had taken his place. Morley and Miller used the optical parts from the classical "interferential refractometer" and carried out observations in a similar manner, with the exception that the light path was made to pass through a static magnetic field. In their results, published in December, 1898, they announced that they had found no displacement more than one-twentieth of a wave length![47]

At the same time, at the new Ryerson Laboratory in Chicago, some peculiar construction was going on in the winter and early spring of 1897. Michelson had been as displeased with Lorentz's critique of his reliance on Stokes's theory as Lorentz had been displeased with Michelson. Therefore, Michelson determined to set up a large-scale vertical interferometer in order to test whether there is or is not an aether gradient corresponding to altitude above the surface of the earth. Michelson believed that Stokes's theory could account for aberration if the relative velocity of light could be shown to be different at the earth's surface and in the aether at a height.

In the hope of detecting a relative motion corresponding to a difference in level, that is, altitude, Michelson had plumbers lay six-inch pipes all around the north wall of the Ryerson Laboratory. These pipes extended lengthwise two hundred feet and vertically fifty feet from ground level to the roof of the laboratory. The rectangle of pipes was constructed in a vertical plane in an east–west direction and covered with a wooden box-sheathing. Adjustable mirrors were placed in each

---

[46] Edwin H. Hall had been a student of H. A. Rowland at Johns Hopkins University in 1879 when he discovered a new action on electric currents exerted by a magnetic field. For details, see Whittaker, *History*, I, 289–291.

[47] E. W. Morley and D. C. Miller, "The Velocity of Light in a Magnetic Field," *Phys. Rev.* 7:283–295 (December, 1898).

corner and the pipes evacuated to one one-hundredth of an atmosphere.

Although there were several difficulties, such as the inefficiency of the wood-box insulation, all observations were carried out in March of 1897, and Michelson without chagrin announced thereafter that "if there is any displacement of the fringes it is less than one-twentieth of a fringe."[48] Since he had found no appreciable differential between the velocity at which light travels at ground level and at fifty feet above ground level, Michelson surmised that "the earth's influence upon the ether [must be] extended to distances of the order of the earth's diameter." Michelson was reticent in announcing this conclusion, saying that it "seems so improbable that one is inclined to return to the hypothesis of Fresnel and to try to reconcile in some other way the negative results" (obtained in his experiment with Morley in 1887). Crediting Lorentz fully for having made this attempt, but still reluctant to grant its plausibility, Michelson continued:

In any case we are driven to extraordinary conclusions, and the choice lies between these three:

1. The earth passes through the ether (or rather allows the ether to pass through its entire mass) without appreciable influence.

2. The length of all bodies is altered (equally?) by their motion through the ether.

3. The earth in its motion drags with it the ether even at distances of many thousands of kilometers from its surface.[49]

It must be emphasized that this experiment of 1897 was not primarily an aether-*drift* test. Michelson was seeking validation for his own (i.e., Stokes's) interpretation of the results of his aether-drift test. Instead of an aether wind, he was looking for an ethereal atmosphere. If the relative motion of earth and aether varies with the vertical height above the terrestrial surface, then the velocity of light might be said to depend upon its distance from the center of the earth. Failing to find any such property, Michelson was still unwilling to let the

---

[48] A. A. Michelson, "The Relative Motion of the Earth and the Ether," *Am. J. Sci.*, 4th ser. 3:477 (June, 1897).

[49] Ibid., p. 478.

FitzGerald-Lorentz hypothesis have the field. He insisted that two other possibilities were at least as plausible as the contraction thesis.

To recapitulate, during the last decade of the nineteenth century four points stand out as major features in the historical development of the aether-drift experiments. First and most obviously, the luminiferous aether had merged into an electromagnetic aether, thence into dielectric aethers of several different sorts; but this metamorphosis had in no way diminished the conceptual need most physicists still felt for a medium. In fact, if anything, the aether concept was never more firmly established than during this period when the marvelous new communication machine called the radio was in its infancy.

Second, Hertz's discovery of wireless waves was more than sufficient to offset, for the time being, any worries that may have lingered about the lack of a consistent theory to explain this medium. Hertzian waves were there for all to perceive as experimental evidence reinforcing belief in an all-pervasive medium. Physical opticians and mathematical physicists became ever more concerned over the theoretical contradictions revealed by radiation studies in the 1890's, but apparently few despaired of eventually finding a solution. Hertz himself was one of the most intense critics of the usual aether theories, but he, like Maxwell and Helmholtz, died without denying the need for a plenum.

Third, insofar as the Michelson-Morley experiment itself was known, it posed a problem indeed, but there were many possible explanations for it, and no direct corroborations of it. Besides the fact that Hertz had seemingly bolstered the same aether which Michelson had undermined, Lodge's test for ethereal viscosity seemed in its failure to cancel out the failure to find an aether wind. Perhaps this double failure simply meant that the Michelson-Morley test was not sensitive enough, after all, to detect so slight an effect. Even though Michelson was trusted beyond question and his measurements were admirably precise, still at least he, Morley, and Miller knew that the seasonal tests had never been made and that there were some extraordinarily delicate possibilities for unnoticed experimental errors.

Fourth, and finally, this chapter has shown by implication that there were *probably* no other experiments performed *on the exact model* of the Michelson-Morley plan or *for that specific purpose* from July, 1887, until the summer of 1902. This is an empirical assertion that may be overthrown by the discovery of some repetitions in Europe or elsewhere, or by a new scholarly consensus about the mathematical or experimental significance of the evidence here presented. But if this interpretation is true, then there is all the more reason to wonder how this uncompleted, unexplained, optical experiment blended into the social fabric of theoretical physics to become a part of the scientific revolution and philosophic renaissance of the twentieth century.[50]

[50] These conclusions, as well as those at the beginning of chapter 12 (below), were tested in my article, "The Michelson-Morley-Miller Experiments Before and After 1905," *Journal for the History of Astronomy* 1, part I:56–78 (February, 1970).

CHAPTER SIX

TRANSFORMATIONS OF AETHER, CIRCA 1900

The end of a century, like the end of a year, is a time for inventory and a time for projections. As the nineteenth century blended into the twentieth, a number of physicists and quasi-physicists recorded their views of the progress of the century. A veritable library of literature of this sort can be mined to evaluate the social situation in physical theory at the turn of the century. This chapter will survey some of this literature in order to understand what the so-called physics of the aether meant to physicists of that era.[1]

[1] The best starting point perhaps is still John Theodore Merz, *A History of European Thought in the Nineteenth Century*, II, 89–199. The problem of adequate historical monographs and a summary list of bibliographic references to "progress-of-the-century" literature are both exposed by I. Bernard Cohen, "Conservation and the Concept of Electric Charge: An Aspect of Philosophy in Relation to Physics in the Nineteenth Century," in Marshall Clagett, ed., *Critical Problems in the History of Science*, pp. 357–376. See also Cohen's valuable bibliographic essay "Some Recent

The date 1900 is notable, of course, for the papers which Max Planck read in that year announcing the necessity for assuming discontinuous energy in the emission or absorption of radiation, thus marking the advent of the quantum theory. Although Planck long resisted the idea that radiation itself is discontinuous, many historians and memoirists have judged the quantum revolution as more fundamentally upsetting to physics than the relativity revolution.[2] However one may side in this evaluation, which, incidentally, depends largely upon how one defines "classical" physics, it is unmistakable that the concept of energy was much less questioned and more unanimously accepted in 1900 than was the concept of the aether. We must remember that the incorporation of a quantum theory into the fabric of physics also took time, and that there were two quantum revolutions, one starting about 1900 and another ending about 1927. When Planck received the [1918] Nobel Prize, in 1920, he warned that from his 1900 discovery of the quantum of action to a truly satisfying comprehensive quantum theory might take as long a time as from Roemer to Maxwell![3] In this prediction he may well have been right, but the majority of physicists in the mid–twentieth century seemed content to

---

Books on the History of Science," in P. P. Wiener and Aaron Noland, eds., *Roots of Scientific Thought: A Cultural Perspective*, pp. 627–656; and Alfred M. Bork, "Physics Just before Einstein," *Science* 152:597–603 (April 29, 1966).

[2] Planck himself drew attention to this distinction in 1931 by pointing out the essentially unifying role of relativity and the disrupting effects of quantum theory. See Max Planck, *The Universe in the Light of Modern Physics*, trans. W. H. Johnston, pp. 18, 24. Banesh Hoffman states the typical case for Planck over Einstein by taking as the "classical" basis for comparison Maxwell's electromagnetic theory rather than Newton's gravitational theory. In this case, relativity is seen as "corroborating" the general form of "classical" (i.e., Maxwellian electrodynamical) theory while quantum theory is seen as "corroding its very fundamentals"! See Banesh Hoffman, *The Strange Story of the Quantum: An Account for the General Reader of the Growth of the Ideas Underlying Our Present Atomic Knowledge*, p. 10. For the best available study, see Max Jammer, *The Conceptual Development of Quantum Mechanics*.

[3] Max Planck, *The Origin and Development of the Quantum Theory: Being the* [1918]*Nobel Prize Address Delivered before The Royal Swedish Academy of Science at Stockholm, 2 June 1920*, trans. H. T. Clark and L. Silberstein, pp. 17–18. See also the basic guide to the literature, Thomas S. Kuhn et al., *Sources for History of Quan-*

view quantum mechanics as having reached a satisfying maturity at least by 1930.

Meanwhile, around 1900, contiguous fields of energy were often conceived as merely local manifestations of the universal aether. Although both concepts—the electromagnetic field and the electromagnetic aether—were presumed continuous in nature, the former could be manipulated by considering fields as isolated systems, whereas the latter apparently could not be. John Trowbridge of Harvard, echoing Paul Drude of Göttingen, characterized physics, at the end of the century, as simply the study of transformations of energy, no longer concerned with *Fernkräfte* but only with *Nahekräfte*.[4] Thomas Preston of Trinity College in Dublin likewise saw the current tendency of physical science as a trend towward regarding "all the phenomena of nature, and even matter itself, as manifestations of energy stored in the ether."[5]

A prime example of this transfer of attention was [Sir] Joseph Larmor's 1898 prize essay entitled *Aether and Matter*. This book has been called an influential statement of the "relations of micro-discontinuity to macro-continuity, of atomic matter to the electromagnetic ether, on a classic pre-1900 basis."[6] When *Aether and Matter* was published, in 1900, the author professed to feel even better about his argument than he had at the time of writing. Subtitled *A Development of the Dynamical Relations of the Aether to Material Systems on the Basis of the Atomic Constitution of Matter, including a Discussion of the Influence*

---

tum Physics: An Inventory and Report, and Kuhn's review-essay, "The Turn to Recent Science," *Isis* 58:409–419 (Fall, 1967).

[4] John Trowbridge, *What Is Electricity?* p. 3. Cf. p. 276: "Perhaps nothing marks so strongly the modern attitude toward physical manifestations as the substitution for action at a distance the action in matter from particle to particle." See also Paul Drude, *Physik des Aethers auf Elektromagnetischer Grundlage*, pp. 8–11.

[5] Thomas Preston, *The Theory of Light*, ed. Charles J. Joly, pp. 28–29.

[6] Lancelot L. Whyte, "A Forerunner of Twentieth Century Physics: A Re-View of Larmor's 'Aether and Matter,'" *Nature* 186:1010 (June 25, 1960). For other examples, see George Basalla, W. Coleman, and R. H. Kargon, eds., *Victorian Science: A Self-Portrait from the Presidential Address of the British Association for the Advancement of Science*, pp. 87–158.

of the Earth's Motion on Optical Phenomena,[7] Larmor's treatise featured an analysis of the Michelson-Morley experiment as a central concern of physics.

Unlike Paul Drude's famous German text of 1894, *Physik des Aethers*, which made no mention of Michelson's experiment, but which did develop the field concept beyond the work of Maxwell and Hertz, Larmor's book quickly became dated. Largely because *Aether and Matter* was written in terms of physical properties rather than mathematical processes, it made Larmor appear to have missed the opportunity to contribute to the rapidly developing field concept. If Larmor's work had relatively little influence on the course of physics after Einstein, it certainly has had great influence on interpretations of the history of physics before Einstein. As a middle link between Lorentz's *Versuch* of 1895 and E. T. Whittaker's *A History of the Theories of Aether and Electricity* of 1910, Larmor's treatise helped to spread the notoriety of the Michelson-Morley experiment.

Larmor's attitude toward the concepts in his main title was summarized in his prefatory note: "Matter may be and likely is a structure in the aether, but certainly aether is not a structure made of matter. This introduction of a suprasensual aethereal medium, which is not the same as matter, may of course be described as leaving reality behind us: and so in fact may every result of thought be described which is more than a record or comparison of sensations."[8] Larmor's essential contribution in this work was to correlate the experimental and theoretical history that led to the view that the basic construction of the microcosmic world is electrodynamic in form. Larmor missed the mainstream of developments in Maxwellian electrodynamics, but he anticipated many of the elements of Einstein's concerns in 1905 and later. He considered the Michelson-Morley experiment fundamental to his subject and supported the FitzGerald-Lorentz contraction as

---

[7] Sir Joseph Larmor, *Aether and Matter*.

[8] Ibid., p. vi (footnote). It should be noted that Drude, too, by 1900 was discusscussing Michelson-Morley at length: see Paul Drude, *The Theory of Optics*, trans. C. R. Mann and R. A. Millikan, pp. 478–482.

st explanation.[9] In advocating the "aethereal constitution of matter, Larmor defended the notions of plenum and continuum at the expense of atomic theory. "All that is known (or perhaps need be known) of the aether itself," he insisted, "may be formulated as a scheme of differential equations defining the properties of a *continuum* in space, which it would be gratuitous to further explain by any complication of structure."[10] The vogue for this kind of thought was short-lived, but Larmor expressed the theoretical temper of his time well, just as Michelson expressed its experimental temper.

In the spring of 1899 Professor Michelson was invited to Massachusetts to give a series of eight lectures at the Lowell Institute on the subject which had won him fame: light waves and their uses.[11] When published, several years later, these lectures in many respects epitomized physical optics in the late Victorian period. All the harshest criticisms of physicalistic positivism can be supported by taking statements made by Michelson in these lectures out of context and contrasting these attitudes with the standards of humility required by mid-twentieth-century science.

Michelson began these lectures with an apology for the difficulties his audience might experience in listening to a scientist's effort to communicate the results of his labor; he ended the lectures with a discussion of the aether, patterned very closely on his paper of 1897; and in between, from first to last, he recounted his own experimental development of confidence in the wave theory and its consequently necessary medium. By modern standards it would seem unforgivable for anyone to say, as Michelson did in his second lecture, that "the more important fundamental laws and facts of physical science have all been discovered, and these are now so firmly established that the possibility of their ever being supplanted in consequence of new discoveries is exceedingly remote." Michelson's supreme confidence that the struc-

---

[9] Larmor, *Aether and Matter*, p. 46. For an example of respect for Larmor, see Max Born, *Einstein's Theory of Relativity*, p. 222.

[10] Larmor, *Aether and Matter*, p. 78. Cf. Larmor's energeticist proclivities as discussed here later, beginning on page 131.

[11] Albert A. Michelson, *Light Waves and Their Uses*.

ture of physical science had finally been revealed was further underscored by his easy repetition of the then current aphorism that "our future discoveries must be looked for in the sixth place of decimals."[12]

Although it is easy to condemn this philosophical attitude as naive realism, historical justice requires that we recognize the context in which such sentiments were uttered. In 1899, as Michelson was delivering these lectures, he unquestionably felt the need to justify his superrefinement of measurement accuracy. Not only to the lay audience but also to some of his fellow scientists in other disciplines, the ultraprecision involved in using interference methods of measurement seemed to be an almost irrational obsession. Why, for instance, should so much time, money, and effort be spent to compare the length of the meter bar at Paris with light waves? What is the object of all this accuracy? Michelson's answer reflects the general loss of confidence by physicists in all material standards at the time of his address: "The standard light waves are not alterable; they depend on the properties of the atoms and upon universal ether; and these are unalterable. It may be suggested that the whole solar system is moving through space, and that the properties of ether may differ in different portions of space. I would say that such a change, if it occurs, would not produce any material effect in a period of less than twenty millions of years, and by that time we shall have less interest in the problem."[13]

In the eighth and last lecture of this series, Michelson discussed the problems involved in the concept of the aether. Reminding his audience of the extraordinary difficulty in trying to comprehend the magnitude of the speed of light, Michelson emphasized the vast gap in velocities between the closest analogy—sound traveling through metal at about 3 miles per second, and light traveling through space at about 186,000 miles per second. For Michelson this enormous gap was far too great to permit any material interpretation of the aether.

On the other hand, his commitment to the undulatory theory allowed him no latitude to conceive of light without a medium. If the

[12] Ibid., pp. 23–24. For a satire on such metaphysics, see A. Boyajian, "A. A. Michelson Visits Immanuel Kant," *Scientific Monthly* 59:438–450 (December, 1944).

[13] Michelson, *Light Waves*, p. 105.

contradictory properties involved in a material view of the ethereal medium could be reconciled, he concluded that it must "be an elastic solid rather than a fluid." But since there was practically no hope for a reconciliation of the contradictory properties required of a material medium, he agreed with FitzGerald and Drude that the aether must be "of an entirely different order" from familiar substances, that it "belongs in a category by itself."[14] Nonetheless, to Michelson it appeared "practically certain that there must be a medium whose proper function it is to transmit light waves." Although he admitted that light had been "fairly well established" as an electromagnetic disturbance, differing only in its restricted range of wavelengths, he felt the need for an electromagnetic medium was not reduced: "The settlement of the fact that light is a magnetoelectric oscillation is in no sense an explanation of the nature of light. It is only a transference of the problem, for the question then arises as to the nature of the medium and of the mechanical actions involved in such a medium which sustains and transmits these electromagnetic disturbances."[15]

Having reviewed the major concepts offered by various physicists in their speculative descriptions of the atom and the aether, Michelson showed his usual restraint by not aligning himself with any of them completely. Most promising for the future of any of the theories offered, he still believed, was the vortex theory of Kelvin. In reviewing his own contribution to the drift problem during the previous decade, Michelson was quite blunt in stating that "the result of the experiment was negative and would, therefore, show that there is still a difficulty in the theory itself; and this difficulty . . . has not yet been satisfactorily explained." [He was] "presenting the case, not so much for the solution, but as an illustration of the applicability of light waves to new problems."[16]

---

[14] Ibid., pp. 147–148. This *sui generis* interpretation of the aether, always more or less implicit, had become eclipsed once again after Hertz and before the failure of Kelvin's last elastic-solid model, about 1894.

[15] Ibid., pp. 159, 161.

[16] Ibid., p. 158. Kelvin's aether-vortex or smoke-ring theory of atomic matter, Michelson noted, "if true, has the merit of introducing nothing new into the hypotheses already made, but only of specifying the particular form of motion required" (p. 161).

The depreciation shown here by Michelson in his own evaluation of the Michelson-Morley experiment was a consistent characteristic throughout his life of his view of that result. He was never enthusiastic about his aether-drift test, primarily because he did not see it as something he had done, but as something he had failed to do. Indeed, his enthusiasm, if one may call it such, was all in the other direction in 1899. Michelson ended his Lowell lectures saying: "Suppose that an ether strain corresponds to an electric charge, and ether displacement to the electric current, these ether vortices to the atoms— if we continue these suppositions, we arrive at what may be *one of the grandest generalizations of modern science*—of which we are tempted to say that *it ought to be true even if it is not*—namely, that *all the phenomena of the physical universe are only different manifestations* of the various modes of motion of one all-pervading substance —*the ether.*[17]

Perhaps the most revealing aspect of the preceding quotation is the explicit aside set off by dashes: "—of which we are tempted to say that it ought to be true even if it is not—." In this statement Michelson clearly reveals the attitude which prevailed among most of his colleagues at the turn of the century. It was an attitude which combined skepticism and confidence, doubt and faith, humility and pride. The temptation to be categorical and to agree with Lord Kelvin that "the ether is the only form of matter about which we know anything at all," had become more and more compelling as a result of the etherealization of matter. Yet to succumb to that temptation and to reify the aether was recognized as being a poetic rather than a scientific tendency. That the aether ought to be true even if it is not, was an attitude which reflected the whole progress of physics in the nineteenth century. Still, the justification for an aether was usually recognized as analogical and heuristic. The aether was, therefore, left deliberately vague, to accommodate the ambiguous answers to the ultimate questions about light.

Few physicists before Planck could have foreseen the possibility of the resuscitation of the particle, or ballistic, theory of light transmission. Had they done so, the Michelson-Morley experiment would have

---

[17] Ibid., p. 162 [italics mine]. See also W. M. Hicks ["Theories of the Aether"], *Br. Ass'n. Adv. Sci. Annual Report* (1895), pp. 595–606.

represented no problem at all, since, according to the emission theory, the Michelson-Morley experiment must give a negative result; the very term *aether drift* would be meaningless.[18]

Although hardly anyone dared in 1900 to advocate the corpuscular theory of light, Planck's quantum of action had introduced an important distinction between the *transmission* or *propagation* and the *emission* or *absorption* of light.[19] A few critics used this for serious attacks on the aether.[20] Generally, however, the recent subsumption of light under electromagnetic phenomena meant that the atomization of energy which Planck now began to advocate (to account for the bunching of spectroscopic energy levels and the "black-body" problem), placed him in the position in which Thomas Young had found himself a full century before. That is, radiant-energy studies would slowly have to prove, as interference studies had done, that their specific concerns were far more general and fundamental than hitherto imagined. Meanwhile, until Einstein reinforced Planck in 1905 and after, quantization of energy was not generally regarded as particularly applicable to the behavior of visible light.[21]

Kelvin's often quoted "two clouds" address in 1900,[22] to the effect that there were on the horizon only two patches of bad weather threat-

[18] An eminent historian of mechanics has reminded us of this fact: "It must be remembered that the Michelson-Morley calculation supposes that the wave theory is accepted or, more accurately, that it is assumed that the velocity of the propagation of light in the sidereal vacuum is independent of the motion of the source. If, on the contrary, the emission theory is taken as the starting-point, it is immediately concluded that the experiment must give a negative result" René Dugas, *A History of Mechanics*, trans. J. R. Maddox, p. 487).

[19] Max Planck, "Ueber das Gesetz der Energieverteilung im Normalspectrum," *Ann. d. Phys.*, 4th ser. 4:553–563; and "Ueber die Elementarquanta der Materie und der Elektricität," pp. 564–566 (January, 1901).

[20] For example, C. Riborg Mann, *Manual of Advanced Optics*, pp. 165–170; W. C. Dampier-Whetham, *The Recent Development of Physical Science*, pp. 246–291.

[21] Again Dugas is worth quoting on this point: "Most physicists of the time, including Planck, were still fundamentally convinced that natural processes are continuous; as Newton had expressed it, *natura non saltus facit*. Indeed, both the ancient faith in the sequence of cause and effect and even the usefulness of the mathematical calculus itself seemed to depend on the proposition that natural phenomena do not proceed by jumps" (Dugas, *History of Mechanics*, p. 581).

[22] Lord Kelvin, "Nineteenth Century Clouds over the Dynamical Theory of Heat and Light," *Am. J. Sci.*, 4th ser. 12:391–392 (November 1901); also reprinted in

ening the fair climate of physical opinion, referred directly to, first, the inexplicable results of Michelson-Morley, and, second, the curious difficulty in radiation studies of explaining the problem of mean kinetic energies. Significantly, to Kelvin the first cloud appeared very dense, a thunderhead in the clear sky, whereas the second cloud seemed much less threatening.

At the turn of the century physical speculation deliberately departed, by two broad avenues, from the experimental evidence regarding the nature of the aether. The first avenue was that of the oversimplification required in popularizations; the second was via the energeticist school of interpretation of microphysical phenomena. Since the energeticist school represented a highly sophisticated and legitimate interpretation of thermodynamics, it will be considered first.

As early as 1854, W. J. M. Rankine had entertained the possibility that thermodynamics might be used as a basis for a unified system of natural science.[23] Rankine, a Scottish engineer, was deeply concerned with the difficulties surrounding the concept of potential energy. The peculiar need for this hypothetical construct had originally grown out of the necessity for some way to conceive of energy as being "stored." Rankine had proposed a "science of energetics," in which the hypothetical concept of energy would be raised to an abstract level where it would enjoy immunity from all conjectural postulates based on indirect evidence. Although Rankine worked with molecular hypotheses and therefore differed greatly from the later school of anti-atomists who adopted his word *energetics*, Rankine did hope to establish the concept of energy with "that degree of certainty which belongs to observed facts," and as the basis for unifying "all branches of physics into one system."[24]

Kelvin's *Baltimore Lectures on Molecular Dynamics and The Wave Theory of Light*, Appendix B, pp. 486–527.

[23] W. J. M. Rankine, "Outlines of the Science of Energetics," *Proceedings of the Philosophical Society of Glasgow*, 3:381, as reprinted in his *Miscellaneous Scientific Papers*. For the chemical continuation of this, see W. H. Brock, ed., *The Atomic Debates: Brodie and the Rejection of the Atomic Theory, Three Studies*.

[24] Rankine, *Miscellaneous Scientific Papers*; see also Ernest Nagel, *The Structure of Science: Problems in the Logic of Scientific Explanation*, p. 126. Cf. Ernst Cassirer, *Substance and Function*, trans. W. C. and M. S. Swabey, pp. 187–203.

Concurrent with the development of confidence in the wave theory, physicists in the late nineteenth century were becoming ever more confident of thermodynamic theory. The spectacular development of the conservation laws, especially the conservation of energy, closely paralleled the development of physical optics.[25] Many physicists who were concerned with thermodynamics tended to view radiation phenomena in statistical terms of heat transfer, in terms of molecular motion rather than wave action. This emphasis on the ultraminute hypermicroscopic world rather than the gross world of sensible matter might lead either to an atomistic or an energistic metaphysics. The dialectic between these two and several other alternative systems before 1900 is a complex chapter of the history of science which has yet to be codified.[26]

Larmor began his article "Energetics" for the eleventh edition of the *Encyclopaedia Britannica* as follows: "The most fundamental result attained by the progress of physical science in the 19th century was the definite enunciation and development of the doctrine of energy, which is now paramount both in mechanics and in thermodynamics."[27] His long article hardly mentions the names of the men now usually tagged as "energeticists": Ernst Mach, Pierre Duhem, Wilhelm Ostwald, Georg Helm. Larmor said only that "of recent years a considerable school of chemists" have insisted "on this procedure as a purification of their science from the hypothetical ideas as to atoms and molecules."[28] The major chemist in this group was Ostwald, but all did have in common an intense desire to purify physical science of all "hypothetical"

---

[25] See Thomas S. Kuhn, "Energy Conservation as an Example of Simultaneous Discovery," and the critiques of this paper by Carl B. Boyer and Erwin Hiebert in Marshall Clagett, ed., *Critical Problems*, pp. 321, 384, 391. Cf. Erwin Hiebert, *Historical Roots of the Conservation of Energy.*

[26] Charles C. Gillispie, *The Edge of Objectivity: An Essay in the History of Scientific Ideas*, p. 496, passim. See also Stephen G. Brush, "Science and Culture in the Nineteenth Century: Thermodynamics and History," *The Graduate Journal* [The University of Texas] 7:477–565 (Spring, 1967); and Andrew G. Van Melsen, *From Atomos to Atom: The History of the Concept Atom*, pp. 160–191.

[27] Sir Joseph Larmor, "Energetics," *Encyclopaedia Britannica*, 11th ed., IX, 390 (1910–1911).

[28] Ibid., p. 397. See also Wilhelm Ostwald, *Die Überwindung des Wissenschaftlichen Materialismus.*

notions. Ernst Mach was the archetypical puritan of this sort. Both the atom and the aether were such notions in 1900.

The significance of the "energeticist movement," for our purposes, is that it placed much greater emphasis on anti-atomism than on anti-aetherism. Hence, simply by having "stormed the wrong door"[29]—the atom instead of the aether—the energeticists were open to accusations of being tolerant of, if not favorable toward, the admission of a universal medium "rigid enough to bear sheer waves and rare enough to pass detectable bodies undetectably through its subtlety."[30]

The energeticist school certainly had no exclusive franchise in seeking recognition for a "physics of the aether" at the turn of the century. While Mach, Ostwald, Helm, and Duhem were propounding anti-atomistic, and, by implication promonistic, philosophic views, several other influences were pressing for recognition on the periphery of physics. From chemistry a group of molecular theorists, supporting Ostwald, but not necessarily following him, were urging the adoption of a colloidal conception of the aether. One such paper, which appeared in 1898, came from Carl Barus at Brown University. In working with the varied and anomalous properties of colloids under pressure, Barus was struck by the similarity between the behavior of his coagulates in their strain, break, and recementation effects and the supposed behavior of a jelly aether. To the physicists Barus offered the immediate analogy from his work that "*the same ether may therefore act, as the case may be, either as a liquid or as a solid.*"[31]

Perhaps the most interesting effort from chemistry to furnish physics with a solution to the aether problem came from the most famous author of the periodic table of the elements. Dmitri Mendeléev

[29] Gillispie, *The Edge of Objectivity*, p. 498. See also Stephen G. Brush, "Mach and Atomism," *Synthese* 18:192–215 (1968).

[30] Gillispie, *The Edge of Objectivity*, p. 499. In view of the important influence of Mach on Einstein, it should be noted that Mach's monism emphatically was not connected with the notions of absolute time or absolute space. Ernst Mach, *The Science of Mechanics: A Critical and Historical Account of Its Development*, trans. T. J. Mc-Cormack; *Supplement to the Third English Edition*, trans. P. E. B. Jourdain, p. xii.

[31] Carl Barus, "The Compressibility of Colloids, with Applications to the Jelly Theory of the Ether," *Am. J. Sci.*, 4th ser. 6:297 (October, 1898). See also Barus's later work, *Interferometer Experiments in Acoustics and Gravitation*.

(1834–1907), the grand old Russian chemist whose marvelous intuition had proven so fruitful in predicting the discovery and order of elements, tried at the turn of the century to fit the electromagnetic aether as an inert gas more subtle than hydrogen in his periodic table. His thesis reached English readers in a 1904 translation entitled *An Attempt Towards a Chemical Conception of the Ether*.[32] This work argued, with all due tentativeness, that "the ether may be said to be a gas, like helium or argon, incapable of chemical combination."[33] Mendeléev's conception of the aether in chemical terms was stillborn, however, since by 1904 the author's very definite preference for chemomechanical models of explanation as well as his dependence on the ideal fluid concept were generally passé.

In addition to the energeticists and positivists of this period, other philosophers of science, Helmholtz, Poincaré, Clifford, Pearson, and, in America, J. B. Stallo and C. S. Peirce,[34] were consulted as influential opinion molders. These were a few of the leaders of a minority of men who were truly scientist-philosophers, men who became at least semiprofessionally involved in both fields. As such, they provided an important leaven to physical theory, and in some cases their criticisms of the aether concept were devastating.[35]

[32] Professor D. Mendeléeff [*sic*], *An Attempt Towards a Chemical Conception of the Ether*, trans. George Kamensky.

[33] Ibid., p. 14. Cf. William Ramsay, *The Gases of the Atmosphere: The History of Their Discovery*, 3rd ed. (1905).

[34] In the light of Poincaré's later importance to the birth of relativity theory, it is interesting to note that C. S. Peirce, the father of pragmatism or "pragmaticism," was heavily critical of Poincaré's early methodological pronouncements outlining a "conventionalist" view of the ether. To Samuel Langley, Peirce wrote in January, 1901: "Now I do not think Poincaré's more than doubtful views ought to be put before the scientific laity with the quasi-sanction of the Smithsonian without the other side of the question, which is the position of all physicists except one little clique of revolutionists (I like the expression, since they propose to put down all mechanical explanations of electricity etc. which are the very crown of our received physics) unless the antidote be administered with the base." Quoted in Carolyn Eisele, "The Scientist-Philosopher C. S. Peirce at the Smithsonian," *J. Hist. Ideas* 18:542 (October, 1957).

[35] One of the most iconoclastic radical relativists was the German-American intellectual Johann Bernhard Stallo (1823–1900), who began in 1873 a campaign to make physicists face their metaphysical theories of cognition. Mach was impressed, and he praised highly Stallo's thorough insistence on the relativity of all knowledge. See J. B. Stallo, "The Primary Concepts of Modern Physical Science," *Popular Science Monthly*

Whatever were the attitudes of more strictly mathematical physicists, like Gibbs, Boltzmann, and Planck, many physicist-philosophers rejected the aether concept, using virtually the same evidence that led many other philosopher-physicists to embrace it. Whereas aether apologists would argue, from the standpoint of physical *dynamics*, that dematerialization demanded an aether as a conceptual necessity, aetherophobes could argue, from the standpoint of *kinematics*, that purely formal analyses of motion deny any such necessity. Henry Crew, writing in 1899, showed how this distinction could be used to avoid the aether problem altogether:

A theory of light may be considered either from a kinematical or from a dynamical point of view. To assume, on experimental grounds, that a ray of light has a different medium, and that it consists in a particular kind of motion, and *thence* to infer the laws of refraction, rectilinear propagation, and diffraction, is to construct a kinematical theory of light. But to assume a certain structure for the luminous body and for the medium, and thence to derive the motions and the different speeds assumed in the kinematical case, is to offer a dynamical explanation of light.

The wave-theory of light is used, nearly always, in the former and narrower sense to mean the kinematical explanation of light; it leaves entirely to one side the dynamical questions hinted at above. It assumes, not without strong experimental evidence, the existence of waves travelling with different speeds in different media, and proposes to explain the cardinal phenomena of optics.[36]

Dissatisfied with such circumlocutions, Arthur Schuster of Manchester complained in 1904 that no theory of optics existed anymore, at least not in the sense of the wave theory of fifty years ago. He advised suspending judgment until the properties of "aether" could be specified more thoroughly. Equally disgusted with the kinematicists' claim to superiority over dynamicists, William G. Hooper of Notting-

---

3:705–716 (October, 1873), and Stallo's book, *The Concepts and Theories of Modern Physics*, edited by P. W. Bridgman. See also Stillman Drake, "J. B. Stallo and the Critique of Classical Physics," in *Men and Moments in the History of Science*, ed. Herbert M. Evans, pp. 25–30.

[36] Henry Crew, ed., *The Wave Theory of Light: Memoirs by Huygens, Young, and Fresnel*, p. vi. See also Merz, *A History of European Thought*, II, 95, and III, 573–574.

ham published in 1903 his *Aether and Gravitation*. This massive qualitative attempt, based squarely on the Michelson-Morley experiment,
to specify the physical properties of "Aether" (as philosophically
identical to "Electricity") argued for matter and energy, too, as modes
of eternal motion. Aether, according to Hooper, was matter, universal,
atomic, gravitative, possessing density, elasticity, inertia, and impressi-
bility. By extending the views of Lodge, Larmor, and Kelvin to their
extremes, Hooper hoped to show that the physical cause of gravitation
is simply that aether is a penultimate atomic and gravitative form of
matter. Apparently without mathematical distinctions he convinced
no one, not even Arthur Schuster who had declared his eagerness to
learn from any quarter more about the physics of the aether.[37]

In 1900, whether or not one felt a necessity for an omnipresent
ethereal medium depended in large measure on whether one considered his own linguistic syntax to be sacrosanct. As obvious as this may
sound today, it was by no means generally recognized in 1900. We
have only to look at some of the popularizations on the progress of
science written at the turn of the century to realize how great has been
the change wrought by physics and by the philosophy of science and
of language since then.[38]

"Gentlemen and fellow physicists of America," began Henry A.
Rowland as he stood before the American Physical Society on October
28, 1899, to take the gavel as president of that body: "We meet today
on an occasion which marks an epoch . . . in the interest of a science
above all sciences which deals with the foundations of the Universe
. . . [and with the] ether of space by which alone the various portions
of matter forming the Universe affect each other."[39] Rowland remind-

---

[37] Arthur Schuster, *An Introduction to the Theory of Optics*, pp. 8, 24; William
G. Hooper, *Aether and Gravitation*, pp. 7, 67, 216–231; Hugh Woods, *Aether: A
Theory of the Nature of Aether and of Its Place in the Universe*, p. 98.

[38] For example, Henry S. Williams, M.D., in *The Story of Nineteenth-Century
Science*, devoted a full chapter to an unequivocal endorsement of the aether as an
undefined plenum. There was no mention of Michelson, Morley, FitzGerald, or Lorentz to mar this picture of the aether.

[39] Henry A. Rowland, "The Highest Aim of the Physicist," *Am. J. Sci.*, 4th ser.
8:401 (December, 1899), reprinted in Nathan Reingold, ed., *Science in Nineteenth
Century America: A Documentary History*, pp. 323–328.

ed his fellow physicists that their fraternity was the aristocracy of the
intellect, his theme being great men on great quests for great entities.
In considering one of the greatest of these entities, the aether of space,
Rowland said: "To detect something dependent on the relative motion of the ether and matter has been and is the great desire of physicists. But we always find that . . . there is always some compensating
feature which renders our efforts useless." He went on to credit the
experiment of Michelson to detect the aether wind with having been
"carried to the extreme of accuracy," but this failure, as well as the
failure of Lodge to detect ethereal viscosity, he supposed might have
been explained if Lodge had used electrified revolving disks.[40]

In 1901 Professor T. C. Mendenhall, a widely respected American
physicist, wrote an article for a newspaper, entitled "Progress in
Physics in the Nineteenth Century," in which he asserted that "the
revival and final establishment of the undulatory or wave theory of
light is one of the glories of the nineteenth century."[41] The demand
for a medium to carry light waves was for him satisfied by "what is
known as the ethereal medium, at first a purely imaginary substance,
but whose real existence is practically established." Mendenhall was
particularly impressed by the molecular theorists: "Waves of light and
radiant heat originate in ether disturbances produced by molecular
vibrations and have impressed upon them all of the important qualities of that vibration."[42]

In the flood of "progress books" published around 1900, there is
ample support for the position that optics had advanced further than
any other branch of science in the century past. "To the sum total of
human knowledge no department has contributed more than that of

[40] Rowland, "The Highest Aim," *Am. J. Sci.*, 4th ser. 8:406–407. Lodge had, incidentally, tried this without a noticeable difference. It is interesting to compare various
presidential addresses, such as this one by Rowland, and those by J. H. Poynting
["On Ether and Atoms"], September 14, 1899, and Joseph Larmor ["Physics of the
Aether"], September 6, 1900, before Section A, *Br. Ass'n. Adv. Sci. Annual Reports*
(1899), pp. 615–624, and (1900) pp. 613–628, respectively. See also George Basalla, William Coleman, and Robert H. Kargon, eds., *Victorian Science: A Self-
Portrait.*

[41] *New York Sun*, February 17, 1901, reprinted in the *Annual Report, . . . Smithsonian Institution, . . . 1900*, pp. 315–331, esp. p. 321.

[42] Ibid., pp. 322, 325.

optics."[43] Regarding the ethereal plenum, "its discovery may well be looked upon as the most important feat of our century."[44] "Among the concepts which have come to stay in scientific thinking, that of the ether must now be included."[45]

Significantly, the last quotation above is followed immediately by this sentence: "It is as real as the concept of 'atom' or 'molecule' but hardly more so." So wrote J. Arthur Thomson in 1902, with an uncommon awareness of the distinctions between concepts and percepts. Less than a decade earlier the most prevalent argument had been something like that of Silvanus Thompson: "If light consists of waves ... it is clear that they must be waves of *something*."[46] By 1902, however, J. Arthur Thomson's attitude was the more general: "That the ether is a necessary conception in modern physics seems to be unanimously admitted by experts, but how exactly the ether is to be conceived of remains quite uncertain." Thomson continued, in the tradition of John Tyndall:

We can well imagine a practical man saying that all this talk of atom and molecule and ether is unreal and unverifiable, and in a certain sense he is undoubtedly right. These molecular and ethereal hypotheses are human imaginings and nothing more; they are constructed in terms of one sense, that of sight; they are attempts to see that which is invisible; to invent a machinery of Nature since the real mechanism is beyond our ken; but it must be observed that these hypotheses are not *vain* imaginings, for they prove themselves yearly most effective tools of research, and they are not *random* guesses, for they are constructed in harmony with known facts.[47]

Antithetical to J. Arthur Thomson's pragmatic position stressing

---

[43] Edward W. Byrn, *The Progress of Invention in the Nineteenth Century*, p. 299.

[44] Henry S. Williams, *Nineteenth-Century Science*, p. 230.

[45] J. Arthur Thomson, *Progress of Science in the Century*, p. 176.

[46] Silvanus P. Thompson, *Light: Visible and Invisible. A Series of Lectures Delivered at the Royal Institution of Great Britain, at Christmas 1896*, p. 108.

[47] Ibid., pp. 177–178. Another, more famous, Thomson (J. J.) was more careful of his language but shared the same attitude: see George P. Thomson, *J. J. Thomson: And the Cavendish Laboratory in His Day*, pp. 37, 155. Two bench marks for the attitude of J. J. Thomson toward discreteness may be found in the following two lectures in pamphlet form separated by twenty years: *The Corpuscular Theory of Matter* (1907) and *Beyond the Electron*, pp. 13–15, (1928).

the functional nature of the aether hypothesis, the realist position with regard to the aether was expressed in the same year (1902) by another British engineer of recent renown. J. Ambrose Fleming, inventor of the diode thermionic valve, argued that the proofs had accumulated to such an extent that "this medium *must* exist."[48] Although he would grant that the aether was not in any way sensible, yet he argued that it is "*a fact* deduced by reasoning from experiment and observation," and continued: "There is abundant proof that it is not merely a convenient scientific fiction, but is as much an actuality as ordinary gross, tangible, and ponderable substances. It is, so to speak, matter of a higher order, and occupies a rank in the hierarchy of created things which places it above the materials we can see and touch."[49]

Since opinion polls and sociological surveys among physicists were as yet nonexistent, it is difficult to say with any degree of confidence whether Thomson's or Fleming's view was the more widely held. When, in 1894, the Marquess of Salisbury and Chancellor of Oxford in a famous address spoke of the aether as a "half-discovered entity" not yet worthy of being called tangible by such words as "body" or "substance," he did indeed catch the essence of the problem of the aether around the turn of the century. Mathematicians could dismiss it as merely *verbal,* but after the Marquess's address, the major problem began to be recognized as *nominal.* Salisbury said it succinctly: "For more than two generations the main, if not the only, function of the word ether has been to furnish a nominative case to the verb 'to undulate.' "[50]

---

[48] J. A. Fleming, *Waves and Ripples in Water, Air and Aether: A Course of Christmas Lectures Delivered at the Royal Institution of Great Britain* [1902], p. 191 [italics mine].

[49] Fleming, *Waves,* p. 192. Perhaps the supreme example of a mechanical model of a hydrodynamical and yet elastic-solid aether was provided by Osborne Reynolds in *The Sub-Mechanics of the Universe,* vol. III of his *Papers on Mechanical and Physical Subjects.* Reynolds's Rede Lecture of 1902 gave pictures and demonstrations to show how discrete grains suitably packed could make up an elastic medium: *On an Inversion of Ideas as to the Structure of the Universe.* Reynolds's work is now classed as a contribution to continuum mechanics rather than to optics; see Clifford A. Truesdell, *Six Lectures in Modern Natural Philosophy.*

[50] Lord Salisbury [Robert Arthur Talbot Gascoyne-Cecil 3rd, Marquess of Salis-

But it is much too easy a solution to dismiss as "merely semantic" the problem of the aether concept. Communities are based on communication, and scientific communication, like any other, must rely on symbol systems that are intelligible largely because of their age and their abstraction from nature. Undoubtedly a few scientists around 1900 still thought, like the Aristotelians, of the aether as the quintessence, the "fifth" or ultimate element that makes up the substratum of the universe. Undoubtedly many more scientists of that era used the symbol of the aether as a synonym for vacuous space across and within which things happened. But surely the historian of mechanics, René Dugas, is correct in saying that the aether "completely bathed" the nineteenth century and that it grew into "the substratum of thought in physics."[51]

To judge the luminiferous aether to have formed "the substratum of thought" at the turn of the century is to generalize from biographical data and from intuitive deduction. More reliable generalizations about the status of the aether concept may be derived from social rather than from individual psychology. The symbol of space, the word *aether*, seems to have suffered no decline in usage in scientific discourse before 1905, although after Lorentz and Poincaré became famous, about 1895, there seems to have been an increase in cautious qualifications.[52]

Young's luminiferous aether at the beginning of the nineteenth century had evolved slowly into an elastic-solid aether and thence into an electromagnetic aether. By the end of the century not only had the aether proliferated into numerous special models, but it was often used

---

bury], *Evolution: A Retrospect*, address to the British Association for the Advancement of Science, pp. 28–29.

[51] "The physics of the XIXth Century was completely bathed in it. That the ether may have ceased after Fresnel and Maxwell, to play the part of the medium and have taken the more abstract one of a system of reference suited to the definition of absolute rest, does not imply that it ceased to function as the substratum of thought in physics" (Dugas, *History of Mechanics*, p. 490). See also the sympathetic article by E. T. Whittaker, "The Aether: Past and Present," *Endeavor* 2:117–120 (July, 1943).

[52] Someone should take a citation index and make a computer program to study the titular frequency of usage in the journal literature of the aether and the field concepts by decades before and after 1905.

interchangeably to express spatial relations and energy transfer. The field concept was still limited mostly to analytical use with Maxwell's equations. Thus, the "physics of the aether" was a descriptive phrase, not an explanatory one. It was used loosely to signify the break with the past and its physics of "matter in motion." With Lorentz and Lodge the phrase "ether of space" gained currency until it became the favorite synonym for Newton's old concept of the "absolute space." In descriptive physics, however, to talk about the aether might be either to talk about space or to talk about radiant energy. Whether microcosmic or macrocosmic space or energy was meant could only be inferred from the context.[53]

The slow transition in preference among physicists for the field concept in place of the aether concept spanned the two decades on either side of 1900. Although Maxwell and Lorentz were the central figures who shaped and characterized this trend, neither was able or willing to ban the use of the aether concept. The ambiguous, ethereal aether, in other words, was still considered necessary, even while the more limited and respectable field theory was superseding it.

Influenced by the prevailing thought of the time, Michelson accepted the aether as something "that ought to be true" and minimized the possibly revolutionary consequences of his own experiments. Meanwhile, H. A. Lorentz continued to wrestle with the problem posed by Michelson's experiment and gradually worked out the famous transformation equations to apply to second-order experiments for aether drift. Lorentz's second-order formulation did not appear in print until 1904, however, and by then all these difficulties with the aether concept plus the growth of confidence in the physical atom were leading toward a new consensus on "what ought to be true."

---

[53] "There was so much talk about lines of force, tubes of force, stresses in the medium, and localized energy that an easy familiarity with the terms began to carry with it a sense of understanding and reality, and curiosity became dulled as the years passed by. The idea of a medium whose state was expressed through the equations of the field was fundamental to the theory, and the idea of action at a distance seemed to retain a historical interest only" (Max Mason and Warren Weaver, *The Electromagnetic Field*, p. xi).

CHAPTER SEVEN

MORLEY-MILLER EXPERIMENTS, 1900–1905

Professors Morley and Miller together attended the Paris Exposition of 1900 when, before the International Congress of Physics, Lord Kelvin spoke of the "two clouds" which overhung the dynamical theory of heat and light. Afterward in conversation with Kelvin, the two Americans were "strongly urged" to repeat the aether-drift experiment with a more powerful apparatus.[1] Without doubt this stimulus was important in setting in motion once again Morley's and Miller's thoughts on the aether-drift tests, but perhaps equally important stimuli were several papers at the turn of the century that purported to vitiate completely the "classic" Michelson-Morley experiment.

In 1898 William Sutherland, an Australian physicist, suggested that

---

[1] Dayton C. Miller, "The Ether-Drift Experiment and the Determination of the Absolute Motion of the Eearth," *Rev. Mod. Phys.* 5:203–242 (July, 1933). This paper is the basic source for the subsequent history and technological detail of Morley and Miller's aether-drift work.

the experimental design of the classic interferometer was not sensitive enough to detect an aether drift. He argued that the lateral shift of the observer's eyes was greater than the fringe shift and suggested that corrected repetitions were in order. In 1902 William M. Hicks, a Fellow of the Royal Society, who had wrestled long with vortex theories of the atom and the aether, published a startling paper which set out to reexamine completely the general theory of the aether-drift interferometer. Hicks argued that the experiment was "not so simple as it may appear" because of "the changes produced by actual reflection at a moving surface." He claimed to show that the effect to be expected from the Michelson-Morley experiment was "the reverse of that hitherto supposed," and he likewise suggested more and better trials of the experiment.[2]

Hicks's announcement especially attracted the attention and provoked the immediate interest of Morley and Miller. Soon Hicks was forced to retract his conclusions because of two mistakes. The first error was pointed out to him by Morley himself: a typographical error in the tabular results of the original paper. The second error was more critical: an algebraic slip in his own analysis had led Hicks to suppose —mistakenly—the FitzGerald-Lorentz contraction to be opposite in sign and direction to that shown by the originators of this hypothesis.[3]

Meanwhile, during 1901 and 1902, H. A. Lorentz was lecturing anew at Leiden on Michelson's experiment, as he reviewed the problems of stellar aberration, mechanical aether models, and molecular attractive and repulsive forces. If Lorentz, Larmor, Hicks, and Kelvin were all so interested in the aether-drift enigma, then clearly the contraction hypothesis itself ought to be tested, Morley and Miller agreed.[4]

---

[2] William Sutherland, "Relative Motion of the Earth and Aether," *Phil. Mag.*, 5th ser. 45:23–31 (January, 1898). William M. Hicks, "On the Michelson-Morley Experiment Relating to the Drift of Ether," *Phil. Mag.*, 6th ser. 3:9–36 (January, 1902).

[3] Hicks, Letter to the Editor (dated January 9, 1902), *Phil. Mag.*, 6th ser. 3:256 (January, 1902); see also p. 555 for still another criticism and reply. Cf. William M. Hicks, Letter to the Editor, *Nature* 65:343 (February 13, 1902).

[4] H. A. Lorentz, "Aether Theories and Aether Models" (1901–1902), in *Lectures on Theoretical Physics*, I, 3–71.

Hicks's paper and Lorentz's lectures were to crop up often in later years to plague interpreters of the Michelson-Morley experiment. Yet apparently Morley and Miller themselves were rather more stimulated by the renewed discussion of the experiment than they were dismayed by its interpretations. Immediately that spring they began making preparations for a new series of interferometric observations to test specifically for the FitzGerald-Lorentz contraction hypothesis. Whether this could be done independently of aether-drift assumptions was not yet at issue.

The question whether any and all types of matter undergo constriction as a result of "absolute velocity" Morley and Miller sought at first to answer by building a much larger interferometer on a white-pine framework. This wood-based cruciform interferometer had optical arms about 430 centimeters long set upon 14-foot planks replacing the sandstone block on the old piers used in 1887. In the summers of 1902 and 1903, they ran series of tests on this wooden interferometer, but they were dissatisfied because of difficulties in trying to keep humidity and temperature constant in order to prevent warping in the wooden supports. By July, 1903, they were convinced that this mundane problem could not be overcome, that no significant results could be obtained by the use of a wood-based apparatus. The path length was more than three times as long as that used in 1887, but precautions were not sufficient to overcome the masking effects of errors introduced by the warping, bending, and expanding wood. Also, their steam-heated basement laboratory was hard on pine.[5]

Meanwhile, in England in 1902 Lord Rayleigh was also testing for the FitzGerald-Lorentz contraction hypothesis by trying to find out if certain substances might become doubly refracting as a result of their motion through the aether. Rayleigh reasoned that if there were a deformation of matter, then this might be accompanied by observable double-refraction effects. Preliminary tests on transparent liquids tra-

[5] For the best of several accounts of this series of tests, see Edward W. Morley and Dayton C. Miller, "Report of an Experiment to Detect the FitzGerald-Lorentz Effect," *Proc. AAAS* (Boston) 41:321–328 (August, 1905).

versed by a light beam in a direction perpendicular to the earth's pre-
sumed motion showed no double refraction of the order to be expected.[6]
But Rayleigh's first answer was as inconclusive as was Morley and
Miller's first answer from their wooden interferometer.

Lacking any trustworthy results on the contraction hypothesis, Morley and Miller in 1903 began to plan a more elaborate and still more
powerful interferometer for the same purpose. They enlisted the help
of a mechanical engineer at Case, F. H. Neff, to design the base for
the new instrument, this time to be made entirely of steel, "in order
to secure structural symmetry and the utmost rigidity."[7] The American
Academy of Arts and Sciences in Boston provided financial aid so that
costs would not inhibit the new experimental design. Although they
could and did use the original cast-iron trough and circular float on the
original piers, most of the 1887 optical parts had been cannibalized in
the course of time for various other purposes. Consequently, the two
experimenters contracted with Michelson's technician, a Chicago artist-
optician, O. L. Petitdidier, to supply them with a beam-splitter, a compensator, and sixteen plain mirrors 10.25 centimeters in diameter.

The general plan of the new steel interferometer was exactly like
that of the classical apparatus. But it, too, was shaped like a Greek
cross with 14-foot crossarms, giving a total light path of 6,406 centimeters. The structural-steel base weighed almost a ton (1,900 pounds)
and floated on a circular bed of about 800 pounds of mercury. The
reflecting mirrors at each end of the cross were arranged in vertical
cast-iron holders in a square form. Two of these frames were bolted
securely to the steel-cross base, while the opposing partner frames were
spring-loaded in order to allow for movement, should there be any
change in path length.

White pine was again used to test for the material contraction. This
time, however, eight long pine dowel rods were used to separate the
mirror frames at opposite ends of the interferometer. These pine rods
were enveloped in brass tubing, trussed in such a way as to guide and

---

[6] Lord Rayleigh, "Does Motion through the Aether Cause Double Refraction?"
Phil. Mag., 6th ser. 4:678–683 (December, 1902). See also Robert J. Strutt [Fourth
Baron Rayleigh], *John William Strutt, Third Baron Rayleigh*, pp. 346–349.

[7] Miller, "The Ether-Drift Experiment," p. 208.

guard the pine rods from external disturbances. Thus "the distances between the opposite systems of mirrors depend upon the pine rods only, while the whole optical system is adequately supported by the steel cross."[8]

Finally, in July, 1904, the refurbished instrument was complete, and the observers were free from classwork. The dates and the time of day for these observations were carefully chosen to coincide with the maximum effects to be expected from the presumed motion of the solar system and the orbital and diurnal motions of the earth. Through a set of 260 turns of the interferometer, Miller walked around the laboratory circuit calling off his estimates of the fringe shift at 16 different azimuth positions, while Morley sat by, recording the yet meaningless data.[9] After having reduced their raw data to comparison figures, they reported that their results did not substantiate the theory then under consideration: "If pine is affected at all as has been suggested, it is affected to the same amount as is sandstone."[10] Considered strictly as an effort to detect the FitzGerald-Lorentz effect, Morley and Miller declared that "the experiment shows that if there is any effect of the nature expected, it is less than the hundredth part of the computed value."[11]

Curiously, however, Morley and Miller did not interpret their experiment exclusively as a test of the contraction hypothesis. Although they admitted Hicks's "profound and elaborate" analysis had prompted the doubts which led to their own reexamination,[12] they

---

[8] Ibid., p. 216.

[9] Morley did most of the planning and calculations for this series of experiments while Miller made most of the observations. See Robert S. Shankland, "Dayton Clarence Miller: Physics across Fifty Years," *Am. J. Phys.* 9:276 (October, 1941). See also Morley and Miller to Lord Kelvin, August 4, 1904 (extract), *Phil. Mag.*, 6th ser. 8:753–754 (December, 1904).

[10] E. W. Morley and D. C. Miller, "On the Theory of Experiments to Detect Aberrations of the Second Degree," *Phil. Mag.*, 6th ser. 9:680 (May, 1905). See also Morley and Miller to H. A. Lorentz, August 5, 1904, in Lorentz Correspondence file, Bohr Library, American Institute of Physics, New York.

[11] E. W. Morley and D. C. Miller, "Report of an Experiment to Detect the Fitz-Gerald-Lorentz Effect," *Phil. Mag.*, 6th ser. 9:685 (May, 1905). Although 1904 was also the year of J. C. Kapteyn's famous analysis of two "star-streams" in our Milky Way galaxy, neither Morley, Miller, nor Michelson seems to have heard the word.

[12] Morley and Miller, *Phil. Mag.*, 6th ser. 9:669 (May, 1905).

believed they had found no support either for the FitzGerald-Lorentz hypothesis or for Hicks's interpretation. Therefore, they reverted to the position that this experiment was simply another, more refined test for aether drift on the classic model: "We assert, then, that the theory of 1887 is correct to terms of the order retained, which were sufficient; that Dr. Hicks's theory agrees with it precisely as to numerical amount and sign of the effect, and that a third examination of the theory gives results differing from those of the two others only by negligible terms of the third order."[13]

Morley and Miller's interpretation of the meaning of their inability to detect any measurable contraction in their basement laboratory in Cleveland was subject to the same objections as were encountered in 1887. Perhaps, after all, Stokes's old aether-drag hypothesis might be correct in one form or another.[14] At any rate, Morley and Miller solved their dilemma temporarily by publishing simultaneously two papers. In the first, they reaffirmed their faith that the theory of the experiment of 1887 had been correct to terms of the second degree, and in the second, they reasserted the final hope (expressed by Michelson and Morley in 1887) that greater height above sea level might show a positive effect. Thus Stokes's drag hypothesis was again invoked as a possible explanation for the null results.

The very next year (1905) Morley and Miller were to carry out their promise and run a series of aether-drift tests on a hilltop. But before I recount that story, it may be well to mention other, concurrent experimental corroborations of the failure to detect the absolute motion of the earth.

In February, 1904, Rayleigh's effort to test for an aether drift by double-refraction experiments had been followed up by D. B. Brace at the University of Nebraska.[15] In 1902 Rayleigh, like Morley and Miller, had set out specifically to test for the FitzGerald-Lorentz contraction in order to remove all doubts regarding that explanation of the

---

[13] Ibid., p. 680.

[14] That is, aether might be entrained by the masonry of the building or by a kind of aether-atmospheric pressure.

[15] D. B. Brace, "On Double Refraction in Matter Moving through the Aether," *Phil. Mag.*, 6th ser. 7:317–329 (April, 1904).

negative results of Michelson and Morley. Brace was deliberately more general in his approch in that he set up two experimental arrangements, one with water and one with glass, in order to test for the production of double refraction in two *states* of matter moving through the aether. He believed that Rayleigh's experimental design could not "be regarded by any means as a conclusive test of the hypothesis" of FitzGerald,[16] but, since it was the only one attempted along this line, he hoped that it could be extended so as to learn more, at least, about the behavior of optical glass. Notably, Brace also cited Hicks's article without realizing it had been retracted. More significantly, Brace reached conclusions very similar to those reached by Morley and Miller in their pine-rod tests: "Hence, if the test is a valid one, the contraction hypothesis cannot explain the negative results of the interference experiments; and with the same reasoning, we also conclude either that the aether moves with embedded matter, or that the effect of the relative motion on the intermolecular forces and the possible consequent relative changes in dimensions are very small."[17]

Concurrently, back in Dublin, an entirely different approach to the problem of the relative motion of earth and aether was being tried. This method had been suggested by FitzGerald, shortly before his death in February, 1901.[18] The idea was to suspend a small charged electrical condenser by a fine wire and to watch for a torque exerted upon it as it was carried through space by the terrestrial motion. The apparatus was very simple, being in principle similar to the oldest of electrical instruments, namely, the compass, or Gilbert's versorium, or Coulomb's torsion balance.[19] But the precautions necessary to rule out extraneous effects were extraordinary. The desired effect to be detected by this experiment was also extraordinarily small; and, like the optical aether-drift tests, it would have to measure a second-order of the dif-

[16] Ibid., p. 317.

[17] Ibid., p. 328.

[18] On FitzGerald's suggestion and its subsequent application, see Sir Edmund Whittaker, *A History of the Theories of Aether and Electricity*, II, 28–38.

[19] For an account of the evolution of this basic electrical instrument, see Duane Roller and Duane H. D. Roller, "The Development of the Concept of Electric Charge: Electricity from the Greeks to Coulomb," in James B. Conant, ed., *Harvard Case Histories in Experimental Science*, II, Case no. 8.

ferential ratio between the velocity of the earth and the speed of light.

In April, 1902, FitzGerald's former student, F. T. Trouton, began trying with small success to carry out FitzGerald's original plan. All kinds of extraneous disturbances had to be eliminated one by one. The next year Trouton was joined by H. R. Noble, and their efforts to refine the very delicate apparatus gradually came to equal their concern with the original relative motion problem. Finally, they announced their assurance that, with regard to the relative motion between the earth and the aether, "there is no doubt that the result is a purely negative one."[20] It is also noteworthy, however, that Trouton and Noble's experiment revealed another unexpected asymmetry in the electrodynamical behavior of moving bodies.

Of central importance to a reconsideration of the meaning of the Michelson-Morley optical experiment were investigations into more strictly electrical phenomena in the years between 1902 and 1905. The experiments of Walther Kaufmann, motivated by Lorentz's electron theory, excited particular interest in 1902 and thereafter,[21] as he began to announce his evidence for the curious fact that the "mass" of a fundamental particle seemed not to be constant but rather to vary with its speed. Those experiments with the beta rays or particles of radium seemed to give definite proof of the variation of mass with velocity since this type of radiation showed measurable deflections in electric and magnetic fields.

Such electronic experiments, by threatening the concept of mass in a way complementary to Ernst Mach's theoretical critique, were severely upsetting to Newtonian dynamics. Rapid advances from Bec-

---

[20] F. T. Trouton and H. R. Noble, "The Mechanical Forces Acting on a Charged Electric Condenser Moving through Space," *Trans. R.S.* 202:181 (1903). From Miller's later point of view, it is significant to note that this apparatus was extremely heavily shielded. Also, "The effect to be looked for was an extremely small one, being a second-order effect only" (p. 166). Miller may also have been influenced by the severe critique by E. H. Kennard, "The Trouton-Noble Experiment," *Bulletin of National Research Council* 4:162–172 (December, 1922).

[21] Whittaker, *History*, II:53 n., gives an extensive list of citations for Kaufmann's reports and calls attention to the difference in interpretations of these experiments before and after relativistic mechanics. Gilbert N. Lewis is credited by Whittaker with making Kaufmann known in English. See especially W. Kaufmann, "Die elektromagnetische Masse des Elektrons," *Physikalische Zeitschrift* 4:54–57 (1902).

querel's discovery of radioactivity in 1896, through the work of Ruther-ford, Crookes, Soddy, the Curies, and others, had proved transmuta-tion in the elements by 1905. So disturbing did the new experiments become that the story of the birth of Special Relativity can be seen with equal validity from this viewpoint.[22] Thus, it came to pass that Kel-vin's second, less threatening, cloud, sighted in 1900, grew into pro-portions of equal magnitude with the Michelson-Morley anomaly by 1904.

In September, 1904, the great French mathematician Henri Poin-caré was in the United States to address the International Congress of Arts and Sciences in St. Louis on "The Principles of Mathematical Physics."[23] He began his address with the rhetorical question, Is a revolution impending in theoretical physics? His answer was an un-equivocal Yes! In giving his catalog of the six general principles of greatest importance, Poincaré listed as number four, "the principle of relativity, according to which the laws of physical phenomena should be the same, whether for an observer fixed, or for an observer carried along in a uniform movement of translation; so that we have not and could not have any means of discerning whether or not we are carried along in such a motion."[24] Poincaré had, of course, for years been concerned with the same kinds of problems as had Lorentz. Recent scholarship has delved deeply into the question of priority for the dis-covery of relativity theory. A central issue in these studies is the ques-tion whether Poincaré's *principle* was as novel as Einstein's *postulate* of relativity. This issue need not detain us, except to note that it must be settled by analysis of the history of philosophy and mathematics.[25]

---

[22] Whittaker's history is, obviously, so organized, but I mean to call attention to the double genetic lineage without analyzing the electrical heritage. For a more ade-quate nontechnical discussion of this genealogy, see J. H. Thirring, *The Ideas of Ein-stein's Theory: The Theory of Relativity in Simple Language*, p. 167. See also Alfred Romer, ed., *The Discovery of Radioactivity and Transmutation*.

[23] This address by Poincaré is most conveniently available in G. B. Halsted's trans-lation, in *The Monist* 15:1–24 (January, 1905).

[24] Ibid., p. 5. For the larger context, see Gerald Holton, "On the Thematic Analysis of Science: The Cases of Poincaré and Relativity," in *Mélanges Alexandre Koyré*, II, 257–268.

[25] Whittaker champions Poincaré and Lorentz; Dugas insists on Poincare's prior-ity; and Einstein himself said that Paul Langevin was also very near the full exposi-

Poincaré was concerned chiefly with Michelson-Morley when he spoke of "the principle of relativity," and not with the electrodynamics of moving bodies. His attitude toward the aether concept can be seen from the following quotations:

We know nothing as to what is the ether, how its molecules are disposed, whether they attract or repel each other; but we know that this medium transmits at the same time the optical perturbations and the electrical perturbations.[26]

.   .   .   .

Indeed, experience has taken on itself to ruin this interpretation [of aether at absolute rest in absolute space] of the principle of relativity; all attempts to measure this velocity of the earth in relation to the ether have led to negative results. This time experimental physics has been more faithful to the principle than mathematical physics; the theorists, to put in accord their other general views, would not have spared it; but experiment has been stubborn in confirming it.

The means have been varied in a thousand ways and finally Michelson has pushed precision to its last limits; nothing has come of it. It is precisely to explain this obstinacy that the mathematicians are forced today to employ all their ingenuity.[27]

Poincaré was using poetic hyperbole, if not flattery, when he spoke of the aether-drift tests as having been performed in "a thousand ways" and when he credited Michelson with having achieved the "last limits" of accuracy. Actually, as we have seen, aether-drift tests were far more limited than was professionally supposed; and Morley and

---

tion of the restricted relativity postulates. If the attitudes of the claimants toward the aether circa 1905 is an adequate criterion, then clearly Einstein is the great in-novator. For an excellent critique of Whittaker's account, see Gerald Holton, "On the Origins of the Special Theory of Relativity," *Am. J. Phys.* 28:633–636 (October, 1960). See also Charles Scribner, Jr., "Henri Poincaré and the Principle of Rela-tivity," *Am. J. Phys.* 32:672–678 (September, 1964); Stanley Goldberg, "Henri Poincaré and Einstein's Theory of Relativity," *Am. J. Phys.* 35:934–944 (October, 1967).

[26] Poincaré, in *The Monist*, p. 5.

[27] Ibid., p. 10. See also H. A. Lorentz, "Electromagnetic Phenomena in a System Moving with Any Velocity Less than That of Light," reprinted from *Proceedings of Academy of Science of Amsterdam* 6 (1904) in Lorentz et al., *The Principle of Rela-tivity*, pp. 11–34.

Miller were even then at work trying to bring greater precision to bear on the elusive aether. Furthermore, Michelson himself had not given up the quest to find a measure of this motion. Only three months later he was to publish another calculation, aiming at another possible way of testing for aether drift with a single pencil of light reflected in opposite directions.[28] Nonetheless, Poincaré's statements are interesting examples of the stature of the enigma presented by Michelson-Morley at this time.

We return now to the year 1905 and the final collaborative effort between Morley and Miller in their mutual quest to find an aether drift. As soon as final examinations for the school year 1905 had been graded, Morley and Miller supervised the moving of the big steel cross to a new location atop a hill a mile or so south of their twin campuses. There, on Euclid (later called Cleveland) Heights, at an altitude of about 300 feet above Lake Erie and 870 feet above sea level, they constructed a temporary housing around some cement piers for their interferometer.[29]

The hut had a panoramic glass window all around, at the plane in which the apparatus, also covered by glass panes, would rotate, so that no opaque obstruction should possibly hinder the free drift of the aether of space through the light paths. The argument that the very location of the previous aether-drift experiments, in basement laboratories surrounded by massive masonry, had prejudiced the results should certainly be obviated by this new arrangement. Even more important, the added height of two hundred feet might reduce the ethereal pressure, if there really were an earthbound ethereal atmosphere with a pressure gradient.

Morley and Miller discarded the pine rods and trussed-brass framework they had used the year before to separate the mirror holders. Ostensibly in their new location they were still testing for an indi-

---

[28] A. A. Michelson, "Relative Motion of the Earth and Aether," *Phil. Mag.*, 6th ser. 8:716–719 (December, 1904). This paper became important fifteen years later as a precursor for Silberstein's test of the general relativity theory. See my later discussion in Chapter 10.

[29] Miller, "The Ether-Drift Experiment," p. 217. Cf. E. W. Morley and D. C. Miller, "Report of Progress in Experiments on Ether Drift," *Science* (new series) 23:417 (March 16, 1906).

cation of the FitzGerald-Lorentz contraction, and this time the structural-steel interferometer base would itself have to be distorted, because all four cast-iron mirror frames were now bolted securely to the crossarms.

The most difficult problem the experimenters could foresee in their new location was the perennial difficulty in interferometric observations of keeping the apparatus under uniform temperature conditions. Indoors this problem was always vexing, but in their flimsy shack on windy Euclid Heights, with no air conditioning and poor insulation, they could only hope that the observations would show such definite and positive fringe shifts as to override the temperature variations. Their hope was supported by one further precaution, however: Miller had constructed a glass-box casing, similar to the wooden and cardboard casings used at various times in the past. This could be set over the optical parts, thus enclosing the whole light path and protecting it from air drafts. With a bicycle acetylene lamp as a light source and the small thirty-five–power observing telescope, the apparatus was complete.

In July, 1905, Miller began to take systematic observations at the new installation, walking around the circuit, peering into the eyepiece, and calling off estimates of fractional fringe shifts against the fiducial mark in his field of view. Morley was sixty-seven years old and only one year from retirement, so he left most of these routine observations to Miller and his students. Morley did watch with interest the daily reports of the raw data obtained, for he was actively engaged in the reduction procedures. In order to plot the aether-drift observations against the theoretical predictions, it was necessary to extract the second-harmonic component that alone represented the second-order, half-period aether-drift effect they sought.

Summer waned and autumn began before Miller and Morley began to accumulate enough reliable data to satisfy themselves that the new experiment was working as well as they could expect under the present circumstances. By November they had reduced 230 turns of the interferometer to tabular figures, with another null result to announce, but one that, as Miller claimed long afterward, showed a "very definite positive effect."[30] This was slightly larger than anything noticed from the basement location, but still far too small to match the theoretical results predicted by Newtonian mechanics.

Then it happened. While Morley and Miller were still struggling to make sense out of their highly refined aether-drift data, a series of three papers was being published by obscure young Albert Einstein in *Annalen der Physik*, the last of which would eventually be recognized as the penultimate solution to the problem of the aether drift. Quietly, unobtrusively it was stated in that paper that: "The introduction of a 'luminiferous ether' will prove to be superfluous inasmuch as the view here to be developed will not require an 'absolutely stationary space' provided with special properties, nor assign a velocity-vector to a point of the empty space in which electromagnetic processes take place."[31] Thus, by eliminating the aether concept and its correlative "absolute space" concept altogether from his considerations on the electrodynamics of moving bodies, Einstein at once made the aether *überflussig* and solved the drift problem by making it meaningless.[32]

By expanding the concept of the electromagnetic field to astronomical dimensions, by extending Galilean insights about relative motion from mechanical into electrical and magnetic phenomena, and by extrapolating the empirically proved regularity of the speed of light

[30] Miller, "The Ether-Drift Experiment." Cf. E. W. Morley and D .C. Miller, "Final Report on Ether Drift Experiments," *Science* (new series) 25:525 (April 5, 1907). See also Dayton C. Miller to Professor [?] Nichols, April 12, 1906, in Lorentz Correspondence file, Bohr Library, American Institute of Physics, New York.

[31] Albert Einstein, "On the Electrodynamics of Moving Bodies," in H. A. Lorentz et al., *The Principle of Relativity*, p. 37.

[32] Although Einstein had published several articles in *Annalen der Physik* before 1905 and therefore may be presumed to have been a regular reader, it is true that Michelson-Morley experiment was not discussed repeatedly in the German journal's pages between 1900 and 1905. There were, however, about five significant articles during these years, any one of which might have stimulated Einstein to reexamine the invariances in electrodynamics: see Egon R. v. Oppolzer, "Erdbewegung und Aether," *Ann. d. Phys.*, 4th ser. 8:898–907 (July 10, 1902); Max Planck, "Ueber die Verteilung der Energie zwischen Aether und Materie," *Ann. d. Phys.*, 4th ser. 9:629–641 (October 21, 1902); W. Wien's three lead articles in the issue for March 8, 1904: "Über die Differentialgleichungen der Elektrodynamik für bewegte Körper" (part I); "Über die Differentialgleichungen der Elektrodynamik für bewegte Körper" (part

into a constant limiting velocity, Einstein grasped a set of steps to a new plateau of physical understanding. He sacrificed the aether for the earth at first, but within a decade and a half he was to retrench somewhat in considering the seriousness of gravitational effects on the behavior of light and electromagnetism. Without much concern for dynamical positional astronomy, he traded traditional presuppositions about the center of the universe and absolute rest and motion for some new insights into space and time as interrelated and into mass and energy as interchangeable. Einstein's own personal and professional growth paralleled that of modern astrophysics and cosmology, and yet there seems to have been surprisingly little direct or immediate exchange of information among struggling professional aspirants in these two young fields.[33]

_____

II); and "Über positive Elektron und die Existenz hoher Atomgewichte," 4th ser., 13; 641–677; and Fritz Hasenöhrl, "Zur Theorie der Strahlung im bewegten Körpern," *Ann. d. Phys.*, 4th ser., 15:344–370 (October 25, 1904).

[33] For some insight into the evolution of "galactocentric" theory, see Thornton and Lou W. Page, eds., *Stars and Clouds of the Milky Way: The Structure and Motion of Our Galaxy*, pp. 17, 50, 96, 110, 118; W. M. Smart, *Stellar Kinematics*, pp. 38–149; G. C. McVittie, *Fact and Theory in Cosmology*, pp. 30, 71, 72; see also my forthcoming study, *Genesis of Relativity*.

CHAPTER EIGHT

THE RISE OF RELATIVITY, 1905–1910

Einstein's famous paper of 1905, like the Michelson-Morley experiment itself, has too often been taken as a "primitive fact," as an irreducible departure from Newtonian mechanics which sprang forth full-grown in 1905, thereafter revolutionizing all physical science. This naive conception of the influence of a single paper ignores science as a social and historical endeavor. Currently there is promise, however, of a greater appreciation for historical continuity in scientific development, combined with a recognition of the discontinuity introduced by Einstein in methodological orientation.[1] As interest

---

[1] This is the aim of a promised full-scale historical study by Gerald Holton as previewed in his paper, "On the Origins of the Special Theory of Relativity," *Am. J. Phys.* 28:628 (October, 1960). See also his "Resource Letter SRT-1 on Special Relativity Theory," *Am. J. Phys.* 30:462–469 (June, 1962). See also several articles by Holton in *Eranos-Jahrbuch*, especially "The Metaphor of Space-Time Events in Science," 34:33–78 (1965); and "Continuity and Originality in Einstein's Special Relativity Theory," *Actes du ixe congrès international d'histoire des sciences*, pp. 499–502.

grows in the history, logic, psychology, and sociology of scientific discovery, we may anticipate from the social sciences more aid applicable to the study of the origins of relativity theory.

But at present we need to overcome the anachronistic tendency to see relativity theory solely in terms of its success. The remainder of this study will be concerned primarily with the viewpoint of the conservative, traditionalist school of "classical" physics, part of which became an antirelativist faction that stoutly resisted the Einsteinian innovations. A slow erosion of confidence in all forms of the aether concept had allowed Einstein to gain a hearing in the first place and then to gain a following. This erosion process did not become an avalanche immediately after 1905. On the contrary, the Einsteinian "revolution" became obvious only as sides were chosen and partisans began to force the issue. This process took about a decade and was accelerated by Einstein's further development of relativity theory. Even then many physicists remained aloof from the controversy, especially if their own particular specialties were not directly involved.

Meanwhile, aether apologists became more vocal than ever as the threat to their intellectual commitment became more powerful. Conversely, the revolutionists became more zealous for the "cause" of relativity as they felt the opposition stiffen and as an iconoclastic brotherhood of initiates began to form. Eventually, Einstein came to find his name being taken in vain by some of those who claimed to be his disciples.[2] Unlike Newton, his illustrious predecessor, Einstein was not a controversialist, being content to go on with his work and to leave the skirmishing to others. Indeed, partly for this reason, he became far better known as a symbol of erudition and sublime profundity than as a human being and natural philosopher.

Einstein's seminal paper on the electrodynamics of moving bodies contains no explicit reference to the Michelson-Morley experiment, nor for that matter does it give any reference to a specific experiment.

[2] See Albert Einstein's position on the aether circa 1920 as found in chapter 10 following. See also the conservatism and generosity exhibited by H. A. Lorentz in his *The Einstein Theory of Relativity: A Concise Statement*, pp. 60, 63. A few of the better biographies are Phillip Frank, *Einstein: His Life and Times*; B. Kuznetsov, *Einstein*; Carl Seelig, *Albert Einstein: Eine Dokumentarische Biographie*; and Ronald W. Clark, *Einstein: The Life and Times*.

This fact has been a source of embarrassment to a number of commentators who have overemphasized the claim that the Special Relativity theory is based directly on experimental difficulties.[3]

Einstein's introductory paragraphs express rather a general concern with the aesthetic problem of theoretical asymmetry arising from Maxwell's electrodynamics as applied to bodies in relative motion. The initial paragraph gives as an example the asymmetric behavior of a magnet and a conductor in the production of electric fields: if the magnet moves and the conductor is at rest, an electric current is produced in the conductor; if the conductor moves and the magnet is at rest, no electric field arises near the magnet, and yet the relative motion does give rise to equal electric currents.

In order to remove this asymmetry and to maintain the beauty and power of the Maxwell-Hertz electrodynamics, Einstein probed deeply into the meanings of and relationships between rigid bodies, clocks, and electromagnetic processes. Based on two postulates, the Principle of Relativity and the Constancy of the Speed of Light, Einstein found a way to make that asymmetry disappear and to preserve the invariance of the electromagnetic field equations.

The first sentence of the second paragraph is the source of the warrant used by popularizers to justify their descriptions of the origins of Special Relativity by beginning with the Michelson-Morley experiment: "Examples of this sort, together with the unsuccessful attempts to discover any motion of the earth relatively to the 'light medium,' suggests that the phenomena of electrodynamics as well as of mechanics possess no properties corresponding to the idea of absolute rest."[4] The phrase in apposition above is as close as Einstein came in this

---

[3] Albert Einstein, "Zur Elektrodynamik bewegter Körper," *Ann. d. Phys.*, 4th ser. 17:891–921 (September 26, 1905). See also the debate between Adolf Grünbaum and Michael Polanyi, "The Genesis of the Special Theory of Relativity," in Herbert Feigl and Grover Maxwell, eds., *Current Issues in the Philosophy of Science*, pp. 43–55.

[4] Albert Einstein, "On the Electrodynamics of Moving Bodies," in H. A. Lorentz et al., *The Principle of Relativity: A Collection of Original Memoirs on the Special and General Theory of Relativity*, p. 37. Most physicists describe the significance of Special Relativity in terms of its denial of the detectability of a preferred reference frame or coordinate system in space, thus elegantly resolving the dilemma posed by aberration and the Fizeau and Michelson experiments.

paper to naming Michelson-Morley as a factor behind his extension of the "Galilean relativity principle" into the domain of electromagnetic phenomena. It is as clear as an implied reference can be, and yet, because it is still ambiguous and so brief, many people interested in scientific discovery asked Einstein at various times throughout his life to place the Michelson-Morley experiment in the development of his thought. Thus there are a number of statements by Einstein himself that attribute different degrees of importance to the role of the aether-drift experiments in the original development of his ideas. From a score of examples which might be collected, let us examine a few.

To Bernard Jaffe on March 17, 1942, Einstein wrote: "It is no doubt that Michelson's experiment was of considerable influence upon my work insofar as it strengthened my conviction concerning the validity of the principle of the special theory of relativity. On the other side, I was pretty much convinced of the validity of the principle before I did know this experiment and its result. In any case, Michelson's experiment removed practically any doubt about the validity of the principle in optics, and showed that a profound change of the basic concepts of physics was inevitable."[5]

In a letter to Michael Polanyi, Dr. N. Balazs reported a personal conversation with Einstein on July 8, 1953: "The Michelson-Morley experiment had no role in the foundation of the theory. He got acquainted with it while reading Lorentz's paper about the theory of this experiment (he of course does not remember exactly when, though prior to his papers), but it had no further influence on Einstein's considerations, and the theory of relativity was not founded to explain its outcome at all."[6]

[5] Reprinted in full in Bernard Jaffe, *Men of Science in America: The Role of Science in the Growth of Our Country*, p. 372. The most exhaustive study of Einstein's debt to Michelson is that of Gerald Holton entitled "Einstein, Michelson, and the 'Crucial Experiment,' " *Isis* 60:133–197 (Summer, 1969).

[6] Reprinted in Michael Polanyi, *Personal Knowledge: Towards a Post-Critical Philosophy*, p. 10 n. This letter represents hearsay evidence, of course, and may be discounted for its editorial comment on a subtle distinction. Polanyi has drawn from Balazs's report and his own subsequent analysis the basis for an interesting, although not unobjectionable, inquiry into scientific epistemology. Having pointed out the error made in physics texts by directly linking Einstein to Michelson-Morley (p. 9), Polanyi states: "The usual textbook account of relativity as a theoretical response

Another and most important example comes from an extensive psychological case study, conducted by Max Wertheimer in person with Einstein, during the war year 1916. Only one paragraph from Wertheimer's lengthy and illuminating firsthand study of Einstein's gestalt is necessary here: "For Einstein, Michelson's result was not a fact for itself. It had its place within his thoughts as they had thus far developed. Therefore, when Einstein read about these crucial experiments made by physicists, and the finest ones made by Michelson, their results were no surprise to him, although very important and decisive. They seemed to confirm rather than to undermine his ideas."[7]

A safe generalization from these and other such statements is that Einstein was always careful to temper his credits to the aether-drift tests with qualifications. When legends persisted in spite of his protestations, Einstein showed some gentle impatience with the insistence that he answer in a few words the question, What caused you to postulate the principle of relativity? Not that he denigrated this kind of question, but he expressly left it to the investigation of others.[8] He had a personal aversion to single-factor explanations, and he was quite aware of the inadequacy of a creator to pass critical judgment on his own work.

Perhaps the point has been belabored, but it should now be clear that the aether-drift experiments formed part of the background for the development of relativity in Einstein's mind, but that they were not by any means the only, and certainly not the most important, source of inspiration for the theoretical advance we call Special Relativity.

To return to the period and document at hand, it is easy to see the

---

to the Michelson-Morley experiment is an invention. It is the product of a philosophical prejudice. When Einstein discovered rationality in nature, unaided by any observation that had not been available for at least fifty years before, our positivistic textbooks promptly covered up the scandal by an appropriately embellished account of his discovery" (p. 11).

[7] Max Wertheimer, *Productive Thinking*, ed. Michael Wertheimer, p. 217. For other important interviews, see I. Bernard Cohen, "An Interview with Einstein," *Sci. Am.* 193:68–73 (July, 1955); and Robert S. Shankland, "Conversations with Albert Einstein," *Am. J. Phys.* 31:47–57 (January, 1963).

[8] Albert Einstein, "Autobiographical Notes," in Paul A. Schilpp, ed., *Albert Einstein: Philosopher-Scientist*, I, 3; cf. II, 665–688. See also Gerald Holton, "The Metaphor of Space-Time Events in Science," *Eranos-Jahrbuch* 34:33–78 (1965).

similarities between Einstein's theoretical discussions and the discussions which had been held on the problems raised by Michelson-Morley. On the other hand, the differences are more subtle, and, given the curious behavior of such colleagues as Lorentz, Planck, and Poincaré toward Einstein's work, their differences are demanding more attention than their likenesses. Einstein's paper certainly stimulated some profound rethinking of the theoretical issues raised by electromagnetic experiments with relative motion. But the correspondence of his theory with Michelson's experiment was less a causal than a sequential relationship.[9]

Einstein's genius was primarily theoretical and not experimental. It lay largely in his ability to combine the essence of experimental evidence with the elegance of theoretical expression and the epitome of epistemological economy. Had Einstein been so concerned in 1905, he certainly could have shown in what specific way his theory might have accounted for the Michelson type of aether-drift anomaly. Instead, he left his postulates to speak for themselves, being content with the following discursive paragraph as the key to his program: "All problems in the optics of moving bodies can be solved by the method here employed. What is essential is, that the electric and magnetic force of the light which is influenced by a moving body, be transformed into a system of coordinates at rest relatively to the body. By this means all problems in the optics of moving bodies will be reduced to a series of problems in the optics of stationary bodies."[10] This passage is now widely recognized as Einstein's most succinct statement of his purpose in this paper: to preserve the invariance of Maxwell's equations so that unlovely electrodynamic asymmetries might be removed.[11] But most

[9] Cf. G. H. Keswani, "Origin and Concept of Relativity," *British J. Phil. of Sci.* (part I), 15:286–306 (February, 1965) and (part II), 16:19–32 (May, 1965); Peter Havas, "Causality Requirements and the Theory of Relativity," *Synthese* 18: 75–102 (January, 1968).

[10] Einstein, in Lorentz et al., *The Principle of Relativity*, p. 59. Aside from the *post hoc, ergo propter hoc* fallacy discussed here, see the interesting distinction made by Hans Reichenbach between "The Lorentz Contraction and the Einstein Contraction," in his *The Philosophy of Space and Time*, pp. 195–202.

[11] Albert Einstein, "Über einen die Erzeugung und Verwandlung des Lichtes betreffenden heuristischen Gesichtspunkt," *Ann. d. Phys.*, 4th ser. 17:132–148 (June 9, 1905). The second paper was entitled, "Über die von der molekularkinetischen

physicists today recognize that the first two of Einstein's three great papers of 1905, published in June and July and concerned largely with reconciling certain contradictions between Newtonian mechanics and Maxwellian electrodynamics, made contributions to physics almost as important as the critique of space and time in the last of those three papers. The first contribution was the "heuristic photon," a suggestion which greatly simplified manipulations in quantum theory and allied Einstein's name with Planck's in the quantum challenge to the wave theory of light. The second paper amplified Brownian motion as an integral feature of kinetic-molecular theory. What is not generally recognized is that the word *heuristic*, introduced into the physical vocabulary with the idea of light quanta, applied with equal force to the postulate of relativity.[12]

Some have averred that Einstein's insistence on the heuristic character of his photon and relativity postulates was primarily a means of defense against the attacks he could foresee on these radical departures. But the tone of confidence, the strength of logic, and Mach's philosophic influence are persuasive evidence that Einstein offered these postulates as "working hypotheses" of real physical promise.[13] Unlike the operationally meaningless "thought experiments" in which he trained his fertile imagination, the heuristic qualifier was added only to those theoretical principles that produced experimentally verifiable or falsifiable predictions. Einstein usually provided invitations to experimentalists to test his models. This is the essential meaning of the twice-translated judgment that "Einstein alone crossed the Rubicon by making a principle of what Poincaré qualified as a *postulate.*"[14] Sim-

---

Theorie der Warme geforderte Bewegung von in ruhenden Flussigkeiten suspendierten Teilchen," *Ann. d. Phys.*, 4th ser. 17:549–560 (July 18, 1905).

[12] Holton, *Am. J. Phys.* 28:630. Note that the word *photon* was not used to refer to the light packets until many years later. See also Holton's "Mach, Einstein, and the Search for Reality," *Daedalus* (Proc. AAAS) 97:636–673 (Spring, 1968). See also Albert Einstein, *Investigations on the Theory of Brownian Movement*, ed. R. Furth.

[13] For a cogent analysis of these postulates and another aspect of Einstein's intellectual development, see Martin J. Klein, "Thermodynamics in Einstein's Thought," *Science* 157:509–516 (August 4, 1967).

[14] René Dugas, *A History of Mechanics*, p. 649. Cf. René Taton, *Reason and Chance in Scientific Discovery*, p. 134.

ilarly, the remark, attributed to Alfred North Whitehead, that Einstein's work provided "a principle, a procedure, and an explanation,"[15] explains in capsule form the scientific importance of the third Einstein paper of 1905.

In the years 1906, 1907, and 1908 Einstein had very little cause to use his heuristic plea as a shield against attack, simply because few men noticed or understood what he was trying to do. Lorentz, for example, in what may be his first public notice of Einstein's relativity postulate, called it "not altogether satisfactory" but credited it as another manifest solution for null results of experiments like Michelson's.[16]

In wrestling with problems of the nature of space and time Einstein had gone beyond the realm of the usual categories of physical thought, and to many if not most of his colleagues it was not precisely clear what his strikingly new viewpoint had to contribute. To call the aether meaningless, and then to go on to substitute the speed of light as a fundamental and absolute constant—in place of Newton's absolute time and space—seemed at the moment equally meaningless from a physical standpoint. Then, too, how should physics interpret the kinematics and extrapolate the dynamics of Einstein's predictions on the behavior of electrons? Were his equations immediately applicable to the far more subtle problems of the aether drift? Could such eclecticism, both philosophically and physically, be trusted? Perhaps most important of all, should mathematical abstraction go beyond Maxwell, Hertz, and Lorentz and be given preference over the still hopeful quest for mechanical models of explanation?

Gradually, as Einstein continued to demonstrate the fertility of his approach, relativity theory began to be taken more seriously. But at first, Einstein's secondary pronouncement that "the mass of the body

---

[15] A. N. Whitehead, quoted by Holton, *Am. J. Phys.* 28:628. It should be noted, however, that Whitehead became a rather severe critic of Einstein after the appearance of the General Theory. In 1920 Whitehead asserted, "To base the whole philosophy of nature upon light is a baseless assumption" (*The Concept of Nature*, p. 195); cf. also Whitehead's alternate theory as expressed in *The Principle of Relativity with Applications to Physical Science*.

[16] H. A. Lorentz, *The Theory of Electrons, and its Applications to the Phenomena of Light and Radiant Heat*, p. 230, from lectures delivered in March and April, 1906. See also Arthur Schuster, *The Progress of Physics . . . 1875–1908*, pp. 106–117.

a measure of its energy content" could not be considered meaningful
until experiments had indeed affirmed that "the theory corresponds to
the facts [and, therefore], radiation conveys inertia between the emit-
ting and absorbing bodies."[17] Einstein's theoretical predictions differed
from those of Abraham and Föppl, leading the experimentalist Walter Kaufmann to announce, rather too hastily, in 1906: "I anticipate
right here the general result of the measurements to be described in
the following: *The measurement results are not compatible with the
Lorentz-Einsteinian fundamental assumption.*"[18]

A. H. Bucherer, following Kaufmann, would provide in the next
three years, by his experiments in deflecting beta particles through
magnetic fields, some of the strongest evidence in direct support of
das Relativitätsprinzip. But even when Bucherer made his major report
in 1909, there was no conviction that the Michelson-Morley experiment had been explained away.[19]

Although the Michelson-Morley results had provided readymade
evidence for the credibility of Einstein's so-called light principle (i.e.,
the postulate that the speed of light is the same regardless of direction
and does not vary with the motion of the source or of the observer),
still one could argue that the experiment had simply not achieved the
necessary precision. The relativity principle does not necessarily *disprove* the existence of aether, although it does negate the plausibility
of aether drift. By disposing of the concept of absolute rest in absolute
space, the new theory stated that only relative motions are measurable.
Thus the journals remained receptive to discussions about the nature of
the aether throughout the decade following 1905. The repugnance to
mathematical abstractions in scientific explanation, the vested interests
of certain aether apologists (like Oliver Lodge and Philipp Lenard),

---

[17] Albert Einstein, "Does the Inertia of a Body Depend upon Its Energy-Con-
tent?" *Ann. d. Phys.*, 4th ser. 17 (1905), reprinted in Lorentz et al., *The Principle
of Relativity*, p. 71.

[18] W. Kaufmann, "Über die Konstitution des Elektrons," *Ann. d. Phys.*, 4th ser.
19:495 (January, 1906) as translated by Holton, *Am. J. Phys.* 28:634. Cf. M. Abraham, *Theorie der Elecktrizität*, and A. Föppl, *Einführung in die Maxwellsche Theorie
der Elektrizität.*

[19] A. H. Bucherer, "Die experimentalle Bestätigung des Relativitätsprinzips,"
Ann. d. Phys., 6th ser. 28:513–536 (1909).

and the lack of acceptance of operational modes of thought among physicists were three of many factors that delayed the acceptance of relativity theory.

In 1906, Morley retired from teaching and moved away from Cleveland, and Miller dropped the whole aether-drift project in order to concentrate on his growing interest in acoustics. Miller's musical tastes and talents were compatible with his scientific tastes and interests, and he soon acquired a reputation as one of the nation's leading authorities on experimental acoustics.[20]

In 1907, Michelson and Morley separately traveled to Europe again to accept different honors from the international scientific community. Michelson became in that year the first American scientist to be honored with a Nobel Prize.[21] From Stockholm he journeyed to London where he met with Morley at the Royal Society's anniversary meeting in December. Morley was awarded the Davy Medal for his chemical determinations of atomic weights, and Michelson, who had been a Foreign Member of the Royal Society since 1902, added the Copley Medal to his awards for "contributions to the science of optics." These honors were given with no mention of the Michelson-Morley experiment or of its relations to the rise of relativity theory. Einstein's papers were not yet widely read or recognized for what they would become. In the speeches accompanying these celebrations the aether concept still figured rather prominently.[22]

The popular legend that grew up in the twenties, based on a remark by Langevin, to the effect that only about a dozen people in the whole world could fully understand Einstein's paper when first published, is

[20] Dayton C. Miller to Professor [?] Nichols, April 12, 1906, in Lorentz Correspondence file, Bohr Library, American Institute of Physics, New York.

[21] Michelson's Nobel citation read: "For his optical experiments of precision and the spectroscopic and metrologic investigations which he carried out by means of them" (Flora Kaplan, *Nobel Prize Winners: Charts; Indexes; Sketches*, p. 21). Cf. Niels H. de V. Heathcote, *Nobel Prize Winners in Physics: 1901–50*, p. 52. See also the ebullient letter showing the effects of the prize on America's first science laureate: A. A. Michelson to T. C. Mendenhall, January 1, 1908, in Mendenhall Collection, Bohr Library, American Institute of Physics, New York.

[22] Presentation and acceptance remarks were printed in *The Times* (London), December 2, 1907, p. 7. See also D. T. McAllister, ed., *The Albert A. Michelson Nobel Prize and Lecture*, Publication of the Michelson Museum no. 2 (1966).

little more than popular mythology. But from the sociohistorical viewpoint the legend is meaningful in the sense that it was indeed an exclusive group of physical theorists who were concerned enough with the same kinds of problems to take Einstein's suggestion seriously.

A majority of astronomers and physicists were still interested in reading articles which held some hope that a satisfactory model of a mechanical aether might be found. For instance, in April, 1908, *Annalen der Physik* published a formidable essay by Hans Witte which examined the status of all the major models of the aether concept in the light of the experimental knowledge of 1908. Witte attempted to show how electromagnetic phenomena generally had come to replace the mechanical theory of light in the consideration of aether theory since the turn of the century.[23]

About the same time, Joseph Larmor was preparing his article "Aether" for the eleventh edition of the *Encyclopaedia Britannica*. Significantly, Einstein's name was not mentioned anywhere in the articles by Larmor, who also wrote "Energetics," "Energy," and "Radiation." Neither "Einstein" nor "Relativity Theory" were even indexed in that classic reference work. Larmor reflected characteristic British bias in saying that theoretical physics had been "largely transformed" into a "science of the aether,"[24] but he was on safer ground in asserting that "we must be content to treat the aether as a *plenum*, which places it in a class by itself."[25]

In 1908 also there appeared one of the most ambitious attempts to

---

[23] Hans Witte, "Weitere Untersuchungen über die Frage nach einer mechanischen Erklärung der elektrischen Erscheinungen unter der Annahme eines kontinuierlichen Weltäthers," *Ann. d. Phys.* 26:235–311 (April, 1908). Even articles critical of hydrodynamical or molecular models for the aether were generally hopeful that future models might be more successful; for example, Luigi d'Auria, "Concerning the Constitution of the Ether," *Pop. Astron.* 15:101–107 (February, 1907).

[24] Joseph Larmor, "Aether," *Encyclopaedia Britannica*, 11th ed., I, 292 (1911). See also Richard C. Maclaurin, *The Theory of Light: A Treatise on Physical Optics*, p. 16.

[25] Larmor, "Aether," p. 293. Even J. J. Thomson, famous for his discrete "electrons," was talking this way in his Adamson Lecture of November 4, 1907: "On the Light Thrown by Recent Investigations on Electricity and on the Relation between Matter and Ether" (pp. 8, 21); see also his "Recent Progress in Physics," *Smithsonian Institution Report for 1909*, p. 191.

explain the Michelson-Morley experiment since Lorentz's theoretical adjustment in 1895. The theory proposed by Walther Ritz audaciously suggested the revival of the old particle theory of light, in conjunction with the photon idea, in order to avoid the bothersome Einsteinian postulate that the velocity of light is independent of the velocity of its source.[26] Ritz's attempted revival of the ballistic hypothesis, in systematic form called the emission theory, attracted considerable attention at the time, because it allowed one to think that the velocity of light and the velocity of the source are additive. According to Ritz, this could be proved if the Michelson-Morley experiment were repeated using some extraterrestrial source for the light beam. Not until the mid-twenties was this test to be tried, to the detriment of the Ritz theory, as we shall see.

Soon after the Ritz emission theory appeared in print, Einstein received, from a flamboyant former teacher in mathematics, a tribute that effectively launched his public career in Germany. On September 21, 1908, shortly before his untimely death, Hermann Minkowski delivered an address to the assembly of German natural scientists and physicians meeting at Cologne. Minkowski's talk, "Raum und Zeit," was widely disseminated, since it was one of the first readable efforts to show the physical meaning of the mathematical fourth dimension. If Minkowski put the hyphen between the two words "space-time," he also showed what this would mean to the future of physics: "The views of space and time which I wish to lay before you have sprung from the soil of experimental physics, and therein lies their strength. They are radical. Henceforth space by itself, and time by itself, are doomed to fade away into mere shadows, and only a kind of union of the two will preserve an independent reality."[27] Minkowski specifically invoked the "famous interference experiment of Michelson" before men-

---

[26] The Ritz Emission Theory was discussed at length by R. C. Tolman, "Some Emission Theories of Light," *Phys. Rev.* 35:136–143 (August, 1912). Cf. Sir Edmund Whittaker, *A History of the Theories of Aether and Electricity*, II, 38–39. Walther Ritz, "Recherches critiques sur l'électrodynamique générale," *Ann. de Chim. et Phys.*, 8th ser. 13:145–275 (February, 1908). See also, more recently, J. G. Fox, "Evidence against Emission Theories," *Am. J. Phys.* 33:1–17 (January, 1965).

[27] Hermann Minkowski, "Space and Time," in Lorentz *The Principle of Relativity*, et al., p. 75.

tioning Einstein's work, and he demonstrated in what respect the Lorentz hypothesis could be considered "completely equivalent to the new conception of space and time, which, indeed, makes the hypothesis much more intelligible."[28] Curiously, Minkowski's interpretation of the difference between Lorentz and Einstein rested with his own emphatic integration of time and space. Lorentz had worked with local time as a special case of absolute time, but, Minkowski said, "the credit of first recognizing clearly that the time of the one electron is just as good as that of the other . . . belongs to A. Einstein. Thus time, as a concept unequivocally determined by phenomena, was first deposed from its high seat. Neither Einstein nor Lorentz made any attack on the concept of space. . . ."[29]

While Minkowski and other mathematical physicists were beginning to inspire appreciation for the elegance of the special theory of relativity, experimentalists generally were much more impressed by the Kaufmann and Bucherer experiments. Lodge, for instance, in a book entitled *Electrons*,[30] in 1907 gave full details of Kaufmann's apparatus without mentioning Einstein. Likewise papers continued to appear scrutinizing more closely the questionable features in the experimental design of Michelson-Morley. One of the more interesting of these pleas for caution in accepting the negative results of the Michelson-Morley tests stressed the importance of using a true monochromatic light source,[31] because all previous aether-drift tests might have been vitiated by incoherent light beams.

[28] Ibid., p. 81. Minkowski's personal attempt to raise the "relativity-postulate" into a "world-postulate" (p. 83) ended with the judgment that here was given "the true nucleus of an electromagnetic image of the world, which, discovered by Lorentz, and further revealed by Einstein, now lies open in the full light of day" (p. 91).

[29] Ibid., pp. 82–83. Minkowski continues by making just such an attack on the concept of space, fully expecting his "violation" to be "appraised as another act of audacity on the part of higher mathematics."

[30] Sir Oliver Lodge, *Electrons: Or the Nature and Properties of Negative Electricity*. On the other hand, see also the important continuation of the Trouton-Noble experiment, in F. T. Trouton and A. O. Rankine, "On the Electrical Resistance of Moving Matter," *Proc. R. S.* (London) 80:420–435 (1908).

[31] Emil Kohl, "Michelson's Ether Research," *Am. J. Sci.*, 4th ser. 27:338 (April, 1909). Closer to the mainstream of interpretation by now were Gilbert N. Lewis and Richard C. Tolman, "The Principle of Relativity and Non-Newtonian Mechanics," *Phil. Mag.* 18:510–523 (October, 1909).

Meanwhile, Poincaré continued to speak and write much on the principle of relativity, always carefully avoiding mention of his German counterpart. Back in 1902 he had overcompensated to show more concern for experimental methodology than for cosmology; Poincaré had then been permissive in regard to the use of the concept of the aether: "Whether the ether exists or not matters little—let us leave that to the metaphysicians; what is essential for us is that everything happens as if it existed, and that this hypothesis is found to be suitable for the explanation of phenomena. After all, have we any other reason for believing in the existence of material objects? That, too, is only a convenient hypothesis; only it will never cease to be so, while some day, no doubt, the ether will be thrown aside as useless."[32] Later and elsewhere he expressed much less patience with those who would allow the aether of space to compete with the relativity of space.[33] This double standard which he found to be current around 1909 was for him methodologically inexcusable: "Whoever speaks of absolute space uses a word devoid of meaning."[34] As a methodologist, however, Poincaré's conventionalism centered on the problem of selection of the most meaningful data and hypotheses. In regard to Michelson's experiment, he apparently never doubted the conventional acceptance of the null results as evidence for the FitzGerald-Lorentz contraction; only the contraction hypothesis itself did he call into question.[35] Thus, although he saw Michelson-Morley as precluding the idea of absolute motion, Poincaré never felt able to outlaw the use of the aether concept: it ex-

[32] Henri Poincaré, *Science and Hypothesis*, pp. 211–212. See also Tobias Dantzig, *Henri Poincaré: Critic of Crisis*, pp. 71–99.

[33] Henri Poincaré, *Science and Method*, p. 10.

[34] Ibid., p. 93. See also a similar critique by Gustave Le Bon, *The Evolution of Forces*, pp. 13, 17, 99.

[35] Poincaré, *Science and Method*, pp. 93–221. This attitude is perhaps best expressed on p. 221: "The hypothesis of Lorentz and FitzGerald will appear most extraordinary at first sight. All that can be said in its favour for the moment is that it is merely the immediate interpretation of Michelson's experimental result, if we *define* distances by the time taken by light to traverse them." See also Stanley Goldberg, "Poincaré's Silence and Einstein's Relativity: The Role of Theory and Experiment in Poincaré's Physics," *Br. J. Hist. Sci.* 5:73–84 (June, 1970).

pressed too well the "natural kinship between all . . . optical phenomena."[36]

"We know nothing as to what the ether is, how its molecules are disposed, whether they attract or repel each other; but we know that this medium transmits at the same time the optical perturbations and the electrical perturbations; we know that this transmission must take place in conformity with the general principles of mechanics, and that suffices us for the establishment of the equations of the electromagnetic field."[37] Poincaré remained skeptical until death that the Michelson-Morley experiment had ever found a satisfactory explanation. He (and certainly many others) would agree with Owen Ely, who wrote in 1910 that the only clear answer to the question, What is the aether? is the reply, "The greatest enigma of all time."[38]

Thus, by 1910 the stage was set for a final duel between those who, like Lodge, believed that "waves we cannot have, unless they be waves in something,"[39] and those who believed with Einstein that the aether was superfluous. For the former group the alternatives by which the Michelson experiment could be explained were four: (1) there is no motion through the aether at all; or (2) there is no aether drift past the earth; or (3) the aether next to the earth is stagnant; or (4) the earth carries the neighboring aether along with it through space.[40]

For the latter group "the principle of relativity is one attempt, and by

[36] Henri Poincaré, *The Value of Science*, p. 140. Cf. Poincaré, *Mathematics and Science: Last Essays*, pp. 89–99.

[37] Poincaré, *Value of Science*, p. 94. This will be recognized as a restatement of Poincaré's earlier position of 1904. See also Stanley Goldberg, "Henri Poincaré and Einstein's Theory of Relativity," *Am. J. Phys.* 35:944 (October, 1967). Cf. Russell McCormmach, "Henri Poincaré and the Quantum Theory," *Isis* 58: 37–55 (Spring, 1967).

[38] Owen Ely, "What is the Ether?" *Pop. Astron.* 18:532 (November, 1910). For another calculation of the enigmatic density of the medium of space (480 × platinum!), see John McKenzie, "The Structure of the Universe," *Minnesota Acad. of Sci. Bulletin* 4:384–405 (1908).

[39] Sir Oliver Lodge, *The Ether of Space*, p. 2.

[40] Ibid., p. 67. For an interesting corroboration of the time lag in acceptance of relativity in one country, see Dennis F. Miller, "The Early Influence of Einstein and the Special Theory of Relativity in America," M.A. thesis, Ohio State University, 1965.

far the most successful attempt as yet, to explain the failure of all experiments designed to detect the earth's motion through space, by its effect on terrestrial phenomena."[41] In the relativist camp we can and indeed do have "waves" which "wave in nothing" save a "field." Furthermore, to those who had accepted the new doctrine, the principle of relativity promised to become at least as far-reaching in its effects as the second law of thermodynamics.[42]

[41] D. F. Comstock, "The Principle of Relativity," *Science* 31:767 (May 20, 1910). Comstock showed an interesting example of personal evolution: see his "Reasons for Believing in the Ether," *Science* (new series) 25:432–433 (March 15, 1907); "A Neglected Type of Relativity," *Phys. Rev.* 30:267 (February, 1910).

[42] H. A. Bumstead, "Applications of the Lorentz-FitzGerald Hypothesis to Dynamical and Gravitational Problems," *Am. J. Sci.*, 4th ser. 26:497 (June, 1909). Cf. *Am. J. Sci.*, 4th ser. 26:493–508 (November, 1908).

CHAPTER NINE

THE FALL OF THE AETHER, 1910–1920

Historians often use World War I as the time of transition between essentially nineteenth- and twentieth-century forms and attitudes. The decade of the Great War may also serve to mark the social transition from essentially classical to relativistic physics. By 1910, the relativity principle had made an impression on the profession as a special meta-theory with promise; by 1920, Einstein had proposed and others had confirmed (in part) so great an extension of the theory that a distinction had to be made between the first Special or restricted theory and the second, the General Relativity Theory.

Quantum theory likewise was being extended and improved enough to be recognized as a promising tool. Planck and Einstein, by mutual reinforcement of each other's productions, accomplished a great deal during this period and were regarded as leaders in the vanguard of physical progress. Neither, however, was satisfied with his own or with the other's quantum theory. One of the reasons for uneasiness was the necessity to admit that somehow *both* the undulatory and the corpuscular theories of light seemed true.[1]

In the three or four years before the outbreak of the war, a number of books and articles were written explaining relativity for physicists who had not kept up with it.[2] Likewise a few textbooks began to include discussions of the meaning of relativity for students. In 1913 the transition from Rutherford's to Bohr's model of the atom made the fraternity of chemists more acutely aware of what was happening in physical theory. Indeed, physical chemistry itself, so greatly stimulated by J. Willard Gibbs's theoretical bases for the synthetic chemical industry, was in some measure a catalyst and a precipitate of World War I.[3]

[1] There is a large and growing body of literature on the history of physics in the early twentieth century. A few of the more pertinent and recent titles are Henry A. Boorse and Lloyd Motz, eds., *The World of the Atom*, II; G. K. T. Conn and H. D. Turner, *The Evolution of the Nuclear Atom*; D. ter Haar, *The Old Quantum Theory*; W. G. V. Rosser, *An Introduction to the Theory of Relativity*, chaps. 1 and 2. On relations between Planck, Einstein, Mach, and Lorentz, see two articles: Gerald Holton, "Mach, Einstein, and the Search for Reality," *Daedalus* 97:636–673 (Spring, 1968), and Stephen G. Brush, "Mach and Atomism," *Synthese* 18:192–215 (1968). See also Martin J. Klein, three articles in *The Natural Philosopher*, edited by Daniel E. Gershenson and Daniel A. Greenberg: "Planck, Entropy, and Quanta, 1901–1906," 1:83–108; "Einstein's First Paper on Quanta," 2:59–86; and "Einstein and the Wave Particle Duality," 3:1–49. Niels Bohr, *On the Constitution of Atoms and Molecules*; Max Jammer, *The Conceptual Development of Quantum Mechanics*; A. d'Abro, *The Decline of Mechanism (in Modern Physics)*; George Gamow, *Thirty Years That Shook Physics: The Story of Quantum Theory*.

[2] As early as 1911 Norman Campbell lamented the misunderstandings and dogmatic rejections of the principle of relativity caused by its mathematical parentage and the failure of its expositors to communicate with physical thinkers: "The Common Sense of Relativity," *Phil. Mag.*, 6th ser. 21:502, 515 (April, 1911). For a bibliography to much of the prewar German literature on relativity, including Max von Laue's comprehensive papers of 1911 and 1913, see Ernst Cassirer, *Substance and Function and Einstein's Theory of Relativity*, pp. 457–460. See also R. D. Carmichael, "On the Theory of Relativity: Philosophical Aspects," *Phys. Rev.* 1:179–196 (March, 1913), and "Mass, Force, and Energy," ibid. 1:161–178 (Feb., 1913); J. W. Nicholson, "On Uniform Rotation, the Principle of Relativity, and the Michelson-Morley Experiment," *Phil. Mag.*, 6th ser. 24:820–827 (1912). See also von Laue, *Die Relativitätstheorie*.

[3] The Haber-Bosch Process for the fixation of atmospheric nitrogen in ammonia, developed semisecretly from 1910 to 1913, relieved Germany of the need for import-

With these new theoretical bases and technological applications for physical science it was inevitable that the nature of the most challenging questions should change. Change they did, and drastically, not only, nor even primarily, because of theory, but also because of changes in instrumentation, in scientific organization, in technical vocational education, and in the feedback from science applied to, in, by, and for industry.

When, in 1914, the first world war since Napoleonic times broke out, the social and economic integration of national societies in Western civilization was so complete that scientific vocations as well as all others suffered a severe redirection. Whether modern wars have done more to stimulate or to impede the progress of science may well be a moot question, but World War I certainly proved as total to scientists as to anyone. Because of the absorption of scientific interests in engines of destruction during the latter half of the decade, 1910–1920, it is not surprising that the four years from 1914 to 1918 should constitute a hiatus for pure science. Most men, scientists included, as the socialist movement discovered to its chagrin, were patriots first and professionals second.[4]

But before 1914 the scientific community was far less divided on nationalistic than on intramural grounds. The physics profession was very far indeed from a unanimous acceptance of the Einsteinian postulates. As a result, these were critical years for the aether hypothesis,[5] as well as for European diplomatic equilibrium. We must now try to see why so many were so reluctant to allow the aether concept to die.

---

ing nitrates from the Atacama beds in Chile. *Ersatz* became a byword after World War I, so common had synthetic materials become as a result of wartime demands. See Muriel Rukeyser, *Willard Gibbs*, pp. 383–402; cf. E. E. Slosson, "Twentieth Century Science and Invention," in Jesse E. Thornton, comp., *Science and Social Change*, p. 178. See also Morris Goran, *The Story of Fritz Haber*.

[4] See I. Bernard Cohen, "American Physicists at War," *Am. J. Phys.* 13:233–235 (August, 1945) and 13:333–346 (October, 1945). See also John U. Nef, *War and Human Progress: An Essay on the Rise of Industrial Civilization*.

[5] For one of several books attempting to analyze the historical nature of the critical impasse over the aether, see Michele LaRosa, *Der Äther, geschichte einer hypothese*. See also Sir Joseph Larmor's article, "Aether," for the famous 11th edition of the *Encyclopaedia Britannica* (1910–1911), I, 292–297, as well as S. Otto Eppenstein's article, "Aberration," 11th ed., I, 54.

In the few years since the turn of the century, physics had changed tremendously. Michelson, at the Ryerson Laboratory at the University of Chicago, had done much to shape these changes by his basic work in experimental optics. Likewise, Miller, at Case School in Cleveland, had done much in his special field of acoustics for the advancement of physics. He, too, was one of the leading experimentalists in America, recognized as a contributor to a renaissance of interest in musical and architectural acoustics.[6]

After his retirement in 1906, Edward Morley moved away from Cleveland and became rapidly less active in any scientific pursuit. Morley was to live on until 1923, but his age and inactivity greatly limited the extent of his interest in the current progress of science. During the war Michelson was called back into the navy to work on an optical range finder, and Miller was commissioned to study the effects of acoustic and shock waves from big-gun explosions. After 1918, Michelson and Miller returned to their separate specialties and separate locations. As a Nobel laureate, Michelson was the more famous scientist, but Miller, too, was growing into an outstanding authority in his field. Neither man was pleased with what was happening in the world of physical theory, ostensibly as a result of the Michelson-Morley experiments. Both were exceedingly reluctant to admit that the aether concept had fallen.

Einstein left his job with the patent office at Bern, and in 1909 he moved to a teaching post at the University of Zürich. Two years later appeared the first of his efforts to extend the restricted relativity theory to a more general application. This was the famous paper of 1911 calling for astronomical experiments to test his deduction that gravitation will cause rays of light to bend. He had treated in a previous memoir the influence of gravitational forces on the propagation of light, but he was not satisfied, "because I now see that one of the most important

[6] Harvey Fletcher, "Dayton Clarence Miller, 1866–1941," *Biographical Memoirs,* National Academy of Sciences, 23:61. Cf. Vern O. Knudsen, "Recent Developments in Architectural Acoustics," *Rev. Mod. Phys.* 6:1–22 (January, 1934). Wallace C. Sabine of Harvard, the acknowledged leader in the field, played host to Miller in 1914 when the latter was invited to give the Lowell Lectures in Cambridge: see D. C. Miller, *The Science of Musical Sounds.* In 1912 Miller had invented a device, called a *phonodeik,* for converting sound waves into visible images.

consequences of my former treatment is capable of being tested experimentally. For it follows from the theory here to be brought forward, that rays of light, passing close to the sun, are deflected by its gravitational field, so that the angular distance between the sun and a fixed star appearing near to it is apparently increased by nearly a second of arc."[7]

At the end of the decade this prediction, verified by a British solar-eclipse expedition to Brazil and Africa, was to be the catapult which would make Einstein and "relativity" front-page news around the world. Meanwhile, and without waiting for experimental confirmation of the bending of light in a gravitational field, the General Theory of Relativity grew into maturity, nurtured by the pacifistic Einstein despite the war. "The Foundation of the General Theory of Relativity,"[8] the famous paper which first presented the full exposition of the extension of the restricted principle of relativity to reference systems in *any* kind of relative motion, was published in 1916. The antipathy to Einstein which concerns us here was directed not at the General Theory but at the Special Theory that had declared the aether superfluous.

To see the aether as it was seen in 1910 requires that we once again remind ourselves of the dangers of anachronisms. Certainly it is clear, as Whittaker writes, that "the theory of relativity had its origin in the theory of aether and electrons. When relativity had become recognised as a doctrine covering the whole operation of physical nature, efforts were made to present it in a form free from any special association with electromagnetic theory, and deducible logically from a definite set of axioms of greater or less plausibility."[9] But with regard to the

[7] Albert Einstein, "On the Influence of Gravitation on the Propagation of Light," in H. A. Lorentz et al., *The Principle of Relativity: A Collection of Original Memoirs on the Special and General Theory of Relativity*, p. 99.

[8] *Ann. d. Phys.*, 4th ser. 49:769–822 (1916). For bibliographic cross-references, see Margaret C. Shields, "Bibliography of the Writings of Albert Einstein to May 1951," in P. A. Schilpp, ed., *Albert Einstein: Philosopher-Scientist*, II, 689; and Nell Boni, Monique Russ, Dan H. Laurence, comps., *A Bibliographic Checklist and Index to the Writings of Albert Einstein.*

[9] Sir Edmund Whittaker, *A History of the Theories of Aether and Electricity*, II, 42–43. H. A. Lorentz himself still argued in 1910–1912 that the "greater plausibility" was his: "If we do not like the name 'aether,' we must use another word as a peg to hang all these things upon. It is not certain whether 'space' can be so extended

aether-drift experiments it is an understatement amounting to historical bowdlerization simply to say, as Whittaker continues: "It should be
mentioned also that when relativity theory had become generally accepted, the Michelson-Morley experiment was rediscussed with a much
more complete understanding and exactitude." That it was "rediscussed," and often certainly with "a much more complete understanding" is not to be denied, but scientific passions were often inflamed
beyond the point of rational understanding.

At least three types of major intellectual opposition arose in response to the challenges of relativity theory in this decade. These three
types of opposition may be called conceptual conservatism, philosophic
relativism, and experimental reversalism.

To begin with responsible conceptual conservatives, we cite first an
honored physicist in anger. On December 28, 1911, William F. Magie
of Princeton addressed the American Physical Society (which had just
elected him president) with a blistering attack on relativity and relativists. Under the title "The Primary Concepts of Physics," Magie
vehemently insisted on retaining the concept of force in spite of Hertz,
and the two separate concepts of time and space in spite of Einstein. He
conceded that the aether was not necessarily indispensable, but in his
opinion the abandonment of the aether concept would be a "great and
serious retrograde step."[10] In his view the radical relativists were attempting Baron Munchausen's feat while trying to escape from a tower
by a rope that was too short: the mythical Munchausen made good his
escape by splicing on some more rope obtained by a cutting from above.

Magie's address demonstrates that the principle of relativity was
well known among his brethren, that the legend of the Michelson-
Morley paternity of relativity was already accepted and that the intra-
fraternal controversy over the usefulness of relativity theory was well

---

as to take care not only of the geometrical properties but also the electric ones. One
cannot deny to the bearer of these properties a certain substantiality . . ." (*Lectures on
Theoretical Physics*, III, 211).

[10] William F. Magie, "The Primary Concepts of Physics," *Science* 35:281–293
(February 23, 1912); quotation from p. 290. See also a similar presidential address,
to the British Association in 1913, published in pamphlet form: Sir Oliver Lodge,
Continuity, esp. pp. 41, 49.

under way. Thinking it unnecssary to give an account of the relativity principle in his address, Magie stated, "it may fairly be said to be based on the necessity of explaining the negative result of the famous Michelson-Morley experiment, and on the convenience of being able to apply Maxwell's equations of the electromagnetic field without change of form to a system referred to moving axes."[11] Magie insisted on the counterevidence that the principle of relativity was not needed to explain the experiments by Fizeau, Mascart, Brace, or those of Kaufmann and Bucherer. Why, then, he asked, should we allow the Michelson-Morley experiment to upset all our primary concepts of physics? "The principle of relativity accounts for the negative result of the experiment of Michelson and Morley, but without an ether how do we account for the interference phenomena which made that experiment possible?"[12]

This last question was perhaps the most embarrassing which the conservative physicist could pose for his assailants. As long as the question of the ultimate nature of light was still unanswered, the desire for waves, or particles, but not both, would seem to demand a division of attitudes based on faith in one theory or the other. Had Magie applied more of his demonstrated historical interests to the origins of Einstein's work, he might have been a little less chauvinistic and a little more critical of the notion, which he evidently accepted, that the Michelson-Morley experiment had been the crucial one.

Nonetheless, he performed a service in showing that the relativists were using the Michelson-Morley experiment in two very different ways. Some were presenting the principle of relativity as a direct inductive conclusion from the negative results of Michelson-Morley; others were deriving from the principle of relativity another postulate of impotence, maintaining that the hopelessness of explaining Michelson-Morley by any theory of the structure of the universe had been demonstrated by the absolute character of the speed of light. Although both these inferences were legitimate, neither was free from objection.

[11] Magie, "Primary Concepts," *Science* 35:287.
[12] Ibid., p. 290. For similar attitudes, see Arthur Schuster, *The Progress of Physics: During 33 Years, 1875–1908*, pp. 106–110; Frederick Soddy, *Matter and Energy*, pp. 184–185.

In the end, Magie conceded that all this discussion was secondary, since the central question remained—could relativity theory ever *explain* natural phenomena? As of 1911 Magie was convinced that relativity theory was "evidently merely descriptive."[13]

Another type of intellectual opposition to relativity theory was exhibited in the 1913 publication of Paul Carus's critical essay on relativity. Carus argued that "applied relativity can neither establish nor refute the principle of relativity. This is true above all of the well-known and most important Michelson-Morley experiment."[14] Carus was a metaphysician who will serve as an example of that philosophic relativism which in the nineteeth century insisted that "all knowledge is relative."

As a rationalist philosopher of science, like Stallo, Carus was a spectator-critic, but, nonetheless, an acute observer, who was also the enormously influential editor of *The Monist* and of the Open Court Publishing Company, which was the American outlet for the works of Poincaré. His commitment to the correspondence theory of truth and his vocational position gave him a unique but limited vantage point from which to view the controversy over relativity.[15]

With the confidence of great conviction Carus claimed that there was nothing novel in the principle of relativity, that indeed "the relativity of time and space, as well as of all real things is a universal and inalienable condition of all existence." Citing Heraclitus, Bradley, and Spencer, Carus chided such relativists as Philipp Frank and Ernst Mach, saying that the upstart demand for a remodeling of all physics was simply indefensible: "I must insist that the principle of relativity has always been subconsciously in the minds of scientists. Only it has lately been forced upon the attention of physicists by the progress of

[13] Magie, "Primary Concepts," p. 292. Among physicists in America this type of argument remained strong for at least a decade. See, for example, Harold A. Wilson, "Gravitation and the Ether," *Rice Institute Pamphlet* 8:23–33 (January, 1921).

[14] Paul Carus, *The Principle of Relativity in the Light of the Philosophy of Science*, p. 66.

[15] See William H. Hay, "Paul Carus: A Case-Study of Philosophy on the Frontier," *J. Hist. Ideas* 17:498–510 (October, 1956). For a mathematician's view, similar to Carus's, see Nicholson, "On Uniform Rotation," *Phil. Mag.*, 6th ser. 24: 820–827.

astronomical measurements." Regarding the aether-drift tests, Carus
said that "there is not one of these so-called experiments, invented to
prove the relativity of time and space, which does not ultimately re-
solve itself into a machine that renders visible aprioristic considerations." Had physicists been better philosophers of science, the relativity problem would never have arisen, for philosophers have always
known, said Carus, that, except for the laws of purely formal thought
perhaps, "there is nothing absolute; everything is relative."[16]

In reflecting on the convulsions in physics which he was witnessing,
Carus relied almost exclusively on the Michelson-Morley experiments
for his understanding of what was at stake. This might have been dangerous were it not for the fact that Carus had been exposed from ado-
lescence to the calculus of extension or geometry of form.[17] This training, together with his semi-professional interest in astronomy, disposed
him throughout life to treat concepts of space with great sophistication.
Because of the lag in the ability of most physicists to understand the
tensor calculus of the mathematical theorists, because of the extraordinary imaginative demands made by Einstein's conceptual critique, and
because of the extremely worrisome trend, as seen by many of the
older generation of physicists, toward a subjectivistic mysticism, or
even possibly a solipsism, the Kantian critique of relativity presented
by Carus attracted widespread attention.

As for the aether concept, Carus was willing to let experiments alone
decide, but he believed that if the luminiferous aether had to be abandoned, then only the energeticists' position would be tenable. Taking
his cue from the cautious reserve of Albert A. Michelson himself,
Carus stated a common puzzlement: "What this famous experiment
has to do with the principle of relativity except in a most general way,
is not yet clear to those who have not joined the ranks of the relativity
physicists; but the relativity physicists insist very vigorously and dog-
matically that it proves, or at least favors, their theory."[18] If so, then

---

[16] Carus, *The Principle of Relativity*, pp. 20, 39, 17, 41.

[17] Carus was tutored in non-Euclidean and projective geometry as a gymnasium
student by the early topologist Hermann Grassmann. See Hay, "Paul Carus," p. 505.

[18] Carus, *The Principle of Relativity*, p. 68. Cf. A. A. Michelson, *Studies in Optics*,
pp. 4, 5.

what are we to do with the astronomical aberration of light? Astronomy, as an observational science, unlike experimental physics, bears a much heavier debt to philosophy. "The question of relativity is a philosophical problem, but the Michelson-Morley experiment is of a purely physical nature, and so we must expect that the last word as to its explanation should be given by the physicists."[19] Carus ended his critique by noting how the mid–nineteenth-century philosophic vogue for the relativity of knowledge had prepared the ground, unconsciously predisposing even physical scientists to accept the tenets of relativity.

Even so, this disposition to accept Einstein and Poincaré uncritically led to what Carus believed was a profoundly disturbing discrepancy: the new relativists, basing their beliefs on the still restricted theory, accepted a finite value for the velocity of light while still assuming without question that gravitation operated infinitely fast. Which is more truly a "universal constant" under these conditions? Carus's philosophic relativism was well expressed by the bit of borrowed doggerel with which he reproved the physical relativists' claim to be expanding scientific horizons:

> Great fleas have little fleas
> Upon their backs to bite 'em,
> And little fleas have lesser fleas,
> And so *ad infinitum*.
>
> And the great fleas themselves, in turn,
> Have greater fleas to go on;
> While these again have greater still,
> And greater still, and so on.[20]

If Carus's kind of criticism was considered impertinent and irrelevant by practicing physicists, a third example of the kinds of opposition which relativity theory encountered in the transitional decade could hardly be so easily ignored. This came from the experimental laboratory itself. In the last months of the year 1913 two reports were

---

[19] Carus, *The Principle of Relativity*, p. 70.

[20] Ibid., pp. 71, 73, 74. This verse was apparently borrowed in part from Augustus de Morgan, who in turn was apparently indebted to Jonathan Swift, and so *ad infinitum*.

given to the Académie française by a French optical physicist, Georges Sagnac, who claimed to have arrived at positive experimental evidence for a luminiferous aether.

Sagnac had combined a number of interferometric principles in building an apparatus which was designed to operate like Michelson's. In a self-contained system he collimated a light beam for greater coherence, then led it into a split lens with a half-silvered inner surface. Here the beam was split into two pencils traversing in opposite directions a polygonal course before being recombined and focused on a photographic plate. The apparatus was mounted on a turntable rotating about once a second, and the experiments were performed with very little possibility of human error—except in the interpretation of the results.

In October, Sagnac announced that he had demonstrated the existence of luminiferous aether by the effects of a relative aether wind on his interferometer in uniform rotation and that "the observed interferential effect is surely the rotational optical effect of the movement of the system in relation to the aether and directly proves the existence of the aether, a necessary support to the light theories of Huygens and Fresnel.[21] Two months later Sagnac was even more positive that he had found experimental evidence for "the reality of the aether": "The result of the procedures showed that, in the surrounding space, the light is propagated with a velocity V_0, independent of the movement of the parts of the light source . . . and of the optical system. This behavior of the space describes experimentally the luminous aether."[22]

As might be expected, Sagnac's experiments received considerable interest, but only on the side of the antirelativists. His results were never really treated fairly or investigated seriously by the opposing faction until after the war. Even then the relativists who looked at this

---

[21] M. G. Sagnac, "L'Éther lumineux démontré par l'effet du vent relatif d'éther dans interféromètre en rotation uniforme," *Comptes Rendus* 157:710 (October 27, 1913) [my translation]. See also John E. Chappell, Jr., "Georges Sagnac and the Discovery of the Ether," *Archives internationales d'histoire des sciences* 18:175–190 (1965).

[22] Sagnac, "Sur la preuve de la réalité de l'éther lumineux par l'expérience de l'interférographe tournant," *Comptes Rendus* 157:1413 (December 22, 1913) [my translation].

adverse evidence generally dismissed Sagnac's experiments by pointing out the numerous possibilities for periodic and systematic errors.
The crux of the misinterpretation was not noticed until 1932.[23]

Sagnac might have been taken more seriously if his original reports
had not been so invidiously worded. The fact remains, however, that
here was a set of experiments, purporting to have accomplished what
the Michelson-Morley experiment was supposed to have done. That
Sagnac's aether-drift tests never got an equivalent consideration might
be attributed to lack of confidence in his trustworthiness. But it would
seem more likely that his misfortune was with historical circumstances
of time and place: in 1913 aether-drift tests of this type had already
been written off as hopeless except by the foremost aether apologists.
And before an interested following could be built up, Sagnac and
his country were smothered in war. The man and his experiments
were forgotten.

There were a number of efforts to "split the difference" between
relativity theory and the aether in the following years.[24] For instance,
in 1914, Leigh Page of Yale submitted a paper that outlined the properties of a type of aether that would be consistent with the principle of
relativity. By positing an "absolutely homogeneous" aether capable of
"transmitting all nonhomogeneities in straight lines with the velocity
of light,"[25] he was able to derive his electrodynamic equations from

[23] Georg Joos in 1932 made clear the distinction between rotational and translational effects to be expected. See his *Theoretical Physics*, p. 449. Still later, however,
Sagnac's interpretation had its champion in H. E. Ives.

[24] For example, Edward V. Huntington, "A New Approach to the Theory of Relativity," *Phil. Mag.*, 6th ser. 23:494–513 (April, 1912).

[25] Leigh Page, "Relativity and the Ether," *Am. J. Sci.*, 4th ser. 38:170 (August,
1914). Very soon hereafter Page joined the relativity camp unequivocally, since he
contributed an important article, "A Century's Progress in Physics," to the anthology
edited by Edward S. Dana, et al., *A Century of Science in America*, with Special
Reference to the *American Journal of Science*, 1818–1918, which gives (p. 379)
an exceptionally clear statement of the status of the aether in the evolving consensus:
"The relativist does not deny the existence of an ether. To him the question has no
more meaning than if he were asked to express an opinion as to the reality of parallels
of latitude on the earth's surface. As a convenient medium of expression in describing
certain phenomena, the ether has justified much of the use which has been made of it.
But to attribute to it a degree of substantiality for which there is no warrant in experiment, is to change it from an aid into an obstacle to the progress of science."

three fundamental assumptions. He relied heavily on Lorentz's work, but was able to reduce by half the number of kinematical hypotheses used by Lorentz.

Rather than straddle the fence between aether and relativity, many physicists preferred to align themselves with Lorentz.[26] The comfort in this procedure lay in retaining the aether concept in the background without allowing it to intrude into considerations of specific analyses. Spectroscopists, especially, seemed to prefer this position. Pieter Zeeman, whose discovery of the broadening of spectral lines in a magnetic field had long since made him famous, joined the support for his mentor and countryman Lorentz in 1915 by repeating Michelson and Morley's repetition of Fizeau's "water-drag" experiment. His results emphasized once again the positive interpretation of that particular type of aether-drag test. As he pushed the degree of accuracy in "what was regarded as one of our most accurate measurements" another step forward, he helped support the conclusion that the Fresnel-Lorentz coefficient for aether drag was definitively confirmed.[27]

By the middle of the decade, confusion was rampant regarding the precise meaning of statements pro and con in the relativity controversy. The tenuous line between physical and metaphysical statements was often breached.[28] Relativists often had to remind their opposition that their contention that it is impossible to measure absolute velocity is not equivalent to the statement that it is impossible to conceive of absolute space.[29] The problem of distinguishing between physical definitions and metaphysical definitions of the principle of relativity was a foremost concern to those in the relativist camp. "The ether is, in truth,

---

[26] Substantial appreciation for this point of view was expressed by E. Cunningham, *Relativity and the Electron Theory.*

[27] Pieter Zeeman, "Optical Investigation of Ether-Drift," *Nature* 96:431 (December 16, 1915).

[28] Even Ernst Mach decried the partisan confusion of issues in his notable Preface of 1913 to *The Principles of Physical Optics: An Historical and Philosophical Treatment*, pp. vii–viii: "I gather . . . that I am gradually becoming regarded as the forerunner of relativity. . . . I must, however, as assuredly disclaim to be a forerunner of the relativists as I withhold from the atomistic belief of the present day. . . . The reason why . . . I discredit the present-day relativity theory, which I find to be growing more and more dogmatical . . . must remain to be treated in the sequel."

[29] Cunningham, *Relativity*, p. 2.

nothing more than the aggregate of the functions which it serves, though we find it difficult to think of it except in some concrete and uniquely existent form."[30]

An Italian philosopher and geometer, Enriques, had as long ago as 1906 pointed out that the Michelson-Morley experiment afforded a confirmation to the first of the two postulates used by Einstein in his Special Theory but not necessarily of the second.[31] Michelson-Morley and Trouton and Noble had given a good basis for the postulate that only relative motions are measurable, but so far the second postulate, that the velocity of light is a universal constant, had a much weaker experimental basis.

Now, in 1918 and 1919, Quirino Majorana in Rome rectified this weakness by publishing his experimental demonstrations, first, of the constancy of the velocity of light reflected from a moving mirror, and second, of the constancy of the velocity of light emitted by a moving source. Majorana's apparatus consisted of a Michelson interfermoeter at one arm of which was located a rotating mirror. In interpreting his results he encountered considerable difficulty with the Doppler-Fizeau effect, but he felt justified in concluding, anyway, that within the limits of significant figures and experimental error he had proven that "the velocity of light does not change by the movement of the source along the direction of propagation."[32]

Another Italian, Augusto Righi, the professor of physics at Bologna who had given Marconi his start toward making wireless ethereal waves detectable, announced repeatedly in 1919–1920 that, contrary to common belief, his own intuition had been confirmed by his analyses

[30] Ibid., p. 4. See also L. Silberstein, *The Theory of Relativity*, pp. 72–87, and 88 n. 2; Richard C. Tolman, *The Theory of the Relativity of Motion*, p. 17.

[31] Federigo Enriques, *Problems of Science*, p. 350.

[32] Quirino Majorana, "On the Second Postulate of the Theory of Relativity: Experimental Demonstration of the Constancy of Velocity of Light Reflected from a Moving Mirror," *Phil. Mag.*, 6th ser. 35:163–174 (February, 1918). Cf. Albert A. Michelson, "Effect of Reflection from a Moving Mirror on the Velocity of Light," *Astrophys. J.* 37:190–193 (April, 1913). The quotation comes from p. 149 of Majorana's second paper, "Experimental Demonstration of the Constancy of Velocity of the Light Emitted by a Moving Source," *Phil. Mag.*, 6th ser. 37:145–149 (February, 1919).

of the Michelson-Morley experiment. He believed he had found a definite difference when the apparatus was turned through ninety degrees.[33]

When World War I was over, and scientists began to return to their civilian pursuits, the work of the negotiators at Versailles was sharing newspaper headlines with reports of a double astrophysical expedition to Brazil and to West Africa. The primary purpose of this scientific expeditionary force was not to settle the controversy still raging over Special Relativity but rather to test one of three new predictions of Einstein's General Theory of Relativity. Led by Arthur Eddington, the astronomers and physicists all looked for evidence of starlight bending around the sun during a solar eclipse to give a definitive answer to the controversy over relativity. By subjecting Einstein's prediction of 1911 to a direct observational test, it was hoped to vault over the Special Theory by confirming General Relativity.[34]

Meanwhile, in both spoken and written words, Sir Oliver Lodge, whose commitment to the dynamic aether concept was congenital, acknowledged that his generation was now in a minority position. With great vigor he cried out against the "extreme relativists" who assert that we do not know and can never know our own motion, who claim "in fact, that motion of matter through ether is a phrase without meaning."[35] As a conservative spokesman, Lodge lamented what he thought to be the extraordinary complexity introduced by relativity. He predicted beforehand that the relativists would claim on their return from

[33] Augusto Righi, "L'Expérience de Michelson et son interprétation," *Comptes Rendus* 168:834 (April 28, 1919); cf. 170:497–501 (March 1, 1920), 1550–1554 (June 28, 1920); 171:22–23 (July 5, 1920). See also Henry Crew, *The Rise of Modern Physics: A Popular Sketch*, p. 285.

[34] See Sir Arthur Eddington, *Space, Time, and Gravitation: An Outline of the General Relativity Theory*, pp. 110–122. For some of A. S. Eddington's preliminary thoughts on light-bending tests, see his two articles entitled "Gravitation—and the Principle of Relativity," in *Scientific American Supplement* no. 2218, I (July 6, 1918): 2–3 and no. 2219, II (July 13, 1918): 22–23. See also C. W. Kilmister, ed., *Men of Physics: Sir Arthur Eddington*.

[35] Sir Oliver J. Lodge, "Aether and Matter: Being Remarks on Inertia, and on Radiation, and on the Possible Structure of Atoms," *Nature* 104:16 (September 4, 1919).

Brazil that the deathblow had been struck to the aether. For him
Michelson-Morley had not done that nor could the eclipse expedition.
The prejudiced contention that it would

is illegitimate, . . . the reality of the aether of space depends on other things,
and . . . the establishment of the principle of relativity leaves it as real as
before, though truly it becomes even less accessible, less amenable to experiment, than we might have hoped. Nevertheless, the aether is needed for
any clear conception of potential energy, for any explanation of elasticity, for
any physical idea of the forces which unite and hold together the discrete
particles of matter, whether by gravitation or cohesion or electric or magnetic
attraction, as well as for any reasonable understanding of what is meant by
the velocity of light.[36]

Lodge felt certain that the exodus of the aether concept would be a
severe handicap to the progress of physics. His own program was quite
simple: "I want people generally to admit that the aether is itself stationary as regards locomotion, and that it is the seat of all potential
energy; and further, at least as a surmise, that it is the medium out of
which matter is probably made, and in which matter is perpetually
moving by reason of its fundamental property called inertia."[37] Insisting on the usage of a word which had now gone out of style,
Lodge's demand for consistency seemed to others to be inconsistent,
since he was obsessed with the idea of continuity without being fair to
the necessity for a field concept to deal with isolated electromagnetic
systems. Insisting on the plenum while simultaneously allowing it to be
synonymous with empty space, Lodge was beginning to appear more
and more eccentric to his comrades.[38] But in historical perspective he

---

[36] Ibid., pp. 16–17. The question of discreteness versus continuity in ultramicro-
scopic phenomena was yet to be recognized as convertible: cf. Sir William H. Bragg,
Electrons And Ether Waves: Being the 23rd Robert Boyle Lecture, given on May 11,
1921.

[37] Lodge, *Nature* 104:17. Lodge's relation to his colleagues may best be judged by
comparing his contribution to the others in the special issue of *Nature* on "Relativity": "The Geometrisation of Physics and Its Supposed Basis on the Michelson-
Morley Experiment," *Nature* 106:795–800 (February 17, 1921).

[38] Lodge had gradually been acquiring a reputation for eccentricity in other ways.
A scurrilous attack on his mystical religious proclivities in 1914 asserted that Lodge
had "reconstructed Christianity with a facility which makes St. Paul, by comparison,

seems to have been a courageous spokesman for a large number of older physicists who sincerely mourned the passing of the aether model. Although nearly all his disciples had outrun him on this point, it is highly significant that Einstein himself by 1920 had undergone a profound reorientation of attitude toward the aether concept. Whereas in 1905 he had iconoclastically declared it unnecessary, by 1920 he was willing to admit publicly that the General Theory of Relativity had forced a reconsideration of the problem of space. Einstein expressed his changed attitude in a neglected speech, "Äther und Relativitäts-theorie," to a Leiden audience in 1920. On the home grounds of his elder friend and conservative colleague H. A. Lorentz, Einstein asserted: "We may say that according to the general theory of relativity space is endowed with physical qualities; in this sense, therefore, there exists an ether. According to the general theory of relativity, *space without ether is unthinkable*; for in such a space there not only would be no propagation of light, but also no possibility of existence for standards of space and time (measuring rods and clocks), nor therefore any space-time intervals in the physical sense."[39] The semantic change in the meaning of the word *aether* as it is exhibited here by Einstein should not be considered in isolation from the gestalt of the General Theory. An aether equivalent only to the kinematical properties of space is a very different idea from any dynamical concept of the aether equivalent to a medium, whether ponderable or imponderable. Nonetheless, the notion of the "aether of space" was here given a new lease on life by Einstein.[40]

---

a clumsy amateur and Jesus Christ a minor prophet" (Joseph McCabe, *The Religion of Sir Oliver Lodge*, p. vii).

[39] Albert Einstein, *Sidelights on Relativity*, p. 23 [Italics mine]. See also Gerald Holton, "On the Origins of the Special Theory of Relativity," *Am. J. Phys.* 28:632 (October, 1960). The original lecture of May 5, 1920, was published as a pamphlet: *Äther und Relativitätstheorie* (15 pages).

[40] Holton quotes this passage from Einstein with the addition of Einstein's next significant sentences in a footnote: "But this ether may not be thought of as endowed with the quality characteristic of ponderable media as consisting of parts which may be tracked through time. The idea of motion may not be applied to it" (*Am. J. Phys.* 28:632 n.). In the popular press Lorentz had recently praised Einstein for cutting loose from "the universal canvas of the ether," saying that "if he had not done so, he probably would never have come upon the idea that has been the foun-

The significant point is that between 1905 and 1916, between the Restricted Relativity Theory and the General Relativity Theory, Einstein had come to extend the field concept so far that verbally the word *aether* might be allowed to express the curvature of the space-time continuum. In the Special Theory he had assumed that there is no such thing as a specially favored coordinate system, therefore no need for an aether concept, and hence no aether drift, and no experiment to measure the same. With the General Theory, however, gravitational fields as well as rotating bodies like the earth ought to affect the behavior of light. Hence, "aether drift" *might* have a physical meaning after all.

Two conclusions emerge clearly from this examination of the literature in the decade following 1910. First, it is clear that the broad, ambiguous, and amorphous concept of an all-pervading luminiferous or electromagnetic aether was becoming less useful in the wake of the successes of atomic physics. In the realm of the atom, the field concept was quite sufficient. On the other hand, for those physicists who were not engaged directly in electronic or electromagnetic researches, but who still were impressed by the achievements and promise of Einstein's work, the simplest way of envisioning and teaching relativity theory was to accept the superficial view that it had grown directly out of the enigma posed by the Michelson-Morley experiment. The adulterated version of Michelson-Morley thus created the same difficulties, within the scientific community, as those often encountered in popularized science. The paradoxes and analogical fallacies so often employed to excite interest in relativity theory created almost as many barriers as bridges to understanding. Second, as a direct consequence of this problem in science education and scientific communications, the Michelson-Morley experiment became in this decade a "celebrated" and a "crucial" experiment. Whereas the Michelson-Morley aether-drift experi-

---

dation of all his examination." Still intermixed, however, was Lorentz's assertion that "it is not necessary to give up entirely . . . the ether" (H. A. Lorentz, *The Einstein Theory of Relativity: A Concise Statement*, pp. 36, 63, 60). See also Stanley Goldberg, "The Lorentz Theory of Electrons and Einstein's Theory of Relativity," *Am. J. Phys.* 37:982–994 (October, 1969).

ment prior to this time had certainly gained great fame as an unsolved problem in optical interpretation, the audience interested in this problem was a small one. With the spreading awareness of the relativity theory, however, Michelson-Morley became an ever more convenient means by which "relativity" could be explained to nonspecialists. Thus Michelson-Morley came to be generally considered the *experimentum crucis* that forcibly caused Einstein to replace Newton.

CHAPTER TEN

MILLER CHALLENGES MICHELSON, 1920–1925

That the aether theory was dead and should be buried was a conviction which seemed, to many of those qualified to assert an opinion and to many more not so qualified, fully confirmed upon the return of Eddington and the 1919 eclipse expeditions. Announcing triumphantly that Einstein's theories had been corroborated by this elaborate astrophysical test of the General Theory of Relativity, Eddington did much to spread Einstein's fame. Even though only one of Einstein's three predictions had herewith passed a test, relativists were exultant that the results were so favorable and decisive.[1]

---

[1] The British eclipse expedition, a fascinating story in itself, was engagingly described by Eddington almost immediately as a central feature in his 1920 publication, *Space, Time and Gravitation: An Outline of the General Relativity Theory*, pp. 110–122. The evidence was not, however, unequivocally in favor of Einstein, although Eddington so interpreted it. See L. Silberstein, "The Recent Eclipse Results and Stokes-Planck's Aether," *Phil. Mag.*, 6th ser. 39:161–170 (February, 1920).

On the other hand, many other physicists were reluctant to be pushed too far too fast. Perhaps for the majority it was enough to accept tentatively the earlier Special Theory without being pushed into full commitment to the more recent and only partially verified General Theory. Others argued that, even though the Restricted Theory had proven fruitful in its limited applications, any viable kind of cosmological physics would have to include the aether, even though it might be renovated almost beyond recognition.

The public obsession with relativity which began in 1919 grew to vast proportions in the few years following. Max Born, a colleague of both Einstein and Planck at Berlin from 1917 to 1921, has related how Einstein was swamped by the "craze" in 1919. Like Woodrow Wilson, Einstein found himself a world celebrity. He had so captured the warweary popular imagination that he saw himself cursed with the King Midas touch.[2] Postwar idealism found in Einstein not only a perfect symbol of the man above the battle, but also the theoretician of a new world order. As disillusionment with idealism in the political sphere grew, there seemed to be almost a transfer of hope that, if not in international relations, then at least in the physics of the cosmos, a new world order could be established. Einstein personally bore his unsought fame with tolerance and humility, but it is not surprising that some of his colleagues felt amusement, chagrin, or even distaste for what so-called popular science was doing with relativity.

One such scientist was Professor Dayton C. Miller of Case School in Cleveland. It was not simply the ballyhoo of 1920 in the yellow press that angered Miller. It was also the half-truths and falsehoods about the Michelson-Morley experiment which had become common currency among his fellow physicists. Innumerable books began to appear about 1920, written by physicists for physicists, by physicists for other scientists, by philosophers for the educated layman, and by the educated layman for the uneducated layman—all kinds of

[2] Max Born, "Physics and Relativity," *Proceedings* [of the Conference on the] *Jubilee of Relativity Theory*, Helvetica Physica Acta, Supplement 4, pp. 255–257. See also the biography by a literary colleague at Berlin, Alexander Moszkowski, *Einstein the Searcher*, pp. 112–114. Daniel J. Kevles, "George Ellery Hale, the First World War, and the Advancement of Science in America," *Isis* 59:427–437 (Winter, 1968).

books for all kinds of people—and each designed to explain relativity theory to its selected audience.[3] Nearly all of these works, insofar as they described the advent and meaning of relativity in terms of Michelson-Morley, were clearly misleading, if not actually false, in Miller's view. Why? Primarily because they so oversimplified the experimental difficulties surrounding the aether-drift tests that the accounts of Michelson-Morley became unrecognizable to one who had worked on them.

In midsummer 1920, George Ellery Hale, the revered director of Mount Wilson Observatory and the astronomical entrepreneur who had sparked the building of two of the world's largest telescopes (the forty-inch refractor at Yerkes and the one hundred-inch Hooker reflector at Mount Wilson), took another initiative by inviting Dayton C. Miller to repeat the Michelson aether-drift experiment atop Mount Wilson, near Pasadena, California. Hale had first broached the idea to Michelson that summer, as the latter was visiting him to arrange the preliminary setup for his velocity-of-light measurements between mountain peaks twenty to twenty-five miles apart. Hale told Miller that Michelson had no objection but doubted whether the effort would be worth the expense, because Morley had last assured him that the small positive residuals of 1905 on Euclid Heights "were almost cer-

[3] Some of the more important literature published in 1920 which will show the spectrum of opinion on Michelson-Morley at that time should be listed: Erwin Freundlich, *The Foundations of Einstein's Theory of Gravitation*, p. 19; J. H. Jeans, *The Mathematical Theory of Electricity and Magnetism*, pp. 593–620; Hermann Weyl, *Space-Time-Matter*, pp. 170–173; Norman R. Campbell, *Foundations of Science: The Philosophy of Theory and Experiment* [formerly titled *Physics: The Elements*], Appendix, pp. 557–559; Moritz Schlick, *Space and Time in Contemporary Physics: An Introduction to the Theory of Relativity and Gravitation*, pp. iii, 7; Alfred North Whitehead, *The Concept of Nature*, pp. 179–195; Benjamin Harrow, *From Newton to Einstein: Changing Conceptions of the Universe*, p. 55. Having perhaps the most influence on the professions of physics was an article by Wolfgang Pauli "Relativitätstheorie," in *Encyklopädie der Mathematischen Wissenschaften* 2: 539–775 (1921), now conveniently available in translation by G. Field as *Theory of Relativity*. Probably most representative of the difficulties in popularization were such essays as those submitted in competition for a $5,000 prize administered by *Scientific American* and judged by Leigh Page of Yale and Edwin P. Adams of Princeton: see J. Malcolm Bird, ed., *Einstein's Theories of Relativity and Gravitation*. Most emphatic in attributing relativity theory directly to Michelson-Morley was Wilhelm Wien, *Die Relativitätstheorie vom Standpunkt der Physik und Erkenntnislehre*.

tainly due to temperature effects." [Hale continued] "In spite of this probability, both he and I would be much interested to see the experiment repeated. You are quite welcome to do the work on Mt. Wilson if you choose, but I am unfortunately not in a position to help financially. . . . Our general facilities are at your service, however . . . and if you do not come during the crowded summer period I can probably give you free board and lodging on the mountain."[4]

Although Miller knew that Morley shared his qualms, he remained unsure whether to attempt a series of repetitions until after some further calculations and consultations with Michelson and Sir Joseph Larmor. Miller paid a hurried visit to Mount Wilson before the fall semester started, but not until January 19, 1921, did he telegraph Hale his decision to accept the invitation and renew the aether-drift experiments. He sent detailed plans by letter a few weeks later. In March, Miller, together with his financial patron, Mr. Eckstein Case, and his technical assistant, Mr. R. T. Haney, was enroute to California by train via the Grand Canyon.[5]

The fact that Einstein himself was seriously reconsidering the usage of the aether concept in terms of the problems of the unified field theory was undoubtedly some inspiration to Miller and even to Michelson. Einstein and his wife arrived in the United States in April, 1921, on a

[4] George E. Hale to Dayton C. Miller, July 19, 1920, in Miller Correspondence file of Director, Mount Wilson Observatory, Pasadena, California. See also Helen Wright, *Explorer of the Universe: A Biography of George Ellery Hale*. Hale may have been stimulated to make this invitation by the great debate on "the scale of the universe" between Harlow Shapley and Heber D. Curtis that Hale had arranged before the National Academy of Sciences on April 26, 1920. See Thornton and Lou W. Page, eds., *Stars and Clouds of the Milky Way: The Structure and Motion of Our Galaxy*, pp. 315–322, and Harlow Shapley, *Through Rugged Ways to the Stars*, chapter 6.

[5] D. C. Miller to G. E. Hale, November 1, 1920; telegram, Miller to Hale, January 19, 1921; Miller to Hale, February 10, 1921; Miller to Hale, March 6, 1921, in Miller Correspondence file of Director, Mount Wilson Observatory. There is some reason to believe that the work of Righi came to Miller's attention late in 1920 and helped him decide to go ahead with the repetitions: see Ernest Merritt (chairman, physics department, Cornell University) to G. E. Hale, November 22, 1920, *idem*. See also D. C. Miller to Prof. Elizabeth R. Laird [Mt. Holyoke College, South Hadley, Mass.], February 8, 1921, Bohr Library, American Institute of Physics, New York.

tour to raise funds for the Zionist movement. During a tight schedule of speeches and appearances before his scientific colleagues as well as the Zionists and their supporters, Einstein came to Cleveland in May, 1921, and made a special point of visiting with Professor Miller at Case, just after his return from California. Einstein had heard that Miller, encouraged by Morley, Michelson, and Hale, had at long last taken on the project of finally completing the aether-drift experiments *for all epochs* of the year.[6]

In February Miller had dismantled the steel-base interferometer last used in 1906 and had shipped it on ahead to California, where Hale had personally supervised its reassembly and transport up to Mount Wilson Observatory. Miller arrived at Mount Wilson at the beginning of April, 1921, and set up his interferometer on four concrete piers next to the eastern edge of the mountaintop, at an elevation of almost six thousand feet. He erected a protective tent made of tar paper and canvas around the apparatus and began his preliminary tests, being careful during his working hours to record all meteorological data. From April 8 to 21 he took systematic readings. And before he had fully reduced the raw data, he found that the readings gave him, as he said in an immediate report to Hale, "strong definite indications of drift four times as large as at Cleveland but accompanied by unexplained disturbance which prevents conclusions till . . . later in year."[7]

Instead, however, of shouting "Eureka!" Miller knew he must anticipate a large array of objections which could be brought against such an announcement. Besides the fact that his flimsy hut at the edge of a precipice afforded inadequate protection against the mountain winds that played havoc with his temperature control, there were five or more

[6] R. S. Shankland, S. W. McCuskey, F. C. Leone, and G. Kuerti, "A New Analysis of the Interferometer Observations of Dayton C. Miller" [page 2 of a mimeographed paper, later revised for publication to exclude most of the purely historical references], published in *Rev. Mod. Phys.* 27:167 (April, 1955).

[7] Telegram, D. C. Miller to G. E. Hale, April 22, 1921, in Miller Correspondence file of Director, Mount Wilson Observatory; see also Dayton C. Miller, "The Ether-Drift Experiment and the Determination of the Absolute Motion of the Earth," *Rev. Mod. Phys.* 5:218 (July, 1933). Cf. Dayton C. Miller, "Ether-Drift Experiments at Mount Wilson Solar Observatory," *Phys. Rev.*, 2d ser. 19:407–408 (April, 1922); a brief note on this paper was also published in *Science* 55:496 (May 5, 1922).

other reasons why his experiment might have gone wrong. So, with considerable encouragement, if not elation, Miller decided to return to Cleveland while workmen built him a more substantial and protective observation post out of corrugated iron and canvas.

Thus, Miller, who had just been elected to the prestigious National Academy of Sciences, was at home in his laboratory on May 25, 1921, when Einstein came out to Case to see the site of the classic Michelson-Morley experiment and to discuss with Miller the unfinished task of testing optically for an aether drift at all seasons of the year. Einstein as yet could hardly speak English, but Miller knew German rather well, and the two apparently spent an enjoyable time together discussing their common interest in the work. Einstein urged Miller to continue the aether-drift trials and to reexamine the earlier results obtained in 1881, 1887, and 1902–1905. Shortly thereafter Miller wrote to T. C. Mendenhall saying that Einstein's visit was most pleasant and that the great theoretician was "not at all insistent upon the theory of relativity."[8]

After Einstein's visit, however, Miller was a bit less enthusiastic about his April findings. He reduced his claim of drift from four to three times as large as found in Cleveland, and described in detail the large periodic effect that appeared once per revolution instead of twice as desired for an aether-drift effect. At Hale's suggestion and with Michelson's endorsement, Miller ordered the Mount Wilson crew to cast in concrete a new base for the interferometer, in order to check out the possibility that magnetostriction might have caused the slight positive fringe shift found in April. To be sure, there were still other possibilities for explaining the April results: they might have been due to radiant heat, to centrifugal force, or to gyrostatic action, perhaps to irregular gravitational effects, or to a yielding of the foundations. Perhaps even magnetic polarization of light could have made it appear that he had found an aether drift of approximately ten kilometers per second at six thousand feet above sea level. But after Einstein's visit, Miller was convinced that the most probable extraneous effect was

[8] Dayton C. Miller to Dr. T. C. Mendenhall, June 2, 1921, in Mendenhall Papers, Bohr Library, American Institute of Physics, New York. Cf. Shankland et al., "A New Analysis" [mimeographed paper, p. 4].

magnetostriction (i.e., some kind of ferromagnetic distortion, corresponding to the FitzGerald contraction or Lorentz transformations).
In hope of eliminating magnetic and heat effects, the concrete cross as a base for another interferometer was cast in the fall of 1921, and all metallic parts were remade of low-expansion materials, either brass or aluminum.[9]

Meanwhile, Miller wrote to H. A. Lorentz, who was then a visiting professor at California Institute of Technology, giving a full status report on his retests and asking for comments or suggestions. Miller concluded this five-page summary by saying: "While these experiments did not promise any new results of importance in connection with Ether Drift, yet I feel that they must be continued until the causes of the fundamental and overtone shifts of the fringes have been fully determined."[10]

Although Lorentz's reaction to this is unknown, perhaps he, like Michelson and Einstein, felt equally beleaguered by another experimental suggestion for an optical test of relativity theory. In a lecture at Princeton University, in mid-May, Dr. Ludwik Silberstein, a Polish-English and now American critic of relativity and of aether, had reissued a proposal for an experiment which, according to the *New York Times*, "may prove Einstein's theory of relativity to be all wrong." Silberstein's suggestion was to test for "the pull of the rotating earth upon the ether to learn whether there is a drag, whole or partial, and [whether] it has several possible results, the most important of which is its effect on the theory of relativity."[11]

Long ago, even before the advent of relativity in 1905, Michelson had wrestled with a similar suggestion for a similar kind of experiment, but at that time he thought it too impractical and too slight a modification of his other work. Now, according to Silberstein's cal-

---

[9] Dayton C. Miller to George E. Hale, June 30, 1921; Hale to Miller, July 28, 1921; Miller to Hale, August 26, 1921, and October 17, 1921: in Miller Correspondence file of Director, Mount Wilson Observatory. See also Miller, "The Ether-Drift Experiment," *Rev. Mod. Phys.* 5:218–219.

[10] Dayton C. Miller to H. A. Lorentz, August 22, 1921: copy in Miller Correspondence file of Director, Mount Wilson Observatory.

[11] *New York Times*, May 13, 1921, p. 7. See also Ludwik Silberstein, "The Recent Eclipse Results," *Phil. Mag.*, 6th ser. 39:164.

culations, if one could arrange a miniature racecourse for light to travel in opposite directions around strong north and south magnetic poles, then one could predict accurately what the fringe shift should be exclusively because of the *rotational* (not *translational*) motion of the earth.[12]

Before making his proposal, Silberstein had prevailed upon Michelson to carry out this experiment. Although reluctant at first, Michelson had agreed to supervise the experiment "in view of the immense interest which there is now in relativity."[13] Einstein, too, was reportedly skeptical at first about the value of this experiment, but gradually he became more interested in this new trial of his General Theory. According to Silberstein, it should be a crucial test: "If, therefore, the experiment which Prof. Michelson will perform gives a full value of the shift, this will harmonize with the general relativity theory, but if the effect is nil, or only a fraction of the full shift of 1.4 per square kilometer, it will be 'a death blow to the relativity theory,' although compatible with the ether theory, testifying simply to a partial drag."[14]

Thus it happened that both Michelson and Miller were drawn into thinking about how to conduct new aether-drift tests atop Mount Wilson. But their experiments were very different kinds of tests, seeking very different effects. Miller, with support from the Carnegie Institute, was in the process of performing a drift test for a ponderable ethereal atmosphere at a high altitude, hoping to prove by optical means the earth's absolute motion, as well as the drag hypothesis of Stokes. Michelson, on the other hand, with some subscription money to be raised by Silberstein, was getting ready to test for an effect of the rotational motion of the earth on the electromagnetic "aether" or earth's

[12] A. A. Michelson, "Relative Motion of the Earth and Aether," *Phil. Mag.*, 6th ser. 8:716–719 (December, 1904). According to this paper, Newcomb and W. Wien had wondered also about the feasibility of this. Apparently a contact between Silberstein and Michelson had been made in January, 1921, when the latter attended the former's lectures at a conference at the University of Toronto. See Ludwik Silberstein, *The Theory of General Relativity and Gravitation*, pp. 30–38.

[13] *New York Times*, June 22, 1921, p. 14. At this time Michelson was also interested in exploiting the use of his new instrument for studying the distribution of intensity in spectral lines: see his "The Vertical Interferometer," *Proc. Nat'l. Acad. Sci.* 6:473–474 (August 15, 1920).

[14] *New York Times*, May 12, 1921, p. 7.

field, as set forth in General Relativity Theory. Miller would simply be rerunning the classic Michelson-Morley experiment with new precautions at a higher elevation, whereas Michelson now would perform a new test which, it was thought, would determine the pull on the aether by the spinning earth and which could corroborate the Einstein theory, or destroy it. Miller was to test for the *translational* motion of the earth; Michelson to test for its *rotational* motion.

In June, 1921, Michelson was still in Paris on a spring semester exchange professorship when his successor as head of the physics department at the University of Chicago, Robert A. Millikan, learned that he would become president of California Institute of Technology later that year. Millikan, who, later still, would also become a Nobel laureate (in 1923, for his experimental determinations of the electronic charge *e* and of Planck's constant *h*), told reporters that Michelson would perform the earth-rotation experiment at Chicago.[15] But Michelson, on his return about midyear, thought it best to go on to Pasadena to conduct the experiment on Mount Wilson at the same level at which Miller's apparatus now rested. On July 28 Michelson reported his plans by letter to the person most interested in their progress:

Dear Dr. Silberstein,

You will be glad to learn that the ether-drift experiment was tried out yesterday with a distance around the triangle of 1,000 feet giving fringes which can be measured to 1/20 fringe.

Tomorrow I go up to Mt. Wilson to select a site for a greater distance, say three or four thousand feet. If this turns out as well, the final test will be made at about ten thousand feet.

It may be possible even to try to measure the displacement, and I should say—contrary to my previous skepticism— that the chances are favorable.

Sincerely yours,
A. A. Michelson[16]

At this same time, of course, Michelson was engaged in several other projects as well as this one. The first love of his life, namely, measuring

---

[15] Ibid., June 22, 1921, p. 14.

[16] Photographic reproduction of this letter is in F. Twyman, "Professor A. A. Michelson, Sc.D., LL.D., Ph.D.," *Physical Society of London: Proceedings* 43:625–632 (September 1, 1931).

with ever-increasing accuracy the fundamental constant *c*, the speed of light in vacuous space, was still his major concern. He was at this time improving apparatus and selecting sites for his plans to reduce the error in that value to not more "than a couple of miles per second."[17] Another of Michelson's projects at Mount Wilson had recently created one of the greatest scientific sensations of the century. Michelson's new 20-foot stellar interferometer mounted on the top of the 100-inch Hooker telescope had made possible the measurement of the diameter of stars. On December 13, 1920, Dr. Francis G. Pease wired Michelson that the diameter of Betelgeuse in the constellation Orion, about 150 light-years away, was 260 million miles.[18] Naturally enough, when Michelson announced the results of these star-size measurements to the American Physical Society on December 29, 1920, scientists and the public for different reasons were alike amazed.[19] To think that certain stars were comparable in size to our solar system and that their widths could actually be measured almost defied comprehension. Biographers, too, have been so concerned with this aspect of Michelson's work in 1920–1921 that hardly any attention has been given to his concurrent interest in the new set of aether-drift tests.[20]

At any rate, in November, 1921, both Michelson and Miller were back on the mountain working on the problems of the luminiferous aether. Both were also quickly disillusioned once again. Miller soon found that his concrete-base interferometer, though considerably less sensitive to changes in temperature than the steel-base apparatus, was far less satisfactory because it was so very much less rigid. Concrete was giving him the same trouble that white-pine boards had in 1902. Nevertheless, from December 4 through December 11, 1921, Miller ran thirteen sets of readings, comprising 153 complete turns of the

---

[17] Quoted in the *New York Times*, November 14, 1921, p. 7.

[18] See John H. Wilson, Jr., *Albert A. Michelson: America's First Nobel Prize Physicist*, p. 159.

[19] Cf. *New York Times*, December 30, 1920, pp. 1, 3; December 31, 1920, p. 10; January 1, 1921, p. 5; February 8, 1921, p. 17.

[20] One would hardly notice this aspect of the aether-drift experiments by relying on the two biographical studies of Michelson: cf. Wilson, *Albert A. Michelson*, p. 172, and Bernard Jaffe, *Michelson and the Speed of Light*. Jaffe almost ignores this part of Michelson's work; see p. 106.

interferometer, under fairly favorable conditions. Michelson, Hale, and the Mount Wilson staff were all interested in the results and helped out with suggestions for improvements in apparatus and design. When the data had been analyzed Miller discovered that the results were almost exactly the same as those obtained with the steel-base interferometer in April: an average periodic amplitude of only four-hundredths of a fringe. Miller was apparently deeply discouraged: he wrote in his research notebook, "all effects are probably due to the instrument. This is the end!"[21]

Apparently Michelson's other interests and commitments at Mount Wilson prevented his hopes, as outlined in the letter to Silberstein, from being fulfilled. For two more years exploratory experiments were conducted around the Cal Tech campus for earth tides and at Mount Wilson in the open air. Finally Michelson became so discouraged with the atmospheric disturbances that he determined to perform the experiment in a specially constructed and evacuated rectangular pipeline on level ground just outside Chicago.[22] Meanwhile, Miller dismantled his apparatus and had it shipped home to Cleveland for extensive laboratory work to iron out the "bugs."

In the early 1920's there was still great ferment in optical theory and great controversy over experimental interpretation among those closest to the problems involved in a decisive search for the aether drift.[23] In 1922, Roy J. Kennedy at Princeton announced a new plan

[21] Miller, quoted by Shankland et al., "A New Analysis" [mimeographed paper, p. 4]. Dr. Edison Pettit of the Mount Wilson Observatory staff was present during this period and recalled, for the author, during a personal interview on August 30, 1961, some of his memories; Miller was considered charming and competent by his associates. Miller apparently made a moving picture in fifteen parts to exhibit this series of aether-drift tests.

[22] A. A. Michelson, "The Effect of the Earth's Rotation on the Velocity of Light," *Astrophys. J.* 61:138 (April, 1925).

[23] It may be well to remember that the award of the Nobel Prize to Einstein in 1921 was not for the General or even the Special Relativity Theory, but rather for his work in theoretical physics notably on the photoelectric effect. Even though astrophysics is not a scientific category recognized by the Nobel Committee, had there been less controversy, Einstein's award might have been more generally stated. Cf. Louis V. de Broglie, "A General Survey of the Scientific Work of Albert Einstein," *Albert Einstein: Philosopher-Scientist*, ed., P. A. Schilpp, I, 123.

whereby decisive evidence might be obtained either for or against aether motion.[24] Miller himself, after having second thoughts about his private discouragement in December, announced that, although he was unable to eliminate a small systematic azimuth disturbance, his results showed "a definite displacement, periodic in each half revolution of the interferometer, of the kind to be expected, but having an amplitude of one tenth the presumed amount."[25]

While none of these experimentalists was so bold as Sir Oliver Lodge, they all had serious doubts about how to explain interference and maintain the undulatory theory of light should the aether concept be abandoned. Lodge told a lay audience in Edinburgh about this time that "an intelligent deep-sea fish would disbelieve in water. . . . Such is our own condition in regard to the aether of space."[26] To him as well as to others, the aether was still conceptually as well as ontologically necessary to physics. Although he did scrupulously change the properties of his conception of the aether in accordance with the latest experimental evidence, Lodge objected vehemently to the abandonment of the word *aether*: "A superstition has recently arisen that the ether is an exploded heresy, and is unnecessary; but that is an absurd misunderstanding. The theory of relativity says nothing of the kind . . . ignoring a thing is not the same as putting it out of existence."[27] Now Lodge was a responsible and a respected physicist, as were also Miller, Sagnac, Righi, and certainly Michelson. But they were of an older generation and woefully in the minority. Their conservatism with respect to the aether concept appeared not only outdated, but to many, by 1923, even reactionary. Ironically, that same year, Arthur H. Compton was

[24] Roy J. Kennedy, "Another Ether-Drift Experiment," *Phys. Rev.*, 2d ser. 20:26–33 (July, 1922).

[25] Miller, "Ether-Drift Experiment," *Phys. Rev.*, 2d ser. 19:407. See also Miller's note in *Science* 55:496 (May 5, 1922).

[26] Sir Oliver Lodge, "Speech through the Aether," *Nature* 108:88 (September 15, 1921). "Let us grant, then," Lodge continued, "that the aether impinges on us only through our imagination; that does not mean that it is unreal. To me, it is the most real thing in the material universe" (p. 89).

[27] Sir Oliver Lodge, "The Ether and Electrons," *Nature* 112:185 (August 4, 1923). Cf. two later Lodge books, *Relativity: A Very Elementary Exposition*, and *My Philosophy: Representing My Views on the Many Functions of the Ether of Space*.

in St. Louis and Chicago doing the experiments that were to put a capstone on wave-particle duality with the "Compton effect" and earn
him a Nobel Prize in Physics in 1927.[28]

Along with the public interest excited by Einstein and the undiffer-
entiated "Theory of Relativity," there began to arise an extraordinary
interest in Einstein and "Relativity" among the lunatic fringe. Just
where this fringe began or ended was difficult to ascertain, since, as
in the aether-drift experiments themselves, there seemed to be no decisive criterion in the social sphere for judging extremism. For instance, Dr. Charles Lane Poor, a professor of celestial mechanics at
Columbia University, was writing and talking about the antithesis between Gravitation and Relativity and "The Errors of Einstein" with
great vigor and relish in 1922–1923: "The Einstein theory, with its
conflicting formulae and inaccuracies, has caused more sins to be committed in the name of relativity than can be attributed to any other
modern scientific postulate. . . . While the relativist uses only approximations in making his observations, once these observations are recorded the approximations are forgotten and results are registered as
being precise."[29] This, too, was responsible criticism. Although less
restrained than Lodge, it was far more restrained than the attacks of
the assorted crackpots, screwballs, anti-Semites, anti-Communists, and
religious fundamentalists who filled out the spectrum of antirelativists.[30]

The decade which produced the Scopes "monkey trial" in Dayton,

[28] See Albert A. Bartlett, "Compton Effect: Historical Background," *Am. J. Phys.*
32: 120–127 (February, 1964).

[29] Charles L. Poor, quoted in the *New York Times*, November 20, 1923, p. 3. See
also Charles L. Poor, *Gravitation Versus Relativity*. Miller admitted to Gano Dunn
that Poor's description of Miller's challenge was accurate and he praised Poor's work
highly, a fact that proved damaging to Miller's reputation.

[30] Martin Gardner has made a beginning, but only a shallow beginning, toward
analyzing this lunatic fringe in his chapter "Down with Einstein!" in *Fads and Fallacies in the Name of Science*, pp. 80–91. Gardner does not allow for the historical
development of relativity, hence he treats Miller as an isolated exception (p. 85).
Also it should be remembered that radio engineers and optometrists, for example,
continued to posit a hypothetical aether with impunity; for example, see Lionel Laurence and H. Oscar Wood, *General and Practical Optics*.

Tennessee, and various colors of shirts among Fascist troopers in Europe was quite likely also to produce demonstrations against intellectuals in other areas. Professionally serious criticism could not always be differentiated from simple prejudice and bigotry. Philipp Lenard's hatred of Einstein was an outstanding example of a personal animosity connected with anti-Semitic bias which resulted in tragedy. In 1921 Nobel laureate Lenard published a paper, later a pamphlet, designed to damn Einstein's dismissal of the aether. It was entitled "Über Äther und Uräther," and by submerging the Michelson-Morley experiment under all other kinds of electromagnetic interference experiments, it purported to show that an aboriginal "Metäther" exists in cosmic space far from all heavenly bodies and that this hyperaether, distinct from space itself, functions as the determinant for the velocity of light.[31] Lenard was a prolific physicist and a capable experimentalist; he, like Johannes Stark, later became a Nazi and prostituted his science to his political creed. Here was a case where an apparently legitimate concern over saving the aether concept, which Lenard believed was threatened by Michelson's experiment and its relativistic explanation, issued into deliberate obscurantism. Aether apologists were therefore in danger of being accused of guilt by association.[32]

The tyranny of majority opinion undoubtedly had some effect on Michelson and Miller in the early twenties, but they were equally free men although unequally famous, and neither was afraid to speak the truth or afraid to pursue the truth as he saw it. In 1923 Michelson told a reporter for the *Chicago Tribune* that he differed with Einstein on the necessity for an aether, but that he was sure Einstein "had changed for the better our view of the world." In reply to a request for his own estimate of his most significant achievement, Michelson said, "I think most people would say that it was the experiment which started the

---

[31] P. Lenard, "Über Äther und Uräther," *Jahrbuch der Radioaktivität und Elektronik* 17:307–356 (July 19, 1921). Cf. Philipp Lenard, *Great Men of Science: A History of Scientific Progress*, pp. 379–390. For a French counterpart see the 950-page tome by Maurice Gandillot, *L'Éthérique: essai de physique experimentale*.

[32] Philipp Frank, *Philosophy of Science: The Link between Science and Philosophy*, pp. 197, 373. "Rather than give up a myth, he preferred to multiply it," said René Dugas of Lenard, in *A History of Mechanics*, p. 490.

Einstein theory of relativity. That experiment is the basis of Einstein. But I should think of it only as one of a dozen of my experiments in the interference of light waves."[33]

Miller was undoubtedly testy during 1922, and, although he had published two small notices of his December trials, he was not eager to follow them up immediately, because he was not confident of his theory, of his data reduction, or of his experimental design.[34] It has been suggested that he was discouraged with his results on top of Mount Wilson, and that "after the 1921 trials at Mt. Wilson, Miller probably would have abandoned further work on the problem except for a visit to Case made by H. A. Lorentz during the following spring."[35] Certainly the contact with Lorentz on April 5, 1922, when he lectured to a packed house in the Case physics lecture hall on the theory of relativity must have been inspiring. If it was primarily Lorentz's influence that induced Miller to reconsider the decision to abandon additional trials, it took about eight months for Miller to react to the influence. Miller must have gained some courage when that grand old man of physics, at that time more venerable than Einstein, told him, after having been shown the white-light fringes in the interferometer, that that was the first time he (Lorentz) had ever seen such fringes![36] Given this divorce between theoreticians and experimentalists, then might it not be the case indeed that relativity theory had gone far beyond its experimental warrant?

At any rate, Miller resumed his laboratory tests on the interferometer with greater vigor than ever. He dared not claim too much

[33] James O'Donnell Bennett, "Superlative Americans. Second Article. Albert Abraham Michelson at 70," the *Chicago Tribune*, rotogravure section, 1923, p. 22 [Fotofax copy, n.d.].

[34] D. C. Miller to G. E. Hale, February 15, 1922, and June 15, 1922, in Miller Correspondence file of Director, Mount Wilson Observatory. The notices are cited earlier in this chapter, in note 7: *Phys. Rev.*, 2d ser. 19:407 (April, 1922), and *Science* 55:496 (May 5, 1922).

[35] Shankland et al., "A New Analysis" [mimeographed paper, p. 5]. Other factors in Miller's discouragement were the resignation of Hale as director of Mount Wilson in March, 1922, and the succession of Walter S. Adams to that post.

[36] Ibid. Lorentz at this time still began and ended his lectures with the "superfluous" ether. See H. A. Lorentz, *Problems of Modern Physics: A Course of Lectures Delivered at the California Institute of Technology* [Spring, 1922], pp. 1, 221.

(from an effect only 20 or 30 percent of the expected amount) before he had satisfied himself that every possible objection could be met. For several months he varied his light sources, he experimented with different observing telescopes, he varied the speed and direction of rotation, and (what was perhaps most important) he determined the order of magnitude of the fringe shifts caused by unequal heating of the air in the optical paths of the interferometer. He satisfied himself that "these experiments proved that under the conditions of actual observation the periodic displacements could not possibly be produced by temperature effects."[37]

In addition to these preparatory trials, Miller made efforts throughout 1923 to keep the observer stationary, but he found that this required too complex an optical system. He tried to photograph the fringe shifts, but found that his pictures were always underexposed. With greater success he found that a bigger telescopic-lens system, without the tube, could be used for direct observation, and that the best source of light was an automobile oxyacetylene headlight. Also, in extending his source of light—finally to the use of sunlight itself—he was able to disprove the Ritz emission, or ballistic, theory which had been advanced in 1908 to explain Michelson-Morley.[38]

Thus, equipped with a much more highly refined instrument than ever before, Miller felt ready to return to a mountain with his interferometer in the summer of 1924. He considered Pikes Peak, but decided to return to Mount Wilson and selected a new site, removed from the canyon edge and reoriented by ninety degrees. There, in a small vale near the solar telescope towers and the one hundred-inch Hooker reflector, carpenters constructed for him a beaverboard-and-canvas observing hut. By mid-August preparations were almost complete and Miller approached this new series of observations confident

---

[37] Miller, "Ether-Drift Experiment," *Rev. Mod. Phys.* 5:220. See also D. C. Miller to Gano Dunn, December 6, 1922; telegram, Miller to Walter S. Adams, January 3, 1923, in Miller Correspondence file of Director, Mount Wilson Observatory.

[38] Shankland et al., "A New Analysis" [mimeographed paper, p. 5]. Significantly, Rudolf Tomaschek, who is usually credited with this accomplishment, was using a Kennedy-type interferometer to do so in Heidelberg at this same time. "Über das Verhalten des Lichtes ausserirdischen Lichtquellen," *Ann. d. Phys.* 73:105–126. (January, 1924).

that all difficulties with his interferometer had been removed: "It was felt that if any of the suspected disturbing causes had been responsible for the previously observed effects, now these were removed, the result would be a true null effect. Such a conclusion would have been accepted with entire satisfaction; and indeed it was almost expected. On the other hand, if the observations continued to give the positive effect, it would certainly have to be considered as real."[39]

On September 4, 5, and 6, 1924, ten sets of readings, making up 136 turns of the interferometer, were taken. The data were reduced once again to a small positive periodic displacement of the interference fringes. Therefore Miller concluded that "the effects were shown to be real and systematic, beyond any further question."[40]

But by what were they caused? If Miller's rerun of the Michelson-Morley experiment had been performed in 1887, the seemingly obvious answer that the fringe shifts were caused by aether drift might still have been puzzling. The value was about seventy percent smaller than expected and any reconciliation of the observed effects with the accepted theories of the aether or of the free space motion of the earth seemed impossible. Since, however, the year-round seasonal tests had never been completed, Miller continued the elaborate reduction of his observations and waited for the spring test.

While Miller was wrestling with the calculations for his solution to the aether-drift problem, Michelson and his staff at the University of Chicago were at last completing the elaborate experiment to test for the effect of the earth's rotation on the velocity of light.[41] At the sub-

---

[39] Miller, "Ether-Drift Experiment," *Rev. Mod. Phys.* 5:221. See also D. C. Miller to Walter S. Adams, May 21, 1924; telegram, Miller to Adams, July 28, 1924, in Miller Correspondence file of Director, Mount Wilson Observatory.

[40] Miller, "Ether-Drift Experiment," *Rev. Mod. Phys.* 5:221. See also D. C. Miller to W. S. Adams, December 8, 1924, in Miller Correspondence file of Director, Mount Wilson Observatory.

[41] The long delay was due in part to Michelson's reluctance to consider this test very important. See H. B. Lemon, "Albert Abraham Michelson: The Man and the Man of Science," *Am. J. Phys.* 4:5 (February, 1936). But it was also due to the difficulty of raising enough money to do the job right (i.e., according to Michelson's perfectionistic demands). A letter from Dean H. G. Gale to Dean J. H. Tufts, dated February 13, 1924, estimated that $7,000 "more" would be needed: "I think you realize that this experiment has roused international interest to a very unusual degree

urb of Clearing, Illinois, on a rectangular tract just over two thousand feet long in an east–west direction, the men from Ryerson Laboratory, supervised by Henry Gordon Gale, had, in the fall of 1924, constructed a pipeline, laid on the surface of the ground. Even with free pipe, furnished by the city, and with a telephone system donated by the Chicago telephone company, the University had to spend $17,000 on this project. Silberstein had been able to raise only $500 more.[42] High-vacuum pumps evacuated the air from this raceway, and a carbon arc at one corner furnished the light source from which Michelson's semi-transparent mirror split a beam into two pencils of light sent in opposite directions around the circuit.

This was supposed to be a test for Einstein's "principle of equivalence," the fundamental postulate of the General Theory from which the equality of the inertial mass and gravitational mass of a system may be deduced. At first this test had been hailed as a crucial experiment for General Relativity Theory, but by January 8, 1925, when the reluctant Michelson announced the provisional results of this experiment, all enthusiasm had worn thin. In recalculating the values of the displacement to be observed, it was discovered that the observed value agreed with the calculated value and that both hypotheses were equally well confirmed: "This result may be explained on the hypothesis of an ether fixed in space, but may also be interpreted as one more confirmation of Einstein's theory of relativity."[43]

This paradoxical announcement was made at the end of a public lecture in Chicago, when Gale came into the downtown theater where a convocation was being held, to inform Michelson of the final returns from this, the second major failure of his career. The next day the *New York Times* carried the story under equally curious headlines:

---

as it is expected to furnish another check on the truth of the Einstein Relativity Theory. Professor Michelson's reputation will insure a very large amount of highly desirable publicity to the University if this experiment is carried out" (A. A. Michelson Papers, The University of Chicago Library).

[42] Michelson, "The Effect of the Earth's Rotation," *Astrophys. J.* 61:139 (April, 1925).

[43] A. A. Michelson, "Light Waves as Measuring Rods for Sounding the Infinite and the Infinitesimal," *The University* [of Chicago] *Record* 11:153 (April, 1925). Cf. W. Ewart Williams, *Applications of Interferometry*, pp. 60–62.

MICHELSON PROVES EINSTEIN THEORY
Experiments Conducted With
5200-Foot Vacuum Tube
Show Light Displacement

ETHER-DRIFT IS CONFIRMED
Rays Found To Travel
At Different Speeds
When Sent In Opposite Directions[44]

However much physicists may have been bothered by these head-lines, they were just as perplexed when they read the final report of the experiment in the *Astrophysical Journal* in April, 1925. There Michelson repeated his previous judgment and attempted no interpretive statements. One of his colleagues remembered him as always convinced that Silberstein's proposal could prove nothing but what Foucault's pendulum had demonstrated, that the earth rotates! But when Michelson and Gale had originally set up their pipeline, a year and a half before, they were (or at least Gale was) still confident that this miniature test of the principle of equivalence would be crucial. To quote the *New York Times* again: "With the present experiment in mind they believe that they have at last struck upon a test which will either controvert the Einstein doctrine or establish it as a physical law on a parity with the law of gravitation."[45]

In the midst of all this confusion, Miller returned to Mount Wilson in late March, seeking a quarter-year confirmation of his "definite consistent effect 30% of expected" and the mathematical advice of Gustaf Strömberg on his reduction procedures. About this time Miller decided to discard all the specific secondary hypotheses that had accompanied Michelson-Morley tests in the past and to embark on "an entirely new quest." Strömberg convinced him that he should try to avoid all assumptions regarding a stagnant, dragged, drifting, or otherwise detectable aether. Miller would ask his apparatus henceforth not to tell him anything about the earth's relative motion due to the

[44] *New York Times*, January 9, 1925, p. 2.
[45] Ibid., October 10, 1923, p. 23. The colleague who remembered Michelson's skepticism was Lemon: see his "Albert A. Michelson," *Am. J. Phys.* 4:5 (February, 1936).

orbital component alone, the FitzGerald-Lorentz contraction, or about magnetostriction, radiant heat, or gravitational deformation. Rather, he would ask only for an answer to a simple question: "If observations were made for the determination of such an absolute motion [of the earth], what would be the result independent of any 'expected' result?"[46]

There was one serious flaw in the preceding rationale. The so-called simple question was not at all simple. Miller never took into serious consideration Mach's and Einstein's intellectual critiques of the Newtonian concept of "absolute" motion." He apparently never really tried to understand the meaning of the relativity of simultaneity, nor had he seriously wrestled with the work of J. C. Kapteyn (1851–1922), Harlow Shapley (1885–), and other statistical astronomers interested in proper motions and in our galaxy's structure and rotation.

At the National Academy of Sciences meeting on April 27, 1925, Arthur H. Compton read a paper that reportedly interpreted the Michelson-Gale experiment as having proved two things: first, that the earth's rotation has no effect on the velocity of light, and, second, that the aether-drag hypothesis has definitely been disproved. Thus, Stokes could no longer be invoked to explain the Michelson-Morley experiment.[47]

At the meeting the next day, the president of the American Physical Society, Dayton C. Miller, made a formal announcement of the small positive result he had found in his Michelson-Morley repetitions. With his commanding stage presence fortified by a tone of positive confidence in his experimental technique, Miller assured his audience that

[46] Miller, "Ether-Drift Experiment," *Rev. Mod. Phys.* 5:222. See also telegram, D. C. Miller to Walter S. Adams, March 9, 1925, in Miller Correspondence file of Director, Mount Wilson Observatory. In an interview with Dr. Gustaf Strömberg at Pasadena on August 1, 1961, the author learned that Strömberg considered the problem still worthy of investigation but recalled that Dr. R. M. Langer was Miller's most influential theoretician. For Strömberg's position, see his article, "Space Structure and Motion," *Science* 76:477–481 (November 25, 1932), and 76:504–508 (December 2, 1932).

[47] *New York Times*, April 28, 1925, p. 20. Other colleagues hailed the fact that Stokes's aether-drag hypothesis, so long a rival to the relativistic explanation of Michelson-Morley, was now removed. See H. N. Russell, "Remarkable New Tests Favor the Einstein Theory," *Sci. Am.* 133:88 (April, 1925).

"there are no corrections of any kind to be applied to the observed values. . . . No assumption has been made as to the expected result. . . . Neither the observer nor the recorder can form the slightest idea as to whether any periodicity is present."[48] After going through the whole history of the development of his renewed interest in aether-drift, Miller went on to conclude that "there is a relative motion of the earth and the ether at this [Mount Wilson] Observatory of approximately nine kilometers per second, being about one-third the orbital velocity of the earth." Miller thought this suggested a partial drag of the aether by the earth and he voiced his expectation that a reassessment of all previous Michelson-Morley experiments would show that they "do not give a true zero result." Finally, he expressed the hope that he would soon have a definite indication of absolute motion of the solar system through space.[49]

Astounding as this announcement must have been, it had the very great virtue, which the Michelson-Silberstein-Gale experiment lacked, of being a positive rather than an equivocal statement. Miller was held in great respect in the profession, as evidenced by the office he served that year. After Miller's announcement, Walter S. Adams, the new director of Mount Wilson, wrote to him, saying: "Your values are very convincing, especially so in the matter of the constancy of the direction indicated. I think everyone will agree that the continuation of this work is one of the most vital problems in modern physics."[50]

[48] Dayton C. Miller, "Ether-Drift Experiments at Mount Wilson," *Proc. N.A.S.* 11:312–313 (June, 1925).

[49] Ibid., p. 314. The committed relativists' view of Miller's claim was well expressed by W. F. G. Swann of Yale in "The Relation of the Restricted to the General Theory of Relativity and the Significance of the Michelson-Morley Experiment," *Science* 62:145–148 (August 14, 1925).

[50] Walter S. Adams to Dayton C. Miller, June 5, 1925, and Miller to Adams, May 21, 1925, in Miller Correspondence file of Director, Mount Wilson Observatory. It must also be remembered that Einstein's own qualms about the reinstatement of the aether concept were not widely known or credited. A common attitude was that expressed by Charles P. Steinmetz in 1923. Steinmetz regarded the aether hypothesis as "finally disproved and abandoned. There is no such thing as the ether, and light and the wireless waves are not motions of the ether" (*Four Lectures on Relativity and Space*, p. 16). See also Ludwik Silberstein's delight in Miller's findings, and Arthur S. Eddington's chagrin, expressed in Letters to the Editor, *Nature*, 115:798 (May 23, 1925), and 115:870 (June 6, 1925).

Physicists throughout the world watched with considerable eagerness for the next pronouncement after seasonal observations were made in August and September of 1925. As promised, on December 29, 1925, at the meeting of the American Association for the Advancement of Science in Kansas City, Dayton C. Miller, as retiring president of the American Physical Society, delivered an address, "The Significance of the Ether-Drift Experiments of 1925 at Mt. Wilson."[51]

Miller began with a review of the history of the luminiferous aether and aberration difficulties; then he went through the history of the Michelson-Morley experiment from 1881 onward to 1925. He made special note of the fact that the classic test of 1887 took into account only the orbital motion of the earth and neither the cosmic motion of the solar system nor the relative motion of the galaxies. He stressed the fact that the whole theory of relativity was related to physical phenomena "largely on the assumption that the ether-drift experiments of Michelson, Morley and Miller had given a definite and exact null result."[52]

Having known all along that the null results were not definite or exact, Miller mentioned his feelings of responsibility as a conscientious physicist. He tried to show how his own recent work differed from the classic Michelson-Morley experiment. The original test had rested on the theory of an absolutely stationary aether against which the relative motion of the earth, or at least its orbital component, might be measured. Now in 1925 he claimed to be asking a different and more general question of the aether-drift experiment, namely, What is the absolute motion, that is, the resultant of all the earth's different motions, of the solar system? While admitting that a few inexplicable azimuth anomalies remained, his final calculations indicated that the absolute motion of the solar system must be about two hundred kilometers per second toward the head of the constellation Draco![53]

---

[51] Miller's address was reprinted in full in *Science* 63:433–443 (April 30, 1926). For background, see Miller's several letters to Adams, especially that of December 19, 1925, in Miller Correspondence file of Director, Mount Wilson Observatory.

[52] Miller, "The Significance of the Ether-Drift Experiments of 1925 at Mt. Wilson," *Science* 63:436 (April 30, 1926).

[53] Ibid., p. 442. Miller avoided a direct statement about reinstituting the aether; rather he said that if an aether-drift and aether-drag is allowed, his data showed an

So impressed were his colleagues with the immediate sensation of Miller's findings that he was awarded the Third American Association prize of $1,000 for this paper. Although his seasonal observations were still incomplete and his distinctions between "absolute motion" of earth and solar system left much to be desired, the AAAS Award Committee, which included Karl T. Compton of Princeton, for physics, tendered this prize to Miller despite the fact that another speaker at that meeting had pleaded for the emancipation of our world geometry from an undue emphasis on special experiments.[54]

---

observed velocity for the "absolute motion" of the earth as 10 kilometers/second toward an apex at right ascension 17 hours and at declination 65°. See also Robert S. Shankland, "Dayton Clarence Miller: Physics Across Fifty Years," *Am. J. Phys.* 9: 279 (October, 1941); H. W. Mountcastle, "Dayton Clarence Miller," *Science* 93: 272 (March 21, 1941).

[54] The prize award committee included W. C. Mendenhall for geology, H. P. Cady for chemistry, H. C. Cowles for botany, and C. E. Seashore for psychology: *Science* 63:105 (January 29, 1926); cf. A. C. Lunn, "Experimental Science and World Geometry," *Science* 63:579–586 (June 11, 1926).

CHAPTER ELEVEN

MICHELSON REAFFIRMS THE NULL, 1925–1930

Almost before Miller had finished speaking, a number of scientists around the world were planning how to go about checking his announcement. Roy J. Kennedy, having moved from Princeton to California Institute of Technology, had already prepared a new experimental design and was fully aware of Miller's challenge to fundamental physical concepts. Robert A. Millikan encouraged him to check that challenge. Kennedy had built a completely closed optical system sealed in a small metal case with helium gas at one atmosphere pressure. He used a plane-parallel, homogeneous light source, polarized in order to eliminate non-interfering rays and to facilitate the adjustment of the reunited beams. Using a divided step-mirror, Kennedy placed the whole apparatus in the constant-temperature room of the Norman Bridge Laboratory. He took precautions to rule out temperature fluctuations and to compensate for the personal equation of observer error.

Kennedy concluded in his report of 1926 that "a shift as small as one-fourth that corresponding to Miller's would be perceived. The result was perfectly definite. There was no sign of a shift depending on the orientation." Anticipating the possible objection of too much shielding, Kennedy transported his apparatus to the top of Mount Wilson, placed it in the observatory building housing the one-hundred–inch telescope, and "here again the effect was null."[1]

Michelson, also in Pasadena from June through September, 1926, was deeply involved with his measurement of the velocity of light between Mount Wilson and Mt. San Antonio. But Miller's challenge had been reaffirmed and strengthened by observations in February, 1926, and by a confirming report (published in April, 1926), made to the American Physical Society while he was serving as its president. Michelson had watched encouragingly at first, but now Miller's challenge required his own attention. As Walter Adams wrote about the progress of the velocity-of-light measurements between mountain peaks, he reminded Michelson that renewed aether-drift tests were "still more important" and that "what the scientific world wants is *your* final word on the subject."[2]

Meanwhile Michelson and his associate Francis Pease, had been designing and preparing to construct the largest interferometer yet built. It was to be an apparatus made of low-expansion structural steel on the same basic form as Miller's steel cross, but about eleven feet in diameter with an effective light path of about fifty-five feet. The observer would ride a counterbalanced bucket seat on one arm of the apparatus as it rotated.

Across the Atlantic, on June 20 and 21, 1926, in eastern France near Switzerland, a free balloon was liberated, carrying in its gondola a miniature interferometer to test at a still higher altitude Miller's claim for an aether drift. Auguste Piccard and Ernest Stahel announced

[1] Roy J. Kennedy, "A Refinement of the Michelson-Morley Experiment," *Proc. N.A.S.* 12:628 (1926).

[2] Walter S. Adams to A. A. Michelson, November 24, 1926, in Michelson Correspondence file of Director, Mount Wilson Observatory, Pasadena, California; Dayton C. Miller, "Ether-Drift Experiments at Mount Wilson in February 1926," *Phys. Rev.* 27:812 (June, 1926).

their preliminary results in *Comptes Rendus* in August: "We have not . . . been able to detect an aether wind."[3] But dissatisfied with several more attempts at telemetered balloon ascents, the next year they took their interferometer mountain climbing to the peak of Mt. Rigi. Piccard and Stahel reported in November that there was no detectable aether wind there either. "We conclude that, in conditions corresponding to the experiment of Miller, the aether wind is not manifested at all. The experimental basis of Einstein's theory therefore remains viable."[4]

Leigh Page and C. M. Sparrow wrote in 1926 that Miller's result was "generally considered to be a crucial difficulty for the theory of relativity," and they proposed to reconcile relativity and Miller's repetition of the Michelson-Morley experiment by assuming that space is anisotropic, that is, that space shows a preferred direction for light to travel.[5] Although the stir that Miller created was greater among the orthodox than among relativists, certainly optical experimentalists had to take seriously the fact that Miller's insistence on the importance of the unexplained residuals was disturbing to relativity theory. In July, 1927, K. K. Illingworth, following in Kennedy's footsteps at California Institute of Technology, used Kennedy's apparatus (reduced in size and with a slight improvement in the step-mirror) in the subbasement constant-temperature room to gain exceptional accuracy of

[3] A. Piccard et E. Stahel, "L'Expérience de Michelson réalisée en ballon libre," *Comptes Rendus* 183:421 (August 17, 1926) [my translation].

[4] A. Piccard et E. Stahel, "L'Absence du vent d'éther au Rigi," *Comptes Rendus* 185:1200 (November 28, 1927) [my translation].

[5] Leigh Page and C. M. Sparrow, "Relativity and Miller's Repetition of the Michelson-Morley Experiment," *Phys. Rev.*, 2d ser. 28:384 (August, 1926). Some measure of the seriousness of the challenge presented by Miller's findings may be derived from an interesting formal debate staged in May, 1926, at Indiana University, prompted largely by Miller's announcement. R. D. Carmichael and H. T. Davis upheld the relativist position; William D. MacMillan and M. E. Hufford argued for Miller and the "classicist" position, but all were agreed that Miller could not be ignored. See R. D. Carmichael et al., *A Debate on the Theory of Relativity*, pp. 75, 89, 109, 124, 135, 137. See also the harsh critique by Hans Thirring, "Kritische Bemerkungen zur Wiederholung des Michelsonversuches auf dem Mount Wilson," *Zeitschrift fur Physik* 35:723–731 (February 8, 1926), abbreviated in English in *Nature* 118:81–82 (July 17, 1926).

observation. He found little more than one-tenth the aether-drift evidence claimed by Miller.[6]

Michelson was by now seventy-five years old, but he was still remarkably active. He maintained a continuous supervisory interest in the progress which Francis Pease was making at the Mount Wilson optical shops in Pasadena in order to carry out the next rigorous test of Miller's results.[7] He was well aware that the General Theory of Relativity had by now been thrice confirmed by observational corroboration of its specific predictions. But his attitude was still ambivalent, as witnessed by the treatment he gives relativity in his second book, published in 1927: "It must therefore be accorded a generous acceptance notwithstanding the many consequences which may appear paradoxical in consequence of the difficulty we find in realizing the unusual conditions of high relative velocities." Furthermore, Michelson recognized how inconsistent the existence of an aether would be with relativity theory, implying as it would the possibility of absolute motion:

But without a medium how can the propagation of light waves be explained? . . . How explain the constancy of propagation, the fundamental assumption (at least of the restricted theory), if there be no medium?

.   .   .   .

It is to be hoped that the theory may be reconciled with the existence of a medium, either by modifying the theory, or, more probably, by attributing the requisite properties to the ether; for example, allowing changes in its

[6] K. K. Illingworth, "A Repetition of the Michelson-Morley Experiment Using Kennedy's Refinement," *Phys. Rev.* 30:692–696 (November, 1927).

[7] In 1926 Sir Oliver Lodge again spoke out for his conservative American friends, saying, "I am told that Morley, before he died, was rather overawed by the consequences deduced from that experiment, and was anxious that it should be repeated with still greater care and under many conditions. . . ." Noting that Miller's results had considerably perturbed the relativist camp but were still *sub judice*, Lodge continued, "I know that Michelson agrees with me (or did in 1920) that there is nothing in the original Michelson-Morley zero result to justify a denial of the existence of the ether; nor is a positive result necessary to reestablish it. The theory of relativity (so-called) has led to the search for *absolute* conditions. . . . On the whole the theory rather tends to strengthen our philosophic conviction of the ether as an integral and absolute ingredient in the universe, although, for purposes of calculation and attainment of results, it—like many other things—can be safely ignored" (Sir Oliver Lodge, *Scientific Worthies*, no. 64: "A. A. Michelson," *Nature* 117:6 [January 2, 1926]).

properties (dielectric constant, for instance) due to the presence of a gravitational field.[8]

In June, 1926, the massive new Michelson-Morley type of "invar" interferometer was ready for operational tests. Although the experiments were not advertised while in progress (with the consequence that they were little known then and are even less well known now), the time, effort, and money spent in building this apparatus and conducting several hundred observations all point to the fact that Miller's reevaluation of the Michelson-Morley experiment was taken quite seriously indeed by the author of that experiment. Dr. Pease, Dr. John A. Anderson of Mount Wilson, and Michelson's personal technical assistant, Fred Pearson, were soon satisfied that mounting the observer on the apparatus made little real difference. In fact, many new difficulties were encountered with the sulky interferometer because it was so big and cumbersome. Preliminary results from observations in the machine shops in Pasadena did not seem encouraging enough to transport this apparatus up the mountain. Nevertheless, through the fall and winter of 1926–1927 they continued to experiment with all the parts, to try out a photographic system, and to increase the accuracy of the tests.[9]

A summit conference on the aether-drift problem at the beginning of February, 1927, assembled all the major characters in the controversy (except Einstein) at Mount Wilson Observatory in Pasadena. Michelson, Lorentz, Miller, and Kennedy were among those who conferred for two days on the theory and practice of aether-drift investigations. Michelson began the conference with a report in which he recalled the genesis and development of his own work in this area: "In 1880 I conceived for the first time the idea that it should be possible to measure optically the velocity v of the earth through the solar sys-

[8] A. A. Michelson, *Studies in Optics*, pp. 161–162.

[9] Walter S. Adams to A. A. Michelson, December 1, 1926, in Michelson Correspondence file of Director, Mount Wilson Observatory. A. A. Michelson, F. G. Pease, F. Pearson, "Repetition of the Michelson-Morley Experiment," *J.O.S.A.* 18: 181–182 (March, 1929). The Mount Wilson solar astronomer, Dr. Edison Pettit, informed me in a personal interview (August 30, 1961) that his own later analysis (made about 1934) of the "invar" used in the 1927–1928 apparatus by Michelson, Pease and Pearson was only 10% instead of the supposed 36% nickel-steel alloy.

tem. . . . Talking in terms of the beloved old ether (which is now abandoned, though I personally still cling a little to it), one might have expected that the aberration of light would be different for a telescope filled with air and with water, respectively."[10] But then, Fresnel's theory had called forth experiments by Airy and others that led by devious routes to a culmination in the Lorentz transformation equations, which "contain the gist of the whole relativity theory." Michelson said nothing that was recorded of his own reruns in progress, but he did credit Roy Kennedy's repetition with the greatest praise."[11]

Lorentz spoke next, beginning with a full tribute to Einstein and admitting his own classical bias toward the notion of absolute time: "There existed for me only this one true time. I considered my time transformation only as a heuristic working hypothesis. So the theory of relativity is really solely Einstein's work."[12] Regarding his own ontological view as to the reality of the contraction hypothesis, Lorentz said, "Asked if I consider this contraction a real one, I should answer 'yes.' It is as real as anything we can observe."[13] This was the year 1927 and Lorentz remembered the period around 1905 as having been in theoretical turmoil quite similar to the present tumultuous activity with regard to quantum mechanics. From the point of view of mid-century nuclear physics, the year 1927 was well past the birth of the "new" quantum theory associated with the names of Louis de Broglie, Max Born, Werner Heisenberg, Erwin Schrödinger, and P. A. M. Dirac.[14]

In the third report of the conference Professor Miller followed very

[10] A. A. Michelson et al., "Conference on the Michelson-Morley Experiment, Held at the Mt. Wilson Observatory, Pasadena, California, Feb. 4, 5, 1927," *Astrophys. J.* 68:341 (December, 1928). Note the first sentence and Michelson's restricted memory of his purpose.

[11] Ibid., p. 343.

[12] Ibid., p. 350.

[13] Ibid., p. 351. See also Robert A. Millikan's concurrence in "The Evolution of Twentieth Century Physics," *Annual Report, . . . Smithsonian Institution, . . . 1927*, pp. 191–199 (1928).

[14] An interesting essay on this historic crisis period between 1926–1928 is R. Kronig, "The Turning Point," in Markus Fierz, ed., *Theoretical Physics in the Twentieth Century: A Memorial Volume to Wolfgang Pauli*, pp. 5–17. See also Gunter Ludwig, *Wave Mechanics*, chapter 2, for "Two Routes to Quantum Mechanics."

closely his presidential address of two years earlier. Miller conceded that some big questions had been answered negatively by the Michelson-Morley experiment, but he insisted that there had always been a small persistent positive effect that could not, he felt, be explained away, except on the basis that it represented an affirmative answer to an even bigger question, absolute motion.[15]

Apparently nothing was settled at the conference on the Michelson-Morley experiment in Pasadena, but certainly the direct confrontation of the principal protagonists offered an opportunity for a thorough reexamination of theory and reinterpretation of experimental findings. The informal discussions must have been more productive than those of which we have any record, but it is certain that all participants left the conference without being persuaded to change their positions on the aether-drift problem. Others, too, were advocating a judicious suspension of verdicts.[16]

Following the Pasadena aether-drift conference in February, 1927, Miller went home to study all his accumulated data once again. It was to take him five years to publish his final results. A few observations were taken in Cleveland, but the main reason he gave for delaying until the fall of 1932 to submit for publication a complete report of his findings was the awesome complexity of his calculations.[17]

By the summer of 1927, Michelson was beginning to think that the whole arrangement of the massive "invar" interferometer was getting too complicated. He wanted to go back to a simpler apparatus. Just then Pease and Anderson fell heirs to a valuable hand-me-down which made possible an entirely new arrangement to corroborate (or refute) Miller's results. In the Pasadena optical shops of the Mount Wilson

[15] Miller in A. A. Michelson et al., "Conference," p. 357.

[16] For example, H. Mineur, "The Experiment of Miller and the Hypothesis of the Dragging Along of the Ether," *Journal of the Royal Astronomical Society of Canada* (Toronto) 21:206–214 (June, 1927).

[17] Dayton C. Miller, "Report on the Ether-Drift Experiments at Cleveland in 1927," *Phys. Rev.* 29:924 (June, 1927); "The Ether-Drift Experiment and the Determination of the Absolute Motion of the Earth," *Rev. Mod. Phys.* 5:232 (July, 1933). See also J. J. Nassau and P. M. Morse, "A Study of Solar Motion by Harmonic Analysis," *Astrophys. J.* 65:73–85 (March, 1927), for another immediate outgrowth of Miller's work.

Observatory lay a tremendous disk of cast iron which had been used a decade earlier as the turntable on which the one-hundred–inch mirror for the Hooker telescope was polished. Anderson and Pease saw the possibility of using this ready-made interferometer base in such a way that the observer would not have to turn with the interferometer as it rotated. The superstructure of the "invar" interferometer might be placed over this bedplate so that an observer could sit on a platform directly over the center of the revolving disk, remaining stationary, while the optical system, including the light source at the center, rotated beneath him, carrying the split beams over a light path of about eighty-three feet. In the fall of 1927 preliminary investigations were begun with this apparatus, but they were plagued by inadequate temperature control, a too-high center of gravity, and asymmetrical strains in the bedplate. These faults introduced inconsistencies in the data recorded, but "still showed clearly that no displacement of the order anticipated was obtained."[18]

Concurrently with these repetitions of the classic aether-drift tests, Michelson was still making headlines with his other experiments in measuring the velocity of light. A leading feature article in the *New York Times Magazine* on November 20, 1927, reveals the reverence in which Michelson was held as "the Apostle of Light." Its author, Waldemar Kaempffert, was effusive: "A whole new physics has been erected on a single experiment of his." And again, "We have to unlearn most of what we thought was true about mass, length, energy, space, time, the universe as a whole—all because the indefatigable Michelson splits a beam of light and races the halves against each other."[19]

Another year went by in which Michelson and his assistants con-

[18] Walter S. Adams to A. A. Michelson, May 31, 1927; Michelson to Adams, June 11, 1927; Adams to Michelson, December 27, 1927, in Michelson Correspondence file of Director, Mount Wilson Observatory. I am especially indebted to D. T. McAllister of the Michelson Museum, U.S. Naval Ordnance Test Station, China Lake, California, for bringing these letters to my attention. See also Michelson et al., "Repetition," *J.O.S.A.* 18:181. For some indication of the growing complexity of instrumentation, see Thomas J. O'Donnell, "An Introduction to Interferometry," *American Machinist* 104:123–146 (June 13, 1960).

[19] Waldemar Kaempffert, "Michelson: The Apostle of Light," *New York Times Magazine*, November 20, 1927, p. 1.

tinued to improve the optical parts mounted on the seven-thousand–pound cast-iron bedplate of the polishing machine for the one-hundred–inch mirror. In their efforts to check Miller's findings by a purely differential method, Pease and Pearson scrupulously followed the time schedule for observations which Miller and Gustaf Strömberg had calculated should show the maximum and minimum effects. But they continued to be troubled by inadequate temperature control and asymmetrical strains. So, in the summer of 1928, they removed the superstructure and put the basic apparatus in a well of the Pasadena Laboratory. Improvising a constant-temperature room, they improved the tank and trough, replaced the mechanical supports with a mercury bath flotation gear, enlarged the path length to eighty-five feet, and built a ceiling over the whole apparatus, above which the observer could sit stationary while the whole sealed optical system rotated beneath him. These and other necessary precautions to eliminate temperature and pressure disturbances as well as observer error were apparently effective. Meanwhile, other evidence corroborating Einstein's General Theory of Relativity continued to accumulate.[20]

During the first three days of November, 1928, a gala celebration was held at the National Bureau of Standards by the Optical Society of America. At this "Michelson Meeting," dedicated to commemorating the life and works of one whose name was said to be "synonymous with optics, especially synonymous with accuracy," Michelson reaffirmed the null results for which he was most famous.

One of the most interesting papers at this Michelson Conference was a review, "The History and Present Status of the Physicists' Concept of Light." Paul R. Heyl, the author, having recounted the struggles of ancients and moderns to make sense of the pecularities and regularities exhibited by optical phenomena, ended his study by asking "Is there still an ether?" and answering: "The concept of the ether has been indeed protean in its nature, all things to all men. Shift-

---

[20] The best account of the evolution of Michelson-Pease-Pearson work was that written by F. G. Pease and published in "Ether Drift Data," *Astron. Soc. Pac.* 42:197–202 (August, 1930). For a classic paper confirming Einstein's predictions regarding the "red shift," see Charles E. St.John, "Evidence for the Gravitational Displacement of Lines in the Solar Spectrum Predicted by Einstein's Theory," *Astrophys. J.* 67: 195–239 (April, 1928).

ing and chameleonlike . . . it has been; in some form or another it is still with us, and bids fair to remain. Paradoxical as it may sound, the ether is becoming etherialized. In this it is but sharing the gradual but general dematerialization of all our physical concepts."[21]

On the last day of that meeting, outside the exhibit which displayed a biographical collection of Michelson's original papers, paintings, and bits and pieces of apparatus, Michelson held a press conference after reporting the latest results of his newest aether-drift tests to the assembly of scientists. The *New York Times* quoted him as saying: "I am merely pointing out a fact and that is that the results of my experiment conducted with greater scientific care, improved apparatus and refined technique, with the intention of eliminating every possible source of error, are again negative. . . . It is for physicists to study and explain these results and reconcile them with the existence of the hypothetical ether."[22]

Miller was also present and was quoted as saying that his experiments had been conducted "in the honest hope of arriving at a negative result also." Miller conceded that the cause of his positive results might have been periodic temperature fluctuations, but he remained adamant: "All I can say is that my own results have been positive."[23]

The two papers announcing formally the results of this series of the originator's repetitions of the Michelson-Morley experiment were not in perfect agreement. The January 19, 1929, issue of *Nature* carried a Michelson note that no displacement as great as one-fifteenth of that expected was found. Later, in March, the *Journal of the Optical Society of America* reported that nothing was observed within one-fiftieth of the expected shift, based on Strömberg's estimate of the solar system's resultant velocity of about three thousand kilometers per second.[24] Was

[21] Paul R. Heyl, *J.O.S.A.* 18:191 (March, 1929). See also Heyl's book, *New Frontiers of Physics*, esp. pp. 2–7, 100–139.

[22] A. A. Michelson in the *New York Times*, November 3, 1928, p. 21. See also Year Books no. 27 and no. 28 of the Carnegie Institution of Washington (D.C.), for official reports of these activities: no. 27, pp. 150–151 (December, 1928); no. 28, p. 143 (December, 1929); cf. Year Book no. 30, p. 171 (December, 1931).

[23] D. C. Miller in the *New York Times*, November 3, 1928, p. 21.

[24] A. A. Michelson et al., "Repetition of the Michelson-Morley Experiment," *Nature* 123:88 (January 19, 1929). Experimental results published in *Nature* for

the difference between one-fifteenth and one-fiftieth simply a refinement of the data? Or did it represent a change in judgment? To most physicists these figures were clearly null results either way, but Miller or his partisans could capitalize on such discrepancies.

Michelson was reported to have said that he had been doubtful about Einstein's dismissal of the aether at the "Michelson Meeting" of the Optical Society in November, but that since the completion of another seasonal check of Miller's results he was "almost convinced" that Einstein was right.[25] But, one may ask, Which Einstein? The man of 1905, of 1915, of 1920—or the man of 1929? Now that several unitary theories (as Einstein himself first called them) had been advanced, Einstein was prevailed upon to write an article for the *New York Times* explaining his present position. In this article, entitled "Field Theories, Old and New," Einstein tried to explain in discursive language the history of the field concept. An extended quotation from his article will not be out of place here.

The physicists of the nineteenth century considered that there existed two kinds of . . . matter, namely ponderable matter and electricity. . . .

Mere empty space was not admitted as a carrier for physical changes and processes. It was only, one might say, the stage on which the drama of material happenings was played.

.   .   .   .

Light waves were, after all, nothing more than undulatory states of empty space, and space thus gave up its passive role as a mere stage for physical events. The ether hypothesis patched up the crack and made it invisible.

The ether was invented, penetrating everything, filling the whole of space, and was admitted as a new kind of matter. Thus it was overlooked that by this procedure space itself had been brought to life. It is clear that this had really happened, since the ether was considered to be a sort of matter which could nowhere be removed. It was thus to some degree identical with space itself; that is, something necessarily given with space. Light was thus viewed as a dynamical process undergone, as it were, by space itself. In this

---

quick exposure to print are often merely preliminary. Cf. Michelson et al., "Repetition," *J.O.S.A.* 18: 182.

[25] A. A. Michelson in the *New York Times*, January 18, 1929, p. 24.

way the field theory was born as an illegitimate child of Newtonian physics, though it was cleverly passed off at first as legitimate.[26]

Then, after recounting the contributions of Faraday, Maxwell, Hertz, and Lorentz, Einstein went on to explain the development of his own thought on the theory of relativity "which had in the last six months entered its third stage of development":

> The characteristics which especially distinguish the general theory of relativity and even more the new third stage of the theory, the unitary field theory, from other physical theories are the degree of formal speculation, the slender empirical basis, the boldness in theoretical construction and, finally, the fundamental reliance on the uniformity of the secrets of natural law and their accessibility to the speculative intellect. It is this feature which appears as a weakness to physicists who incline toward realism or positivism, but is especially attractive, nay, fascinating, to the speculative mathematical mind. Meyerson in his brilliant studies on the theory of knowledge justly draws a comparison of the intellectual attitude of the relativity theoretician with that of Descartes, or even of Hegel, without thereby implying the censure which a physicist would read into this.
>
> However that may be, in the end experience is the only competent judge.
>
> Yet in the meantime one thing may be said in defense of the theory. Advance in scientific knowledge must bring about the result that an increase in formal simplicity can only be won at the cost of an increased distance or gap between the fundamental hypothesis of the theory on the one hand and the directly observed facts on the other hand. Theory is compelled to pass more and more from the inductive to the deductive method, even though the most important demand to be made of every scientific theory will always remain: that it must fit the facts.[27]

Is this the Einstein who had "almost convinced" Michelson, a month before? Perhaps, but probably not. Michelson, the experimentalist, was well aware that Miller, also an experimentalist, still had

[26] Albert Einstein, "Field Theories, Old and New," reprinted from the *New York Times*, February 3, 1929. Hereafter page numbers are supplied for this unpaginated pamphlet.
[27] Ibid., pp. 3, 4. Cf. the new essay added to the 15th edition of Albert Einstein, *Relativity: The Special and General Theory*, entitled "Relativity and the Problem of Space," pp. 147–157. See also Einstein's widely reprinted article "Space, Ether and the Field in Physics," *Forum Philosophicum* I:173–184 (1930).

a valid argument by which to negate all the negative results of these newest repetitions of the Michelson-Morley experiment. Michelson could foresee that Miller would argue: "It is considered very important that the interferometer should not be enclosed in a metallic casing, nor even in an opaque covering; also that it should not be placed inside a room with heavy walls such as are required for a constant-temperature room. The apparatus should be, as nearly as possible, in the open so that there is no possibility of entrainment of the ether in massive materials surrounding it."[28]

Therefore, notwithstanding the fact that the elaborate interferometer in Pasadena had shown no displacements worthy of notice, in the summer of 1930 Pease and Pearson had Michelson's tremendous instrument transported up the mountain and installed in the base of the one hundred-inch telescope, where some control at least could be maintained over temperature effects. If this new location was not in the open air, nor even in a glass house, it was at least at the right elevation and housed in a less massive (though more metallic) structure.[29]

Meanwhile, from across the ocean in Germany came the report in September, 1930, that an automated (i.e., completely automatic), Michelson interferometer had been at work for a year without finding any significant aether drift.[30] The combined talents and resources of Georg Joos and of the Zeiss Optical Manufacturing Company had been brought to bear on the problem of checking out Dayton C. Miller's positive results in the most thorough manner possible. An instrument almost twelve feet high, with arms almost twenty feet in diameter and completely enclosed in a cylindrical cross, with the light source above and the objective cameras below the center of rotation, had been constructed and operated, in Jena, in direct response to Miller's 1925 an-

[28] Miller, "The Ether-Drift Experiment," *Rev. Mod. Phys.* 5:212.

[29] Pease, "Ether Drift Data," *Astron. Soc. Pac.* 42:202. Correspondence between Henry G. Gale, Michelson's colleague at Chicago, and Walter S. Adams, during January, 1930, indicates that Michelson had lost interest by then and that the latest observational data was not suitable for publication: Michelson Correspondence file of Director, Mount Wilson Observatory. Again, I am indebted to D. T. McAllister of the Michelson Museum for locating and recognizing the value of these seven letters.

[30] Georg Joos, "Die Jenaer Wiederholung des Michelsonversuchs," *Ann. d. Phys.*, 5th ser. 7:385–407 (1930).

nouncement. The summary of the result follows: "A description is given of a recording Michelson-interferometer of 21 m light path. Microphotometric measurements of the photographs taken by it show that, assuming the presence of an aether wind, any effect is less than 1/1000 of a fringe width, an amount of aether-wind that must be less than 1.5 km/sec."[31]

Thus the simultaneous conclusion from the Zeiss plant in Jena (to the effect that if there were an aether drift, then it would be less than 1/1000 of one fringe width), with the conclusions from the Mount Wilson optical shops in California, tended to clinch the matter of aether drift permanently. Together, the independent reports and experimental authority of Michelson and of Joos seemed conclusively to outweigh Miller's interpretation of the Michelson-Morley experiment.

Miller might still insist, as he would three years later[32] that there was too much shielding in both of these most elaborate reruns of the Michelson-Morley experiment. But this single objection could be seen as grasping at a straw. To those scientists who had no vested interest to protect, Michelson's steadfast interpretation of his own experiment plus Joos's independent corroboration appeared decisive. The so-called whispering campaign against Einstein and relativity theory should be dispelled once and for all by these combined results. As Joos himself wrote in his next textbook: "Experiment has thus decided against our acoustic analogy and against the existence of a stationary medium carrying light; i.e., the existence of a cosmic ether or absolute space is disproved."[33]

A social gathering of scientists at the California Institute of Technology in mid-January, 1931, brought together in person for the first and only time the two men so widely honored as the founders, in experiment and theory, of physical relativity. Albert Michelson, barely four months before his death, and Albert Einstein met publicly under

[31] Ibid., p. 407 [my translation].

[32] Miller, "The Ether-Drift Experiment," *Rev. Mod. Phys.* 5:240. See also Dayton C. Miller, "The Absolute Motion of the Solar System and the Orbital Motion of the Earth Determined by the Ether-Drift Experiment," *Phys. Rev.* 43:1054 (June 15, 1933).

[33] Georg Joos, *Theoretical Physics*, p. 227.

the auspices of Cal Tech's president Robert A. Millikan at a grand banquet on January 15 in the new Atheneum. There Millikan introduced Michelson first as the great experimentalist who laid the empirical foundations of special relativity. Michelson rose to express his gratitude for Einstein's appreciation of his aether-drift work begun a full half century earlier, then said: "I may recall the fact that in making this experiment there was no conception of the tremendous consequences brought about by the great revolution which Dr. Einstein's theory of relativity has caused, a revolution in scientific thought unprecedented in the history of science."[34]

After this, W. W. Campbell spoke for Charles E. St. John, R. C. Tolman, E. P. Hubble, and others among the distinguished company present who had played leading roles in corroborating general relativity. Then with his characteristic bias, Millikan introduced the other guest of honor, Einstein, as the great theoretician who had made modern science "essentially empirical." Whereupon Einstein arose to praise these and other famous men with special attention to the frail and aging Michelson:

You, my honored Herr Michelson, began [this work] when I was only a small boy, not even a meter high. It was you who led the physicists into new paths, and through your marvelous experimental labors prepared for the development of the relativity theory. You uncovered a dangerous weakness in the ether theory of light as it then existed, and stimulated the thoughts of H. A. Lorentz and FitzGerald from which the special theory of relativity emerged. This latter, in turn, led the way to the general theory of relativity and to the theory of gravitation. Without your work this theory would today be scarcely more than an interesting speculation; your verifications furnished the real basis for this theory.[35]

[34] A. A. Michelson, quoted in the *New York Times*, January 16, 1931, p. 3. For an edited transcript of these proceedings, see "Professor Einstein at the California Institute of Technology: Addresses at the Dinner in His Honor," *Science* 73:375–381 (April 10, 1931).

[35] Albert Einstein, quoted in the *New York Times*, January 16, 1931, p. 3. For the full text in German and in English of this formal address, see "Dr. Einstein's Address," *Proceedings Am. Phil. Soc.* 93:544–545 (1945). For an elaborate study of the meaning of this meeting, see Gerald Holton, "Einstein, Michelson, and the 'Crucial Experiment,'" *Isis* 60:186–190 (Summer, 1969). The translation is Holton's from ibid., p. 190 n. 162.

CHAPTER TWELVE

AETHER EXPERIMENTS AND PHYSICAL THEORY

The final verdict in the search for the ethereal aether was not given by Michelson and Joos in 1930, although it would be comforting and convenient to think so. Almost all theoretical physicists and most experimental physicists had long since been convinced that the concept of the aether was indeed superfluous, if not actually obscurantist, but there were yet a few who, for one reason or another, preferred to retain the notion of an aether, if only as a synonym for the plenum or the curvature of interstellar space. Repetitions and refinements of the Michelson-Morley experiment, so thoroughly stimulated by Miller after 1925, did not abruptly cease in 1930. Therefore, this concluding chapter will attempt two tasks: first, to risk historical judgment on the significance of Michelson-Morley to modern physical science, and second, to look briefly beyond Michelson's life and into the near past.

The story recounted here, it should be emphasized, is only one aspect of the grand intellectual reorientation of modern science symbolized by Einstein and his theories. In the latter chapters this history has stressed the viewpoint of those scientists who failed to win acceptance rather than of those who succeeded. I have deliberately chosen to emphasize here, not the birth and development of relativity theory, but rather the gradual denouement or recession of the aether theory. The story followed here bypasses for the most part the influence of the development of quantum theory on the gradual demise of the aether concept. Such self-imposed strictures severely limit the value of any general conclusions to be drawn from this study.

Nevertheless, I must attempt to come to terms with the question asked at the beginning of this study: to what extent can it be said legitimately that Einstein's Special Theory of Relativity was a lineal outgrowth of the Michelson-Morley experiment? This question was posed historically, and it must be answered historically.

There are, of course, already, hallowed pat answers to this question. Max Planck himself, in denouncing strictly operationalist modes of thought that would allow physical meaning only to specifiably reproducible actions, said that "if physicists had always been guided by this principle, the famous experiment of Michelson and Morley undertaken to measure the so-called absolute velocity of the earth would never have taken place, and the theory of relativity might still be non-existent."[1] Elsewhere Planck made a similar remark with the added observation that "the problem of the earth's absolute velocity has for some time been seen to be somewhat insignificant: Yet the trouble spent upon it has proved extremely useful in physics."[2]

On the other hand, a careful student of the aether-drift experiments, Thomas W. Chalmers, has asserted that "had the ether-drift experiments never been performed, the theory of relativity would have arisen in the way it did, but it would have lacked one of its several sources of

[1] Max Planck, *Scientific Autobiography and Other Papers*, p. 139. The father of operationalism, P. W. Bridgman, on the other hand, insisted that Einstein was hardly influenced by Michelson-Morley at all: see his *A Sophisticate's Primer of Relativity*, pp. 53, 135, 160.

[2] Max Planck, *Philosophy of Physics*, p. 70. For a similar judgment, see Hermann Bondi, *Relativity and Common Sense: A New Approach to Einstein*, p. 60.

experimental confirmation."[3] It is certainly true that this experiment furnished Einstein, through Lorentz, with ready-made experimental evidence supporting the new theory, but the difficulty overlooked by Chalmers is that Einstein's debt to Lorentz was in turn a direct debt to Michelson.

Einstein himself after 1916 contributed to an oversimplified historical view by various didactic accounts of relativity theory.[4] Furthermore, as we have seen, Einstein's "first step away from naive visualization" was retracted in a sense at one point, but the movement he had led moved on, "sometimes even against the resistance of its creator, [giving] birth to further fruitful development, following its own autonomous course."[5]

The fact is that neither Planck's implied historical determinism nor Chalmers's overcompensating judgment that Michelson-Morley was significant only as ready-made evidence for relativity will satisfy the historical evidence presented in this study. The idea that Michelson-Morley was the *experimentum crucis* for the luminiferous aether and therefore the empirical basis for relativity implies a logical tidiness in human affairs which seldom exists and which certainly is not exhibited by this history. Conversely, science is itself that human enterprise above all others that is designed to overcome such untidiness, and this history of the interferometric tests for an aether drift certainly should have shown more mutual interdependence between the experiments and the theory than mere sequential juxtaposition.

---

[3] Thomas W. Chalmers, *Historic Researches: Chapters in the History of Physical and Chemical Discovery*, p. 81. For a similar judgment, see Martin Gardner, *Relativity for the Million*, p. 34; cf. p. 25.

[4] To see this change in Einstein's attitude toward the receding past, compare the level of sophistication in the treatment of Michelson-Morley in 1916–1920 in Einstein's own early popularization, *Relativity: The Special and General Theory*, pp. 63–64, with his later collaborative popularization with Leopold Infeld, *The Evolution of Physics: The Growth of Ideas from Early Concepts to Relativity and Quanta*, p. 183. The latter denies any recognition at all to Dayton C. Miller's claim, and in fact Einstein is on record regarding this book as representing "not at all the way things happened in the process of actual thinking": Einstein in Max Wertheimer, *Productive Thinking*, ed. Michael Wertheimer, p. 228.

[5] Wolfgang Pauli, *Theory of Relativity*, p. vi. See also Max Born, *Einstein's Theory of Relativity*, p. 224.

Perhaps the question whether the special relativity and aether theories were truly mutually exclusive contributed most to the confusion and controversy over the proper interpretation to be given to Miller's challenge of Michelson's celebrated null result. When Michelson died, on May 9, 1931, still reluctant to admit the irreconcilable loss of the "beloved" aether, yet firmly convinced that an optical drift had not yet been detected, he left an unresolved legacy which others could exploit. Throughout the 1930's and until his death in 1941, Miller maintained the hope that a reconciliation might be accomplished, because he was convinced that he had detected the aether drift and the absolute motion of the earth.[6] Other advocates for the aether during that decade, among them O. C. Hilgenburg, L. Zehnder, H. E. Ives, and Hermann Fricke,[7] may not have agreed on the degree of reconciliation to be expected, but the hypothesis of an imponderable aether of some impalpable sort did seem necessary to them. Perhaps it is true, as Max Planck often contended, that "a new scientific truth does not triumph by convincing its opponents and making them see the light, but rather because its opponents eventually die, and a new generation grows up that is familiar with it."[8]

The new generation, or the regenerated older generation, of physicists after 1930 no longer needed a visualizable aether. As astrophysical cosmology became fully reinstated to the status of a legitimate science, and as conceptual analysis became a recognized adjunct of physical theory, the boundary between physics and metaphysics shifted. Aether became a metaphysical concept in the pejorative sense, whereas cosmogony and cosmology became respectable once again with the

[6] D. C. Miller, "The Ether-Drift Experiment and the Determination of the Absolute Motion of the Earth," *Nature* 133:162–164 (February 3,1934). See also Harvey Fletcher, "Dayton Clarence Miller, 1866–1941," in *Biographical Memoirs*, National Academy of Sciences 23:63; cf. G. W. Stewart, "Dayton Clarence Miller," *Journal of Acoustical Society of America* 12:479 (April, 1941).

[7] See Carl F. Krafft, *Ether and Matter*, p. 5. One notable discussion of Miller's results that took him seriously was written by Robert B. Lindsay and Henry Margenau in 1936: *Foundations of Physics*, pp. 319–326, 354, 377.

[8] Planck, *Scientific Autobiography*, pp. 33–34. Cf. Planck, *Philosophy of Physics*, p. 97. See also Stephen E. Toulmin, "The Evolutionary Development of Natural Science," *American Scientist* 55:456–471 (December, 1967), esp. p. 470.

work of Harlow Shapley, Edwin P. Hubble (1889–1953), and many other astrophysicists. One of the first of the new cosmologists following Einstein's lead, Willem de Sitter, made this point quite clear in 1931:

This giving of names is not so innocent as it looks. It has opened the gate for the entrance of hypotheses. Very soon the field is materialized, and is called aether. From the mathematical point of view, of course, "aether" is still just another word for "field," or, perhaps better, for "space"—the absolute space of Newton—in which there may or may not be a field. From the point of view of physical theory . . . however, the "aether of space" . . . is not simply space, it is something substantial, it is the carrier of the field, and mechanical models consisting of racks and pinions and cog wheels are devised to explain how it does the carrying.[9]

Maxwell's aether, or Kelvin's, of course, had not been everyone's model, but de Sitter was quite right in insisting that aether apologists define exactly what they meant by the word and show to what it referred. Inability to do this meant that the aether apologists were left ever farther behind, until they found themselves in the position of speaking a nonphysical language.

The relativist, or majority, position here was aptly illustrated in 1932 by Georg Joos's thorough exposition of the latest experiments on the aether-drift anomalies: "The fact that the stationary ether concept leads to correct results for rotation systems in optical experiments also is no more a proof of existence of the ether than are the mechanical phenomena of Coriolis and centrifugal forces."[10] Then, after pointing out how the Michelson-Gale experiment served as an optical analog to Foucault's pendulum, Joos hammered home the major objection revealed by that experiment. The Michelson-Gale experiment, wrote Joos, "is of importance from the fact that it makes all attempts to

[9] Willem de Sitter, *Kosmos: A Course of Six Lectures on the Development of Our Insight into the Structure of the Universe*, delivered at the Lowell Institute, Boston, in November, 1931, p. 105. See also Harlow Shapley, *Flights from Chaos: A Survey of Material Systems from Atoms to Galaxies*, and (ed.) *Source Book in Astronomy*; Edwin P. Hubble, *The Realm of the Nebulae* (1936), and *The Observational Approach to Cosmology* (1937). For an excellent historical and philosophical analysis, see John David North, *The Measure of the Universe: A History of Modern Cosmology*.

[10] Georg Joos, *Theoretical Physics*, p. 449.

explain the negative result of Michelson-Morley by convection of the ether seem futile, for it seems absurd that the ether could be carried along by the earth, completely in translation and not at all in rotation."[11]

What seemed manifestly absurd to Joos and to most other physicists remained, then and now, far less absurd than the relativists' discard of "common sense" seemed to those whose physical imagination worked dynamically rather than kinematically. There continues to be a nonprofessional literature demanding that "science must leave something for waves to wave in."[12] And not few have been the attempts, often scurrilously personal, to discredit Einstein and relativity and to reinstate some kind of an aether.[13]

Here then must be our verdict: The classic Michelson-Morley aetherdrift experiment derived its importance far more from what it *suggested* than from what it *imposed*. Only by distant interpretations of it and after several delayed repetitions of it was the experiment regarded as crucial. Oversimplified descriptions, by relativists and by aether apologists, gave it a mythic life that made it seem to have been a crucial test for the existence of aether. But there is a severe danger of anachronism if we attribute this historical interpretation to Michelson's experiment without compensating for its temporal development and for the concurrent development of relativity theory and of physics as a whole. The present is never quite so simple as the past usually seems. Historical accuracy requires us to recognize that the past was also once the present.

The sociohistorical conflict between relativity and the aether

[11] Ibid.

[12] Eigil Rasmussun, *Matter and Gravity: Logic Applied to Physics in a Common-Sense Approach to the Classical Ether Theory*, p. 7. Cf. H. L. Speiro, *The Fluid Ether*. See also a grandfather of the polemical literature edited by Hans Israël, Erich Ruckhaber, and Rudolf Weinmann, *Hundert Autoren gegen Einstein*.

[13] A few of the most interesting examples of antirelativist literature are (in chronological order): Arthur Lynch, *The Case Against Einstein*; George de Bothezat, *Back to Newton: A Challenge to Einstein's Theory of Relativity*; Alfred O'Rahilly, *Electromagnetics: A Discussion of Fundamentals*; Blamey Stevens, *The Psychology of Physics*; Charles A. Muses, *An Evaluation of Relativity Theory: After a Half Century*; Arthur S. Otis, *Light Velocity and Relativity*; Dewey B. Larson, *New Light on Space and Time*; Michael M. Hare, *Microcosm and Macrocosm*.

TABLE 1*

EXPERIMENTAL BASIS FOR THEORY OF SPECIAL RELATIVITY

Theory		Light propagation experiments							Experiments from other fields			
		Aberration	Fizeau convection coefficient	Michelson-Morley	Kennedy-Thorndike	Moving sources and mirrors	De Sitter spectroscopic binaries	Michelson-Morley, using sunlight	Variation of mass with velocity	General mass-energy equivalence	Radiation from moving charges	Meson decay at high velocity	Trouton-Noble
Ether theories	Stationary ether, no contraction	A	A	D	D	A	A	D	D	N	A	N	D
	Stationary ether, Lorentz contraction	A	A	A	D	A	A	A	A	N	A	N	A
	Ether attached to ponderable bodies	D	D	A	A	A	A	A	D	N	N	N	A
Emission theories	Original source	A	A	A	A	A	D	D	N	N	D	N	N
	Ballistic	A	N	A	A	D	D	D	N	N	D	N	N
	New source	A	N	A	A	D	D	A	N	N	D	N	N
Special theory of relativity		A	A	A	A	A	A	A	A	A	A	A	A

Legend: A, the theory agrees with experimental results.
D, the theory disagrees with experimental results.
N, the theory is not applicable to the experiment.

* Panofsky and Phillips, *Classical Electricity and Magnetism*, Table 15–2 (Cambridge: Addis Wesley, Second Edition, 1962).

TABLE 2*

COMPARISON OF THEORETICAL FEATURES

	Emission theory	Classical ether theory	Special theory of relativity
Reference system	No special reference system	Stationary ether is special reference system	No special reference system
Velocity dependence	The velocity of light depends on the motion of the source	The velocity of light is independent of the motion of the source	The velocity of light is independent of the motion
Space-time connection	Space and time are independent	Space and time are independent	Space and time are interdependent
Transformation equations	Inertial frames in relative motion are connected by a Galilean transformation	Inertial frames in relative motion are connected by a Galilean transformation	Inertial frames in relative motion are connected by a Lorentz transformation

* Panofsky and Phillips, *Classical Electricity and Magnetism*, Table 15–3 (Cambridge: Addison-Wesley, 1955).

was more complex than physicists generally could afford to recognize. The abandonment of the aether concept meant the loss of mechanical analogy, while the victory of relativity was another victory of mathematical formalism. The force of traditional Newtonian absolutism was only slowly overcome by the tide of success accompanying Einstein's relativism. Chalmers ended with an apt judgment: "In a way, the choice may be said to lie between ether and no ether. In a more refined way it can be said to lie between freedom to admit the play of the imagination into our efforts to interpret Nature and its rigorous exclusion on a basis which refuses to recognize dogma and redefines reality in terms solely of mathematical truth."[14]

Tables 1 and 2 (above), first published in 1955, conveniently summarize the theoretical issues most meaningful to most physicists after 1930. For a generation or so, lack of concern for historical chronology contributed to the acceptance of this paradigm. Let us

[14] Chalmers, *Historic Researches*, p. 83. See also David Bohm, *The Special Theory of Relativity*, pp. 41–53.

return to a chronological trace for a glimmer of a thread that runs into contemporary history.

In 1930, Roy J. Kennedy and Edward M. Thorndike deliberately set out to establish experimentally the relativity of time. Considering the classical Michelson-Morley tests to have concerned only the nature of space and not of time, Kennedy and Thorndike used a special arrangement of the Michelson-Morley interferometer, with deliberately unequal arm lengths and a highly sophisticated light source, all enclosed in a high-vacuum chamber, to test for the denial of significance for absolute time. They concluded two years of work in 1932 by publishing what they believed to be the first true complement to Michelson-Morley, showing the relativity of time by confirming the Lorentz transformation equations, which by implication confirmed Einstein's famous "clock axiom."[15]

As we have seen, not until 1933 did Dayton C. Miller publish his major paper on the aether drift. This long paper was something of a secret history of the developments in the 1920's which Miller had precipitated.[16] Undoubtedly, most physicists agreed with Planck that Miller's concern over the absolute motion of the earth was "somewhat insignificant." Although papers on "relativistic cosmology" excited

[15] Roy J. Kennedy and Edward M. Thorndike, "Experimental Establishment of the Relativity of Time," *Phys. Rev.* 42:400–418 (November, 1932). Cf. Adolf Grünbaum, "Logical and Philosophical Foundations of the Special Theory of Relativity," in Arthur Danto and Sidney Morgenbesser, eds., *Philosophy of Science*, p. 417. Grünbaum has often urged that philosophical mastery of Special Relativity Theory is required for decoding its history: see his "The Bearing of Philosophy on the History of Science," in *Science* 143:1406–1412 (March 27, 1964). See also J. J. C. Smart, *Problems of Space and Time.*

[16] Dayton C. Miller, "The Ether-Drift Experiment and the Determination of the Absolute Motion of the Earth," *Rev. Mod. Phys.* 5:203 (July, 1933). Cf. an abstract of this paper in *Phys. Rev.*, 2d ser. 43:1054 (June 15, 1933). Georg Joos wrote a letter to the editor of *Physical Review* defending his own work and attacking Miller's paper; Miller defended his complaint about shielding with a counterattack, insisting upon his own regularly periodic variations: see *Phys. Rev.*, 2d ser. 45:114 (January 15, 1934). Curiously, Miller never took full advantage of the possibility of correlating his data with that of such astronomers as W. W. Campbell, J. H. Moore, and V. M. Slipher, who were most concerned with the proper-motion problem. See Robert G. Aitken, *Our Journey through Space*, Leaflet no. 43, *Astron. Soc. of Pac.* 1:175–178 (1932).

greater interest, Miller continued to publish his message in the hope that some would see it as more than a mere curiosity.[17]

In 1937–1938 a series of six papers was published by Herbert E. Ives of the Bell Telephone Laboratories which analyzed in minute detail the problems in experimental design raised by the Michelson-Morley experiment.[18] Ives ran a long series of new tests to check specific points and compare them with the predictions of relativity theory. His immediate stimulus had been the Kennedy-Thorndike experiment and Kennedy's subsequent claim to have simplified the theory of the Michelson-Morley experiment.[19] Ives accepted the claim of Sagnac that he had produced positive experimental evidence for the existence of the luminiferous aether, and he tried to save the aether by attacking Einstein's light principle: "Calling upon the Michelson-Morley experiment as proof of the invariance of light signal phenomena therefore carries with it the acceptance of the luminiferous ether here assumed."[20] Together with G. R. Stilwell, Ives did another experiment in 1938 confirming the transverse Doppler effect which, it was asserted, gave positive results that "may hence be claimed to give more decisive evidence for the Larmor-Lorentz theory."[21] This statement of interpretation illustrates that Miller was not the last of the mavericks.

[17] For example, H. P. Robertson, "Relativistic Cosmology," *Am. J. Phys.* 5:62–90 (January, 1933). A condensed version of Miller's "The Ether-Drift Experiment," cited above, apeared under the same title in *Nature* 133:162–164 (February 3, 1934). See also Dayton C. Miller to G. S. Fulcher, February 12, 1938, in Fulcher Collection, Bohr Library, American Institute of Physics, New York.

[18] These six papers by H. E. Ives all appeared in *J.O.S.A.* and will be cited here with abbreviated titles: "Graphical Exposition of the Michelson-Morley Experiment," 27:177 (May, 1937); "Light Signals on Moving Bodies," 27:263 (July, 1937); "Abbreviation of Clocks and the Clock Paradox," 27:305 (September, 1937); "Apparent Lengths and Times," 27:310 (September, 1937); "Doppler Effect and Michelson-Morley," 27:389 (November, 1937); "Light Signals Sent around a Closed Path," 28:296 (August, 1938).

[19] Roy J. Kennedy, "Simplified Theory of the Michelson-Morley Experiment," *Phys. Rev.* 47:965–968 (June, 1935).

[20] Ives, "Light Signals," *J.O.S.A.* 27:271; "Light Signals," *J.O.S.A.* 28:296. See footnote 18, above.

[21] Herbert G. Ives and G. R. Stilwell, "An Experimental Study of the Rate of a Moving Atomic Clock," *J.O.S.A.* 28:226 (July, 1938). Cf. Grünbaum in Danto and Morgenbesser, eds., *Philosophy of Science*, p. 418. See also Grünbaum's *Philosophical Problems of Space and Time*, pp. 341–409.

The next year, in direct reply to Ives's experiments, Herbert Dingle gave an unusual relativist answer to Ives and Stilwell's claim that a fixed aether was still preferable to Special Relativity: "On the contraction hypothesis we change the length of a body by changing its state of motion: on the relativity hypothesis we change the form of our equations by changing our mind."[22] Dingle, having first disposed of Miller's experiments as irrelevant to the present controversy, proceeded to chastise Ives for thinking that FitzGerald, Larmor, or Lorentz could have given an adequate explanation for the Kennedy-Thorndike experiment. Michelson-Morley had been an effort to detect the undetectable, but Kennedy-Thorndike was an experiment of a different kind, not a test for aether drift in free space, but for the time of travel of two light pencils over closed paths of different lengths.[23] Dingle criticized Ives for introducing too many new, independent hypotheses, then remarked about the impossibility of "proving" a universal negative statement: "The principle of relativity is of precisely the same character as the second law of thermodynamics—a negative statement based on all the relevant evidence available. . . . Either hypothesis might be disproved by a single observation."[24]

That such an observation must be positive and must rest firmly on no more independent hypotheses than were advanced by Michelson-Morley was understood, but Dingle's argument led him later to deny categorically any significance to Dayton C. Miller's work: "If we accept Miller's results we do not merely destroy a theory; we destroy belief in the rationality of our experience, in what we used to call the 'uniformity of nature.' It is much to be hoped that a satisfactory explanation of the apparent contradiction will be found."[25] Dingle's defense of the majority position was thought extreme in some circles, and

[22] Herbert Dingle, "The Interpretation of the Michelson-Morley and Kennedy-Thorndike Experiment," *Phil. Mag.*, 7th ser. 27:695 (June, 1939).

[23] For a more detailed technical exposition of this difference, see H. P. Robertson, "Postulate *versus* Observation in the Special Theory of Relativity," *Rev. Mod. Phys.* 21:378–382 (July, 1949).

[24] Dingle, "The Interpretation," *Phil. Mag.,* 7th ser., 27:702.

[25] Herbert Dingle, *The Special Theory of Relativity*, p. 22. Cf. W. H. McCrea, *Relativity Physics*, pp. 6, 12 n.5. While Dingle and McCrea continue to do battle, the former also argues that, contrary to popular opinion, no time was measured in

in later years his interpretation, too, became that of a minority. One early critic wrote in reply: "When it is reflected what a simple idea a fixed medium is, as well as how indispensable it seems to be for nearly all astronomical, optical, and physical facts, it seems rather perverse to think that because we have so far failed to find the earth's velocity through it there can be no such medium."[26] Dingle's point, that relativists in 1939 regarded the search for aether drift as equivalent to the search for perpetual-motion machines, indicates how far Dayton Miller and his comrades had fallen behind current physical thought.

After another world war had come and gone one might expect that the aether would be gone forever. The atomic bomb should have annihilated the notion of aether, along with everything else in its way, if Special Relativity Theory, symbolized grossly by mass-energy equivalence, were truly the complete antithesis to the aether theory, as prewar partisans of relativity had so boldly assumed. But the power of survival of the aether concept was demonstrated in November, 1951, when P. A. M. Dirac, one of the most respected men in all theoretical physics because of his role in founding quantum electrodynamics, wrote a letter to the editor of *Nature*. Published under the headline, "Is There an Aether?" this letter began Dirac's advocacy, from cosmological considerations, of the position that: "Aether is no longer ruled out by relativity, and good reasons can now be advanced for postulating an aether. . . . With the new theory of electrodynamics we are rather forced to have an aether."[27]

Dirac's startling proposition for a revival of the aether concept was based on the need to explain, cosmologically, preferred motion and the

Michelson-Morley and therefore the question of additive velocities is still unsettled; see Herbert Dingle, "A Re-Examination of the Michelson-Morley Experiment" in Arthur Beer, ed., *Vistas in Astronomy* 9:97–100.

[26] A. Eagle, "Letter to the Editor," *Phil. Mag.*, 7th ser. 28:595 (November, 1939). Dingle himself has continued to goad the complacency of those who believe in the positive experimental support for Special Relativity Theory; see his "Reason and Experiment in Relation to the Special Relativity Theory," *British Journal for Philosophy of Science* 15:49 (May, 1964); and his "The Case against Special Relativity," with reply by W. H. McCrea, "Why the Special Theory of Relativity is Correct," *Nature* 216:113–125 (October 14, 1967).

[27] *Nature* 168:906 (November 24, 1951). See also P. A. M. Dirac, "The Evolution of the Physicist's Picture of Nature," *Sci. Am.* 208:45–53 (May, 1963).

hypotheses of continual creation, put forward by the cosmologists Hermann Bondi and Thomas Gold. These men, as well as Leopold Infeld, answered Dirac in subsequently published letters.[28] Although Dirac admitted that the necessary formulae could be purified and that the cosmological creation of electric charges must be very small, he insisted that the quantum electrodynamical need for avoiding action-at-a-distance was still great enough to make feasible the resuscitation of the aether.[29]

In the wake of Dirac's suggestion, English physicists returned to a new generation of interferometers to toy with the latest technology and perchance to improve on or to explain away Dayton C. Miller's anomalous results. In 1952, R. W. Ditchburn and O. S. Heavens, using a commercial type Michelson interferometer with one arm inclined forty-five degrees upward from the horizontal, reported that no fringe shift was observed.[30] Two years later H. L. Furth proposed and Louis Essen accepted a challenge to try the Michelson-Morley experiment once again, this time with standing rather than progressive waves.[31]

Essen, of Britain's National Physical Laboratory, made use of a new device called a "cylindrical cavity resonator" to repeat the Michelson-Morley experiment with short radio waves instead of with light. His technique consisted in observing the variation of resonant radio frequencies rather than the deviation of optical interference fringes. Essen's report in 1955 began as follows: "The special theory of relativity is so well established by indirect evidence that the significance of

[28] *Nature* 169:146 (January 26, 1952); 169:702 (January 31, 1952). For a readable explanation of this English ferment see Sir Harrie Massey, *The New Age in Physics*, pp. 127–138.

[29] Dirac in *Nature* 169:702. See also G. H. F. Gardner, "Rigid-Body Motions in Special Relativity"; and J. L. Synge, "Gardner's Hypothesis and the Michelson-Morley Experiment," *Nature* 170: 243–244 (August 9, 1952).

[30] "Relativistic Theory of a Rigid Body," *Nature* 170–705 (October 25, 1952); cf. pp. 243–244. More recently Dirac seems to have been joined by another famous theoretician, Louis de Broglie, who returned to an earlier position in favor of some sort of "subquantized medium": see Broglie, *The Current Interpretation of Wave Mechanics: A Critical Study*, pp. viii, 43.

[31] H. L. Furth, "Proposal for a New Aether Drift Experiment," *Nature* 173:80 (January 9, 1954); L. Essen, Letter to the Editor, *Nature* 173:734 (April 17, 1954); cf. H. L. Furth, Letter to the Editor, *Nature* 174:505 (September 11, 1954).

the original aether-drift experiments has decreased, and the experimental discrepancies in the results have caused few misgivings. Such discrepancies should, however, be examined if further developments in experimental technique enable new and possibly more precise methods to be employed."[32] Dirac's suggestion remained in the background, but Miller's 1933 conclusions about the aether drift were the direct "discrepancies" at issue. Essen's verdict in the paper above was that Miller's "conclusions cannot be accepted; the effect, if any, is shown to be not more than one-tenth of that reported by him, which must probably be ascribed to some systematic error."

The related development of microwave technology aroused another search for a significant problem, and the aether-drift anomaly was made to order for a test by the amplification and stimulation of emitted radiation. Like Michelson's interferometer, the evolving device now called the "maser" or the optical "laser" was applied in its infancy to the problem of the ethereal aether. Charles H. Townes and his colleagues in 1963 obtained null results at a new order of negative magnitude.[33]

Perhaps the definitive explanation of why Miller's interferometer seemed to him to give results at variance with all other experiments of this nature was put forward in 1955, a half century after Einstein's original paper, by the publication of "A New Analysis of the Interferometer Observations of Dayton C. Miller."[34] The senior author of this work was Robert S. Shankland, who, as a student and assistant

[32] L. Essen, "A New Aether-Drift Experiment," *Nature* 175: 793–794 (May 7, 1955). For an account of the operational use of Special Relativity Theory in modern high energy particle accelerators, see R. E. Peierls, *The Laws of Nature*, pp. 121–145.

[33] Charles H. Townes, interview with author, July 1, 1964, Houston, Texas; and Townes, "Quantum Electronics, and Surprise in Development of Technology," *Science* 159: 699–703 (February 16, 1968). See also James P. Gordon, "The Maser," *Sci. Am.* 199:42–50 (December, 1958); and T. S. Jaseja, A. Javan, C. H. Townes, "Frequency Stability of He-Ne Masers and Measurements of Length," *Physical Review Letters* 10:165–167 (March, 1963), which claims to verify the FitzGerald-Lorentz contraction due to earth's orbital velocity to one part in one thousand.

[34] Robert S. Shankland et al., "A New Analysis of the Interferometer Observations of Dayton C. Miller," *Rev. Mod. Phys.* 27: 167–178 (April, 1955). See also André Mercier, "Fifty Years of the Theory of Relativity," *Nature* 175:919–921 (May 28, 1955).

TABLE 3

SIGNIFICANT TRIALS OF THE MICHELSON-MORLEY EXPERIMENT

Observer	Year	Place	Arm Length	Expected Shift	Observed Shift	Rat
Michelson	1881	Potsdam	120 cm	0.04 fringe	0.0200	2
Michelson & Morley	1887	Cleveland	1100	0.40	0.0100	40
Morley & Miller	1902–04	Cleveland	3200	1.13	0.0150	75
Miller	1921	Mt. Wilson	3200	1.12	0.0800	14
Miller	1923–24	Cleveland	3200	1.12	0.0300	37
Miller (sunlight)	1924	Cleveland	3200	1.12	0.0140	80
Tomaschek (starlight)	1924	Heidelberg	860	0.30	0.0200	15
Miller	1925–26	Mt. Wilson	3200	1.12	0.0880	13
Kennedy	1926	Pasadena & Mt. Wilson	200	0.07	0.0020	35
Illingworth	1927	Pasadena	200	0.07	0.0004	175
Piccard & Stahel	1927	Mt. Rigi	280	0.13	0.0060	22
Michelson *et al.*	1929	Mt. Wilson	2590	0.90	0.0100	90
Joos	1930	Jena	2100	0.75	0.0020	375

Legend:

Arm length or l = optical path length of one interferometer arm.

Expected shift or

$2l/\lambda)\ (v/c)^2$ = maximum anticipated fringe displacement, assuming a 30 km/sec aeth drift velocity, when apparatus rotates through 90°.

Observed shift or $(2A)$ = actually "observed" shift of fringes or the amplitude of second harmon found after data reduction.

Ratio = comparison of Expected shift with Observed shift.

Adapted from R. S. Shankland's Table 1, *Rev. Mod. Phys.* 27:168, 178 (April, 1955).

of Miller's in 1932, had participated in the final analysis of Miller's voluminous accumulation of data.

Twenty years later Professor Shankland had inherited both Miller's papers and his position as head of the physics department at the former Case School of Applied Science, known after 1947 as Case Institute of Technology (a merger in 1967 created Case Western Reserve University). His personal interest in his former teacher resulted in several biographical articles after Miller died in 1941. A decade later, Dirac's suggestion appeared, and Shankland, after extensive consultation with Einstein, decided to subject Miller's observations to a thoroughgoing review by the use of the latest statistical techniques and by programming these data for computer analysis. He enlisted the aid of three colleagues, S. W. McCuskey, a mathematician, F. C. Leone, an astronomer, and G. Kuerti, a mechanical engineer, all of whose talents were essential to the solution of the problem. Shankland and his colleagues concentrated on the small periodic residuals of the Mount Wilson experiments, finishing their analysis in 1954. As a team of scientists reviewing the half-century history of the Michelson-Morley-Miller problem, they began with a chronological assessment of the relevant data points. As shown in the accompanying table entitled "Significant Trials of the Michelson-Morley Experiment," which is slightly modified, with Shankland's permission, for easier interpretation, the comparisons of expected to observed fringe shifts over the years of increasing optical sophistication still left little, though something, to be explained.

In July, 1954, the authors mailed a mimeographed preprint of their work to some twenty-five readers for comment. Einstein saw the final draft before his death in March of that year and wrote a personal letter of appreciation to Professor Shankland for having finally explained the small periodic residuals from the Mount Wilson experiments.[35] The concluding portion of the paper Einstein received from Shankland's team is as follows:

The Mt. Wilson experiments as reinterpreted by the present analysis of Miller's original readings are now believed to contain no effect of the kind predicted by the ether theory and, within the limitations imposed by local disturbances, are entirely consistent with a null result at all epochs during a year.

We believe that the Mt. Wilson experiments of Dayton C. Miller are entirely consistent with the null results obtained in all other trials of the Michelson-Morley experiment.[36]

The above extract from the concluding summary in the draft paper illustrates the general conclusion of Shankland and his team, but more

---

[35] Interview with R. S. Shankland, professor of physics, Case Institute of Technology, July 14, 1960. See also Shankland, "Conversations with Albert Einstein," *Am. J. Phys.* 31:47–57, (January, 1963).

[36] R. S. Shankland et al., "A New Analysis of . . . Dayton C. Miller" [mimeographed paper dated July, 1954], p. 31.

cogently the version published a year later shows vividly the purported closing of a ring which had begun in 1880:

We believe that this discussion of the effect of temperature permits the following inference: Under the most favorable experimental circumstances the second harmonics in the Mt. Wilson data remain essentially constant in phase and amplitude through periods of several hours and are then associated with a constant temperature pattern in the observation hut. This, together with the statistical and mechanical analyses, forces us to conclude that the observed harmonics in the fringe displacements are not due to cosmic phenomenon (ether drift), nor to magnetostriction, nor to mechanical causes, but rather to temperature effects on the interferometer. These disturbances were much more severe at Mt. Wilson than those encountered by other observers in their repetitions of the Michelson-Morley experiment performed in laboratory rooms.[37]

It will be remembered that at the very beginning of Michelson's conception of a second-order aether-drift measurement he had been warned by Helmholtz in Berlin that the difficulty of keeping a constant temperature might be insuperable. Michelson had written: "I had quite a long conversation with Dr. Helmholtz concerning my proposed method for finding the motion of the earth relative to the ether, and he said he could see no objection to it, except the difficulty of keeping a constant temperature."[38] To say that we have come full circle by noting that the temperature problem was recognized as paramount at the beginning and at the end of the Michelson-Morley experiments is artistically satisfying but scientifically inaccurate. Reevaluations of both the experiment and the theory behind it continue into the specious present to reduce the null result.[39]

[37] Shankland et al., "A New Analysis," *Rev. Mod. Phys.* 27:178.

[38] A. A. Michelson to Simon Newcomb, November 22, 1880, from Berlin, in Miscellaneous Letters Received, vol. 3, Records of the Naval Observatory, NA, RG 78.

[39] See J. P. Cedarholm, C. F. Bland, B. L. Havens, and C. H. Townes, "New Experimental Test of Special Relativity," *Phys. Rev. Ltrs.* 1:342–343 (November 1, 1958). Recent experimental work on gravitation and General Relativity Theory has also renewed interest in the foundations of the Special Theory to some degree: see Robert H. Dicke, *The Theoretical Significance of Experimental Relativity*; and Louis Witten, ed., *Gravitation: An Introduction to Current Research*; J. P. Cedarholm and C. H. Townes, "A New Experimental Test of Special Relativity," *Nature* 184:1350–1351 (October 31, 1959); A. Zajac and H. Sadowski, S. Licht, "Real Fringes in the

Space and time and space-time, as abstract concepts about the universe in which we live, are not now, and may never be, *completely* understood. But if "like all science [the problem] of space must be classified as unfinished business,"[40] we may still say without qualification that the work of Michelson, Morley, and Miller in struggling with the ethereal aether of space advanced our understanding greatly. Their progress may have been inadvertent, and the aether may be a vacuous concept now, but *space* has just arrived in the form of a new age.[41] The faltering but persistent steps taken by Michelson, Morley, and Miller were a vital part of the thought and experience that have brought us to this age.

Sagnac and Michelson Interferometers," *Am. J. Phys.* 29:669–673 (October, 1961); Carl E. Ockert, "Speed of Light," *Am. J. Phys.* 36: 158–161 (February, 1968); K. C. Turner and H. A. Hill, "New Experimental Limit on Velocity-Dependent Interactions of Clocks and Distant Matter," *Phys. Rev.* 134:252–256 (1964); S. A. Bludman and M. A. Ruderman, "Possibility of the Speed of Sound Exceeding the Speed of Light in Ultradense Matter," *Phys. Rev.* 170:1176–1184 (June 25, 1968). George A. Articolo, "The Michelson-Morley Experiment and the Phase Shift upon a 90° Rotation," *Am. J. Phys.* 37:215–218 (February, 1969); D. K. Sen, *Fields and/or Particles*; and John David North, *The Measure of the Universe*. For accounts of sequelae to Michelson-Morley by Charles H. Townes and others, see especially T. S. Jaseja, A. Javan, J. Murray, and Townes, "Test of Special Relativity or of the Isotropy of Space by Use of Infrared Masers," *Physical Review* 133:1221–1225 (March 2, 1964); and E. K. Conklin, "Velocity of the Earth with Respect to the Cosmic Background Radiation," *Nature* 222:971–972 (June 7, 1969).

[40] Max Jammer, *Concepts of Space: The History of Theories of Space in Physics*, p. 196. See also John A. Wheeler, "Our Universe: The Known and the Unknown," *American Scholar* 37:248–274 (Spring, 1968); Nannielou H. Dieter and W. Miller Goss, "Recent Work on the Interstellar Medium," *Rev. Mod. Phys.* 38:256–297 (April, 1966); Hong-Yee Chiu and William F. Hoffmann, eds., *Gravitation and Relativity*, p. xxiii, and chapters by J. A. Wheeler, R. F. Marzke, and R. H. Dicke, especially pp. 40, 121; and V. V. Fedynskii, ed., *The Earth in the Universe*, p. 15.

[41] For a juxtaposed exposition of Michelson-Morley with an essay by Thomas Gold, "The First Five Years of Space Research," see S. T. Butler and H. Messel, eds., *The Universe of Time and Space: A Course of Selected Lectures in Astronomy, Cosmology, and Physics*. For the beginnings of the space age, see the bibliography in L. S. Swenson, Jr., James M. Grimwood, and Charles C. Alexander, *This New Ocean: A History of Project Mercury*.

APPENDIX A

Michelson's 1881 Paper

Art. XXI.—*The relative motion of the Earth and the Luminiferous ether;* by Albert A. Michelson, Master, U. S. Navy.

The undulatory theory of light assumes the existence of a medium called the ether, whose vibrations produce the phenomena of heat and light, and which is supposed to fill all space. According to Fresnel, the ether, which is enclosed in optical media, partakes of the motion of these media, to an extent depending on their indices of refraction. For air, this motion would be but a small fraction of that of the air itself and will be neglected.

Assuming then that the ether is at rest, the earth moving through it, the time required for light to pass from one point to another on the earth's surface, would depend on the direction in which it travels.

Let V be the velocity of light.

v = the speed of the earth with respect to the ether.

D = the distance between the two points.

d = the distance through which the earth moves, while light travels from one point to the other.

d_1 = the distance earth moves, while light passes in the opposite direction.

Suppose the direction of the line joining the two points to coincide with the direction of earth's motion, and let T = time required for light to pass from the one point to the other, and T_1 = time required for it to pass in the opposite direction. Further, let T_0 = time required to perform the journey if the earth were at rest.

Then $T = \dfrac{D+d}{V} = \dfrac{d}{v}$; and $T_1 = \dfrac{D-d}{V} = \dfrac{d_1}{v}$

From these relations we find $d = D\dfrac{v}{V-v}$ and $d_1 = D\dfrac{v}{V+v}$ whence $T = \dfrac{D}{V-v}$ and $T_1 = \dfrac{D}{V+v}$; $T-T_1 = 2T_0\dfrac{v}{V}$ *nearly,* and $v = V\dfrac{T-T_1}{2T_0}$.

If now it were possible to measure $T-T_1$ since V and T_0 are known, we could find v the velocity of the earth's motion through the ether.

In a letter, published in "Nature" shortly after his death, Clerk Maxwell pointed out that $T-T_1$ could be calculated by measuring the velocity of light by means of the eclipses of Jupiter's satellites at periods when that planet lay in different directions from earth; but that for this purpose the observations of these eclipses must greatly exceed in accuracy those which have thus far been obtained. In the same letter it was also stated that the reason why such measurements could not be made at the earth's surface was that we have thus far no method for measuring the velocity of light which does not involve the necessity of returning the light over its path, whereby it would lose nearly as much as was gained in going.

The difference depending on the square of the ratio of the two velocities, according to Maxwell, is far too small to measure.

The following is intended to show that, with a wave-length of yellow light as a standard, the quantity—if it exists—is easily measurable.

Using the same notation as before we have $T = \dfrac{D}{V-v}$ and $T_1 = \dfrac{D}{V+v}$. The whole time occupied therefore in going and returning $T + T_1 = 2D\dfrac{V}{V^2 - v^2}$. If, however, the light had traveled in a direction at right angles to the earth's motion it would be entirely unaffected and the time of going and returning would be, therefore, $2\dfrac{D}{V} = 2T_0$. The difference between the times $T + T_1$ and $2T_0$ is

$$2DV\left(\frac{1}{V^2 - v^2} - \frac{1}{V^2}\right) = \tau\,; \ \tau = 2DV\frac{v^2}{V^2(V^2 - v^2)}$$

or nearly $2T_0\dfrac{v^2}{V^2}$. In the time τ the light would travel a distance $V\tau = 2VT_0\dfrac{v^2}{V^2} = 2D\dfrac{v^2}{V^2}$.

That is, the actual distance the light travels in the first case is greater than in the second, by the quantity $2D\dfrac{v^2}{V^2}$.

Considering only the velocity of the earth in its orbit, the ratio $\dfrac{v}{V} = \dfrac{1}{10\,000}$ approximately, and $\dfrac{v^2}{V^2} = \dfrac{1}{100\,000\,000}$. If $D = 1200$ millimeters, or in wave-lengths of yellow light, 2 000 000, then in terms of the same unit, $2D\dfrac{v^2}{V^2} = \dfrac{4}{100}$.

If, therefore, an apparatus is so constructed as to permit two pencils of light, which have traveled over paths at right angles to each other, to interfere, the pencil which has traveled in the direction of the earth's motion, will in reality travel $\dfrac{4}{100}$ of a wave-length farther than it would have done, were the earth at rest. The other pencil being at right angles to the motion would not be affected.

If, now, the apparatus be revolved through 90° so that the second pencil is brought into the direction of the earth's motion, its path will have lengthened $\frac{4}{100}$ wave-lengths. The total change in the position of the interference bands would be $\frac{8}{100}$ of the distance between the bands, a quantity easily measurable.

The conditions for producing interference of two pencils of light which had traversed paths at right angles to each other were realized in the following simple manner.

Light from a lamp a, fig. 1, passed through the plane parallel glass plate b, part going to the mirror c, and part being

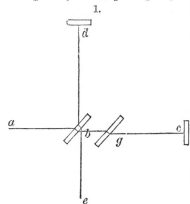

reflected to the mirror d. The mirrors c and d were of plane glass, and silvered on the front surface. From these the light was reflected to b, where the one was reflected and the other refracted, the two coinciding along be.

The distance bc being made equal to bd, and a plate of glass g being interposed in the path of the ray bc, to compensate for the thickness of the glass b, which is traversed by the ray bd, the two rays will have traveled over equal paths and are in condition to interfere.

The instrument is represented in plan by fig. 2, and in perspective by fig. 3. The same letters refer to the same parts in the two figures.

The source of light, a small lantern provided with a lens, the flame being in the focus, is represented at a. b and g are the two plane glasses, both being cut from the same piece; d and c are the silvered glass mirrors; m is a micrometer screw which moves the plate b in the direction bc. The telescope e, for observing the interference bands, is provided with a micrometer eyepiece. w is a counterpoise.

In the experiments the arms, bd, bc, were covered by long paper boxes, not represented in the figures, to guard against changes in temperature. They were supported at the outer ends by the pins k, l, and at the other by the circular plate o. The adjustments were effected as follows:

The mirrors c and d were moved up as close as possible to the plate b, and by means of the screw m the distances between a point on the surface of b and the two mirrors were made approximately equal by a pair of compasses. The lamp being lit, a small hole made in a screen placed before it served as a point of light; and the plate *b*, which was adjustable in two planes, was moved about till the two images of the point of light, which were reflected by the mirrors, coincided. Then a sodium flame placed at *a* produced at once the interference bands. These could then be altered in width, position, or direction, by a slight movement of the plate *b*, and when they were of convenient width and of maximum sharpness, the

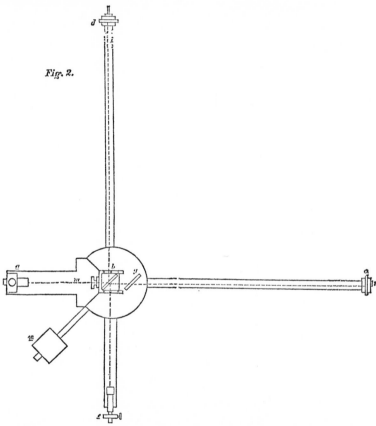

Fig. 2.

sodium flame was removed and the lamp again substituted. The screw *m* was then slowly turned till the bands reappeared. They were then of course colored, except the central band, which was nearly black. The observing telescope had to be focussed on the surface of the mirror *d*, where the fringes were most distinct. The whole apparatus, including the lamp and the telescope, was movable about a vertical axis.

It will be observed that this apparatus can very easily be

made to serve as an "interferential refractor," and has the two important advantages of small cost, and wide separation of the two pencils.

The apparatus as above described was constructed by Schmidt and Hænsch of Berlin. It was placed on a stone pier in the Physical Institute, Berlin. The first observation showed, however, that owing to the extreme sensitiveness of the instrument to vibrations, the work could not be carried on during the day. The experiment was next tried at night. When the mirrors were placed half-way on the arms the fringes were visible, but their position could not be measured till after twelve o'clock, and then only at intervals. When the mirrors were moved out to the ends of the arms, the fringes were only occasionally visible.

It thus appeared that the experiments could not be performed in Berlin, and the apparatus was accordingly removed

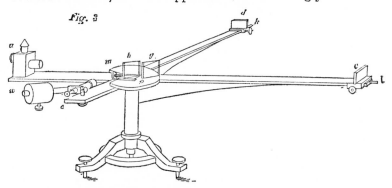

Fig. 3

to the *Astrophysicalisches Observatorium* in Potsdam. Even here the ordinary stone piers did not suffice, and the apparatus was again transferred, this time to a cellar whose circular walls formed the foundation for the pier of the equatorial.

Here, the fringes under ordinary circumstances were sufficiently quiet to measure, but so extraordinarily sensitive was the instrument that the stamping of the pavement, about 100 meters from the observatory, made the fringes disappear entirely!

If this was the case with the instrument constructed with a view to avoid sensitiveness, what may we not expect from one made as sensitive as possible!

At this time of the year, early in April, the earth's motion in its orbit coincides roughly in longitude with the estimated direction of the motion of the solar system—namely, toward the constellation Hercules. The direction of this motion is inclined at an angle of about +26° to the plane of the equator, and at this time of the year the tangent of the earth's motion in its orbit makes an angle of $-23\frac{1}{2}°$ with the plane of the equator; hence we may say the resultant would lie within 25° of the equator..

The nearer the two components are in magnitude to each other, the more nearly would their resultant coincide with the plane of the equator.

In this case, if the apparatus be so placed that the arms point north and east at noon, the arm pointing east would coincide with the resultant motion, and the other would be at right angles. Therefore, if at this time the apparatus be rotated 90°, the displacement of the fringes should be *twice* $\frac{8}{100}$ or 0·16 of the distance between the fringes.

If, on .the other hand, the proper motion of the sun is small compared to the earth's motion, the displacement should be $\frac{6}{10}$ of ·08 or 0·048. Taking the mean of these two numbers as the most probable, we may say that the displacement to be looked for is not far from one-tenth the distance between the fringes.

The principal difficulty which was to be feared in making these experiments, was that arising from changes of temperature of the two arms of the instrument. These being of brass whose coefficient of expansion is 0·000019 and having a length of about 1000 mm. or 1 700 000 wave-lengths, if one arm should have a temperature only one one-hundredth of a degree higher than the other, the fringes would thereby experience a displacement three times as great as that which would result from the rotation. On the other hand, since the changes of temperature are independent of the direction of the arms, if these changes were not too great their effect could be eliminated.

It was found, however, that the displacement on account of bending of the arms during rotation was so considerable that the instrument had to be returned to the maker, with instructions to make it revolve as easily as possible. It will be seen from the tables, that notwithstanding this precaution a large displacement was observed in one particular direction. That this was due entirely to the support was proved by turning the latter through 90°, when the direction in which the displacement appeared was also changed 90°.

On account of the sensitiveness of the instrument to vibration, the micrometer screw of the observing telescope could not be employed, and a scale ruled on glass was substituted. The distance between the fringes covered three scale divisions, and the position of the center of the dark fringe was estimated to fourths of a division, so that the separate estimates were correct to within $\frac{1}{12}$.

It frequently occurred that from some slight cause (among

others the springing of the tin lantern by heating) the fringes would suddenly change their position, in which case the series of observations was rejected and a new series begun.

In making the adjustment before the third series of observations, the direction in which the fringes moved, on moving the glass plate *b*, was reversed, so that the displacement in the third and fourth series are to be taken with the opposite sign.

At the end of each series the support was turned 90°, and the axis was carefully adjusted to the vertical by means of the foot-screws and a spirit level.

	N.	N.E.	E.	S.E.	S.	S.W.	W.	NW.	Remarks.
1st revolution	0·0	0·0	0·0	−8·0	−1·0	−1·0	−2·0	−3·0	Series 1, footscrew
2d "	16·0	16·0	16·0	9·0	16·0	16·0	15·0	13·0	marked B, toward
3d "	17·0	17·0	17·0	10·0	17·0	16·0	16·0	17·0	East.
4th "	15·0	15·0	15·0	8·0	14·5	14·5	14·5	14·0	
5th "	13·5	13·5	13·5	5·0	12·0	13·0	13·0	13·0	
	61·5	61·5	61·5	*x*	58·5	58·5	56·5	54·0	
	S. 58·5	W. 56·5			N.E 61·5		S.E. 60·0		
	120·0		118·0		120·0		114·0		
	118·0				114·0				
Excess,	+2·0				+6·0				
1st revolution	10·0	11·0	12·0	13·0	13·0	0·0	14·0	15·0	Series 2, B toward
2d "	16·0	16·0	16 0	17·0	17·0	2·0	17·0	17·0	South.
3d "	17·5	17·5	17·5	17·5	17·5	4·0	18·0	17·5	
4th "	17·5	17·5	17·0	17·0	17·0	4·0	17·0	17·0	
5th "	17·0	17·0	17·0	17 0	16·0	3·0	16·0	16·0	
	78·0	79·0	79·5	81·5	80·5	*x*	82·0	82·5	
	S. 80·5	W. 82·0			N.E 79·0		S.E. 81·5		
	158·5		161·5		160·0		164·0		
	161·5				164·0				
Excess,	−3·0				−4·0				
1st revolution	3·0	3·0	3·0	3·0	2·5	2·5	2·5	10·0	Series 3, B toward
2d "	18·0	17·5	17·5	18·0	18·5	19·0	19·5	26·0	West.
3d "	11·0	·11·0	13·0	12·0	13·0	13·5	13·5	21·0	
4th "	1·0	0·0	0·5	0·5	0·5	0·0	0·0	14·0	
5th "	4·0	4·0	5·0	5·0	5·0	5·5	5·5	16·0	
	37·0	35·5	39·0	38·5	39·5	40·5	71·0	*x*	
	S. 39·5	W. 41·0			N.E. 35·5		S E 38·5		
	76·5		80·0		76·0		79·5		
			76·5				76·0		
Excess,			+3·5				+3·5		
1st revolution	14·0	21·0	15·5	17·0	14·0	14·5	14·5	16·0	Series 4, B toward
2d "	10·0	20·0	12·0	12·0	13·0	13·0	13·0	13·5	North.
3d "	14·0	25·0	15·0	16·0	16·0	16·0	16·0	17·0	
4th "	18·0	27·0	18·5	18·5	18·5	19·0	20·0	21·0	
5th "	15·0	24·0	15·0	15·0	15·0	16·0	16·0	16·5	
	71·0	*x*	76·0	78·5	76·5	78·5	79·5	84·0	
	S. 76·5	W. 79·5			N.E. 73·5		S·E. 78·5		
	147·5		155·5		152·0		162·5		
			147·5				152·0		
Excess,			+8·0				+10·5		

The heading of the columns in the table gives the direction toward which the telescope pointed.

The footing of the erroneous column is marked x, and in the calculations the mean of the two adjacent footings is substituted.

The numbers in the columns are the positions of the center of the dark fringe in *twelfths* of the distance between the fringes.

In the first two series, when the footings of the columns N. and S. exceed those of columns E. and W., the excess is called positive. The excess of the footings of N.E., S.W., over those of N.W., S.E., are also called positive. In the third and fourth series this is reversed.

The numbers marked "excess" are the sums of ten observations. Dividing therefore by 10, to obtain the mean, and also by 12 (since the numbers are twelfths of the distance between the fringes), we find for

	N.S.	N.E., S.W.
Series 1	+0·017	+0·050
" 2	−0·025	−0·033
" 3	+0·030	+0·030
" 4	+0·067	+0·087
	4) 0·089	0·137
Mean =	+0·022	+0·034

The displacement is, therefore,

In favor of the columns N.S. .... +0·022
"   "   "   N.E., S.W. .... +0·034

The former is too small to be considered as showing a displacement due to the simple change in direction, and the latter should have been zero.

The numbers are simply outstanding errors of experiment. It is, in fact, to be seen from the footings of the columns, that the numbers increase (or decrease) with more or less regularity from left to right.

This gradual change, which should not in the least affect the periodic variation for which we are searching, would of itself necessitate an outstanding error, simply because the sum of the two columns farther to the left must be less (or greater) than the sum of those farther to the right.

This view is amply confirmed by the fact that where the excess is positive for the column N.S., it is also positive for N.E., S.W., and where negative, negative. If, therefore, we can eliminate this gradual change, we may expect a much smaller error. This is most readily accomplished as follows:

Adding together all the footings of the four series, the third and fourth with negative sign, we obtain

N.	N.E.	E.	S.E.	S.	S.W.	W.	N.W.
31·5	31·5	26·0	24·5	23·0	20·8	18·0	11·0

or dividing by 20×12 to obtain the means in terms of the distance between the fringes,

N.	N.E.	E.	S.E.	S.	S.W.	W.	N.W.
0·131	0·131	0·108	0·102	0·096	0·086	0·075	0·046

If x is the number of the column counting from the right and y the corresponding footing, then the method of least squares gives as the equation of the straight line which passes nearest the points x, y—

$$y = 9·25x + 64·5$$

If, now, we construct a curve with ordinates equal to the difference of the values of y found from the equation, and the actual value of y, it will represent the displacements observed, freed from the error in question.

These ordinates are:

N.	N.E.	E.	S.E.	S.	S.W.	W.	N.W.
—·002	—·011	+·003	—·001	—·004	—·003	—·001	+·018

N.	—·002	E.	+·003	N.E.	—·011	N.W.	+·018
S.	—·004	W.	—·001	S.W.	—·003	S.E.	—·001
Mean=	—·003		+·001	Mean=	—·007		+·008
	+·001				+·008		
Excess=	—·004			Excess=	—·015		

The small displacements —0·004 and —0·015 are simply errors of experiment.

The results obtained are, however, more strikingly shown by constructing the actual curve together with the curve that should have been found if the theory had been correct. This is shown in fig. 4.

4.

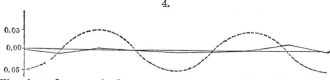

The dotted curve is drawn on the supposition that the displacement to be expected is one-tenth of the distance between the fringes, but if this displacement were only $\frac{1}{100}$, the broken line would still coincide more nearly with the straight line than with the curve.

The interpretation of these results is that there is no displacement of the interference bands. The result of the hypothesis of a stationary ether is thus shown to be incorrect, and the necessary conclusion follows that the hypothesis is erroneous.

This conclusion directly contradicts the explanation of the phenomenon of aberration which has been hitherto generally accepted, and which presupposes that the earth moves through the ether, the latter remaining at rest.

It may not be out of place to add an extract from an article published in the Philosophical Magazine by Stokes in 1846.

" All these results would follow immediately from the theory of aberration which I proposed in the July number of this magazine; nor have I been able to obtain any result admitting of being compared with experiment, which would be different according to which theory we adopted. This affords a curious instance of two· totally different theories running parallel to each other in the explanation of phenomena. I do not suppose that many would be disposed to maintain Fresnel's theory, when it is shown that it may be dispensed with, inasmuch as we would not be disposed to believe, without good evidence, that the ether moved quite freely through the solid mass of the earth. Still it would have been satisfactory, if it had been possible to have put the two theories to the test of some decisive experiment."

In conclusion, I take this opportunity to thank Mr. A. Graham Bell, who has provided ·the means for carrying out this work, and Professor Vogel, the Director of the *Astrophysicalisches Observatorium*, for his .courtesy in placing the resources of his laboratory at my disposal.

APPENDIX B
Michelson-Morley Paper of 1886

ART. XXXVI.—*Influence of Motion of the Medium on the Velocity of Light;* by ALBERT A. MICHELSON and EDWARD W. MORLEY.†

THE only work of any consequence, on the influence upon the velocity of light of the motion of the medium through which it passes, is the experiment of Fizeau. He announced the remarkable result that the increment of velocity which the light experienced was not equal to the velocity of the medium, but was a fraction x of this velocity which depended on the index of refraction of the medium. This result was previously obtained theoretically by Fresnel, but most satisfactorily demonstrated by Eisenlohr,‡ as follows:

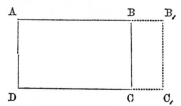

Consider the prism AC in motion relatively to the ether in direction AB with velocity θ. Suppose the density of the external ether to be 1 and of the ether within the prism, $1+\Delta$. In the time dt the prism will advance a distance $\theta dt =$ BB, At

---

† This research was carried on by the aid of the Bache Fund.
‡ Verdet. Conferences de Physique, ii, 687.

the beginning of this time the quantity of ether in the volume BC, (if S=surface of the base of the prism,) is $S\theta dt$. At the end of the time the quantity will be $S\theta dt(1+\varDelta)$. Hence in this time a quantity of ether has been introduced into this volume equal to $S\theta dt\varDelta$.

It is required to find what must be the velocity of the ether contained in the prism to give the same result. Let this velocity be $x\theta$. The quantity of ether (density$=1+\varDelta$) introduced will then be $Sa\theta dt(1+\varDelta)$ and this is to be the same as $S\theta dt\varDelta$, whence $x=\dfrac{\varDelta}{1+\varDelta}$. But the ratio of the velocity of light in the external ether to that within the prism is n, the index of refraction, and is equal to the inverse ratio of the square root of the densities, or $n=\sqrt{1+\varDelta}$ whence $x=\dfrac{n^2-1}{n^2}$ which is Fresnel's formula.*

* The following reasoning leads to nearly the same result; and though incomplete, may not be without interest, as it also gives a very simple explanation of the constancy of the specific refraction.

Let l be the mean distance light travels between two successive encounters with a molecule; then l is also the "mean free path" of the molecule. The time occupied in traversing this path is $t=\dfrac{a}{v_{\prime}}+\dfrac{b}{v}$, where a is the diameter of a molecule, and $b=l-a$, and $v_{\prime}$ is the velocity of light within the molecule, and v, the velocity in the free ether; or if $\mu=\dfrac{v}{v_{\prime}}$ then $t=\dfrac{\mu a+b}{v}$. In the ether the time would be $t_{\prime}=\dfrac{a+b}{v}$, hence

$$n=\dfrac{t}{t_{\prime}}=\dfrac{\mu a+b}{a+b}. \tag{1}$$

If now the ether remains fixed while the molecules are in motion, the mean distance traversed between encounters will no longer be $a+b$, but $a+a+b+\beta$; where a is the distance the first molecule moves while light is passing through it, and β is the distance the second one moves while light is moving between the two. If θ is the common velocity of the molecules then $d=\dfrac{-\theta}{v_{\prime}}a$, and $\beta=\dfrac{\theta}{v-\theta}b$.

The time occupied is therefore $\dfrac{a}{v_{\prime}}+\dfrac{b}{v-\theta}$ or $\dfrac{\mu a}{v}+\dfrac{b}{v-\theta}$. The distance traversed in this time is $a+b+\left(\dfrac{\mu a}{v}+\dfrac{b}{v-\theta}\right)\theta$; therefore the resulting velocity $v=\dfrac{a+b}{\dfrac{\mu a}{v}+\dfrac{b}{v-\theta}}+\theta$.

Substituting the value of $n=\dfrac{\mu a+b}{a+b}$ and neglecting the higher powers of $\dfrac{\theta}{v}$, this becomes

$$v=\dfrac{v}{n}+\left(1-\dfrac{1}{n^2}\dfrac{b}{a+b}\right)\theta. \tag{2}$$

But $\dfrac{v}{n}$ is the velocity of light in the stationary medium; the coefficient of θ is therefore the factor

$$x=\dfrac{n^2-1}{n^2}+\dfrac{1}{n^2}\dfrac{a}{a+b}. \tag{3}$$

It seems probable that this expression is more exact than Fresnel's; for when the particles of the moving medium are in actual contact, then the light must be accelerated by the full value of θ: that is the factor must be 1, whereas $\dfrac{n^2-1}{n^2}$ can

Fresnel's statement amounts then to saying that the ether within a moving body remains stationary with the exception of the portions which are condensed around the particles. If this condensed atmosphere be insisted upon, every particle with its atmosphere may be regarded as a single body, and then the statement is, simply, that the ether is entirely unaffected by the motion of the matter which it permeates.

It will be recalled that Fizeau* divided a pencil of light, issuing from a slit placed in the focus of a lens, into two parallel beams. These passed through two parallel tubes and then fell upon a second lens and were re-united at its focus where they fell upon a plane mirror. Here the rays crossed and were returned each through the other tube, and would again be brought to a focus by the first lens, on the slit, but for a plane parallel glass which reflected part of the light to a point where it could be examined by a lens.

At this point vertical interference fringes would be formed, the bright central fringe corresponding to equal paths. If now the medium is put in motion in opposite directions in the two tubes, and the velocity of light is affected by this motion, the two pencils will be affected in opposite ways, one being retarded and the other accelerated; hence the central fringe would be displaced and a simple calculation would show whether this displacement corresponds with the acceleration required by theory or not.

Notwithstanding the ingenuity displayed in this remarkable contrivance, which is apparently so admirably adopted for eliminating accidental displacement of the fringes by extraneous causes, there seems to be a general doubt concerning the results obtained, or at any rate the interpretation of these results given by Fizeau.

never be 1. The above expression, however gives this result when the particles are in contact—for then $b=0$ and $x=\dfrac{n^2-1}{n^2}+\dfrac{1}{n^2}=1$.

Resuming equation (1) and putting $a+b=l$ we find $(n-1)l=(\mu-1)a$. But for the same substance μ and a are probably constant or nearly so; hence $(n-1)l$ is constant.

But Clausius has shown that $l=\kappa\dfrac{\sigma}{\rho}a$, where κ is a constant, σ the density of the molecule; ρ, that of the substance; and a, the diameter of the "sphere of action." σ and a are probably nearly constant, hence we have finally $\dfrac{n-1}{\rho}=$ constant.

Curiously enough, there seems to be a tendency towards constancy in the product $(n-1)l$ for different substances. In the case of 25 gases and vapors whose index of refraction and "free path" are both known, the average difference from the mean value of $(n-1)l$ was less than 20 per cent. though the factors varied in the proportion of one to thirteen; and if from this list the last nine vapors (about which there is some uncertainty) are excluded, the average difference is reduced to 10 per cent.

* Ann. de Ch. et de Ph., III, lvii, p. 385, 1859.

This, together with the fundamental importance of the work must be our excuse for its repetition. It may be mentioned that we have tried to obtain formulated objections to these experiments but without success. The following are the only points which have occurred to us as being susceptible of improvement.

1st. The elimination of accidental displacement of the fringes by deformation of the glass ends of the tubes, or unsymmetrical variations of density of the liquid, etc., depends on the assumption that the two pencils have traveled over identical (not merely equivalent) paths. That this is not the case was proven by experiment; for when a piece of plate glass was placed in front of one of the pencils and slightly inclined, the fringes were displaced.

2d. The arrangement for producing the motion of the medium necessitated very rapid observation—for the maximum velocity lasted but an instant.

3d. The tubes being of necessity of small diameter and only their central portion being available (since the velocity diminishes rapidly toward the walls) involved considerable loss of light—which, having to pass through a slit was already faint.

4th. The maximum velocity (in the center of the tube) should be found in terms of the mean velocity. (Fizeau confessedly but guesses at this ratio.)

These are the suggestions which determined the form of apparatus adopted, a description of which follows:

The Refractometer.—After a number of trials, the following form was devised and proved very satisfactory. Light from a source at a (fig. 5) falls on a half silvered surface b, where it divides; one part following the path $b c d e f b g$ and the other the path $b f e d c b g$. This arrangement has the following advantages: 1st, it permits the use of an extended source of light, as a gas flame; 2d, it allows any distance between the tubes which may be desired; 3d, it was tried by a preliminary experiment, by placing an inclined plate of glass at h. The only effect was either to alter the width of the fringes, or to alter their inclination; *but in no case was the center of the central white fringe affected.* Even holding a lighted match in the path had no effect on this point.

The tubes containing the fluid were of brass, 28^{mm} internal diameter; and, in the first series of experiments, a little over 3 meters in length, and in the second series, a little more than 6 meters. The ends of these tubes were closed with plane parallel plates of glass which were not exactly at right angles but slightly inclined so as to reflect the light below the telescope, which would otherwise be superposed on that which passed through the tubes. The tubes were mounted on a wooden support entirely disconnected from the refractometer which was mounted on brick piers.

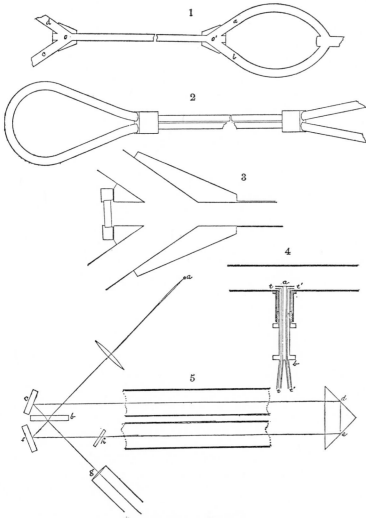

EXPLANATION OF FIGURES.

FIG. 1.—Vertical section through tubes. FIG. 2.—Plan of tubes. FIG. 3.— One end of tubes, showing glass plate inclined to axis. FIG. 4.—Gauge for velocity at different points. FIG. 5.—Plan of refractometer.

The flow of water was obtained by filling a tank four feet in diameter and three feet high, placed in the attic, about 23 meters above the apparatus, with which it was connected by a three inch pipe. The latter branched into two parts, and each branch again into two; the two pairs being joined each to one of the tubes. The branches were provided with large valves, by turning which the current was made to flow in either direction through the tubes and into a large tank, from which it was afterward pumped up to the upper tank again. The flow lasted about three minutes, which gave time for a number of observations, with the flow in alternating directions.

Method of observation.—In the first series of observations a single wire micrometer was used in the eyepiece of the observing telescope, but afterward a double wire micrometer was employed. The tubes being filled with distilled water, the light from an electric lamp was directed toward the central glass of the refractometer and the latter adjusted by screws till the light passed centrally down both tubes, and then the right angled prism at the further end adjusted till the light returned and was reflected into the telescope, where generally two images were observed. These were made to coincide, and the fringes at once appeared. They could then be altered in width or direction by the screws, till the best result was obtained. A slight motion of one of the mirrors produced an inclination of the fringes, and the horizontal wire of the micrometer was placed at the *portion of the fringes which remained fixed*, notwithstanding the movement of the mirror. This adjustment was frequently verified, and as long as it was true, no motion of the tubes or distortion of the glasses could have any effect on the measurements. During this adjustment it was found convenient to have a slow current of water, to avoid distortions on account of unequal density.

The signal being given the current was turned on, and the micrometer lines set, one on each of the two dark bands on either side of the central bright fringe, and the readings noted. The difference between them gave the width of the fringe, and their mean, the position of the center of the central white fringe. This being verified the signal was given to reverse the current; when the fringes were displaced, and the same measurements taken; and this was continued till the water was all out of the upper tank. Following is a specimen of one such set of observations.

No. 63.

Direction of current, Micrometer wire,	+		—	
	l.	*r.*	*l.*	*r.*
	11	34	80	93
	13	35	71	88
	10	40	73	90
	13	38	67	92
	14	40	65	89
	10	35	61	94
Means	11·8	37·0	69·5	91·0

Width of fringe_____	48·8	60·5
Mean width_____		54·6 + (3·0 = index error)
Displacement _____	57·7	46·0
Mean displacement _____		51·8

$$\varDelta = \frac{51\cdot8}{57\cdot6} = \cdot899.$$

(Long tube, vertical fringes, full current.)

Velocity of water.—The velocity of the water in the tubes was found by noting the time required to fill a measured volume in the tank, and multiplying by the ratio of areas of tank and tube. This gave the mean velocity. In order to find from this the maximum velocity in the axis of the tube the curve of velocities for different radii had to be determined. This was done as follows: a tight fitting piston *ab* (fig. 4) containing two small tubes *tt, t, t,* was introduced into the tube containing the water. The ends of the tubes were bent at right angles in opposite ways, so that when the water was in motion the pressure would be greater in one than in the other. The other ends of the small tubes were connected to a U tube containing mercury, the difference in level of which measured the pressure. The pressures were transformed into velocities by measuring the velocity corresponding to a number of pressures. Following is the table of results :—

Pressures.	Velocities.	$\dfrac{v}{\sqrt{p}}$
26	393	77·1
108	804	77·1
190	1060	76·9
240	1190	76·8

It is seen from the approximate constancy of the last column that within limits of error of reading, the square roots of the readings of the pressure gauge are proportional to the velocities.

To find the curve of velocities along a diameter of the tube, the piston was moved through measured distances, and the corresponding pressures noted. The diameter of the tube was about 28^{mm}, while that of the small tubes of the gauge was but 2^{mm}, so that the disturbance of the velocity by these latter was small except very close to the walls of the tube. The portion of the piston which projected into the tube was made as thin as possible, but its effect was quite noticeable in altering the symmetry of the curve.

In all, five sets of observations were taken, each with a different current. These being reduced to a common velocity all gave very concordant results, the mean being as follows: x = distance from the axis in terms of radius; v = corresponding velocity in terms of the maximum.

$x.$	$v.$
0·00	1·000
·20	·993
·40	·974
·60	·929
·80	·847
·90	·761
·95	·671
1·00	·000

The curve constructed with these numbers coincides almost perfectly with the curve

$$v=(1-x^2)^{.165}.$$

The total flow is therefore $2\pi\int_0^1 (1-x^2)^{.165} x\,dx = \dfrac{\pi}{1\cdot165}$. The area of the tube being π, the mean velocity $= \frac{1}{1\cdot165}$ of the maximum; or the maximum velocity is $1\cdot165$ times the mean. This, then, is the number by which the velocity, found by timing the flow, must be multiplied to give the actual velocity in the axis of the tube.

Formula.

Let l be the length of the part of the liquid column which is in motion.

$u =$ velocity of light in the stationary liquid.

$v =$ velocity of light in vacuo.

$\theta =$ velocity of the liquid in the axis of tube.

$\theta x =$ acceleration of the light.

The difference in the time required for the two pencils of light to pass through the liquid will be $\dfrac{l}{u-\theta x} - \dfrac{l}{u+\theta x} = \dfrac{2l\theta x}{u^2}$ very nearly. If $\varDelta$ is the double distance traveled in this time in air, in terms of λ, the wave-length, then

$$\varDelta = \frac{4l\theta n^2 x}{\lambda v} \text{ whence } x = \frac{\lambda v}{4ln^2\theta}\,\varDelta.$$

λ was taken as ·00057 cm.

$v =$ 30000000000 cm.

$n^2 =$ 1·78.

The length l was obtained as follows: The stream entered each tube by two tubes a, b (figs. 1, 2) and left by two similar ones d, c. The beginning of the column was taken as the intersection, o, of the axes of a and b, and the end, as the intersection, o', of the axes of d and c. Thus $l = oo'$. $\varDelta$ is found by observing the displacement of the fringes; since a displacement of one whole fringe corresponds to a difference of path of one whole wave-length.

Observations of the double displacement Δ.

1st Series. $l = 3{\cdot}022$ meters.

$\theta = 8{\cdot}72$ meters per second.

$\Delta =$ double displacement; $w =$ weight of observation.

Δ.	w.	Δ.	w.	Δ.	w.	Δ.	w.
.510	1·9	·521	0·9	·529	0·6	·515	2·5
·508	1·6	·515	·9	·474	2·0	·525	2·7
·504	1·7	·575	·6	·508	1·4	·480	·8
·473	1·4	·538	2·1	·531	·8	·493	10·6
·557	·4	·577	·6	·500	·5	·348	2·8
·425	·6	·464	1·7	·478	·6	·399	5·7
·560	2·8	·515	1·2	·499	1·0	·482	2·1
·544	·1	·460	·4	·558	·4	·472	2·0
·521	·1	·510	·5	·509	2·0	·490	·8
·575	·1	·504	·5	·470	2·1		

2d Series. $l = 6{\cdot}151, \theta = 7{\cdot}65.$

Δ.	w.	Δ.	w.	Δ.	w.	Δ.	w.
·789	4·9	·891	1·7	·909	1·0	·882	6·6
·780	·3·5	·883	2·5	·899	1·7	·908	5·9
·840	4·6	·852	11·1	·832	4·3	·965	2·0
·633	1·1	·863	1·5	·837	2·1	·967	3·3
·876	7·3	·843	1·1	·848	1·9		
·956	3·6	·820	3·4	·877	4·7		

3d Series. $l = 6{\cdot}151, \theta = 5{\cdot}67.$

Δ.	w.	Δ.	w.	Δ.	w.	Δ.	w.
·640	4·4	·626	11·9	·636	3·1	·619	6·5

If these results be reduced to what they would be if the tube were 10^{m} long and the velocity 1^{m} per second, they would be as follows:

Series.	Δ.
1	·1858
2	·1838
3	·1800

The final weighted value of Δ for all observations is $\Delta = {\cdot}1840$. From this, by substitution in the formula, we get

$$x = {\cdot}434 \text{ with a possible error of} \pm{\cdot}02.$$
$$\frac{n^2-1}{n^2} = {\cdot}437.$$

The experiment was also tried with air moving with a velocity of 25 meters per second. The displacement was about $\frac{1}{100}$ of a fringe; a quantity smaller than the probable error of observation. The value calculated from $\frac{n^2-1}{n^2}$ would be ·0036.

It is apparent that these results are the same for a long or short tube, or for great or moderate velocities. The result was also found to be unaffected by changing the azimuth of the fringes to 90°, 180° or 270°. It seems extremely improbable that this could be the case if there were any serious constant error due to distortions, etc.

The result of this work is therefore that the result announced by Fizeau is essentially correct; and that *the luminiferous ether is entirely unaffected by the motion of the matter which it permeates.*

APPENDIX C
Michelson-Morley Paper of 1887

THE

AMERICAN JOURNAL OF SCIENCE.

[THIRD SERIES.]

————•♦•————

ART. XXXVI.—*On the Relative Motion of the Earth and the Luminiferous Ether ;* by ALBERT A. MICHELSON and EDWARD W. MORLEY.*

THE discovery of the aberration of light was soon followed by an explanation according to the emission theory. The effect was attributed to a simple composition of the velocity of light with the velocity of the earth in its orbit. The difficulties in this apparently sufficient explanation were overlooked until after an explanation on the undulatory theory of light was proposed. This new explanation was at first almost as simple as the former. But it failed to account for the fact proved by experiment that the aberration was unchanged when observations were made with a telescope filled with water. For if the tangent of the angle of aberration is the ratio of the velocity of the earth to the velocity of light, then, since the latter velocity in water is three-fourths its velocity in a vacuum, the aberration observed with a water telescope should be four-thirds of its true value.†

---

* This research was carried out with the aid of the Bache Fund.

† It may be noticed that most writers admit the sufficiency of the explanation according to the emission theory of light; while in fact the difficulty is even greater than according to the undulatory theory. For on the emission theory the velocity of light must be greater in the water telescope, and therefore the angle of aberration should be less; hence, in order to reduce it to its true value, we must make the absurd hypothesis that the motion of the water in the telescope carries the ray of light in the opposite direction !

On the undulatory theory, according to Fresnel, first, the ether is supposed to be at rest except in the interior of transparent media, in which secondly, it is supposed to move with a velociy less than the velocity of the medium in the ratio $\dfrac{n^2-1}{n^2}$, where n is the index of refraction. These two hypotheses give a complete and satisfactory explanation of aberration. The second hypothesis, notwithstanding its seeming improbability, must be considered as fully proved, first, by the celebrated experiment of Fizeau,[*] and secondly, by the ample confirmation of our own work.[†] The experimental trial of the first hypothesis forms the subject of the present paper.

If the earth were a transparent body, it might perhaps be conceded, in view of the experiments just cited, that the inter-molecular ether was at rest in space, notwithstanding the motion of the earth in its orbit; but we have no right to extend the conclusion from these experiments to opaque bodies. But there can hardly be question that the ether can and does pass through metals. Lorentz cites the illustration of a metallic barometer tube. When the tube is inclined the ether in the space above the mercury is certainly forced out, for it is incompressible.[‡] But again we have no right to assume that it makes its escape with perfect freedom, and if there be any resistance, however slight, we certainly could not assume an opaque body such as the whole earth to offer free passage through its entire mass. But as Lorentz aptly remarks: "quoi qui'l en soit, on fera bien, à mon avis, de ne pas se laisser guider, dans une question aussi importante, par des considérations sur le degré de probabilité ou de simplicité de l'une ou de l'autre hypothèse, mais de s'addresser a l'expérience pour apprendre à connaitre l'état, de repos ou de mouvement, dans lequel se trouve l'éther à la surface terrestre."[§]

In April, 1881, a method was proposed and carried out for testing the question experimentally.[‖]

In deducing the formula for the quantity to be measured, the effect of the motion of the earth through the ether on the path of the ray at right angles to this motion was overlooked.[¶]

* Comptes Rendus, xxxiii, 349, 1851; Pogg. Ann. Ergänzungsband, iii, 457, 1853; Ann. Chim. Phys., III, lvii, 385, 1859.

† Influence of Motion of the Medium on the Velocity of Light. This Journal, III, xxxi, 377, 1886.

‡ It may be objected that it may escape by the space between the mercury and the walls; but this could be prevented by amalgamating the walls.

§ Archives Néerlandaises, xxi, 2ᵐᵉ livr.

‖ The relative motion of the earth and the luminiferous ether, by Albert A. Michelson, this Jour., III, xxii, 120.

¶ It may be mentioned here that the error was pointed out to the author of the former paper by M. A. Potier, of Paris, in the winter of 1881.

The discusssion of this oversight and of the entire experiment forms the subject of a very searching analysis by H. A. Lorentz,* who finds that this effect can by no means be disregarded. In consequence, the quantity to be measured had in fact but one-half the value supposed, and as it was already barely beyond the limits of errors of experiment, the conclusion drawn from the result of the experiment might well be questioned; since, however, the main portion of the theory remains un-questioned, it was decided to repeat the experiment with such modifications as would insure a theoretical result much too large to be masked by experimental errors. The theory of the method may be briefly stated as follows:

Let *sa*, fig. 1, be a ray of light which is partly reflected in *ab*, and partly transmitted in *ac*, being returned by the mirrors *b* and *c*, along *ba* and *ca*. *ba* is partly transmitted along *ad*,

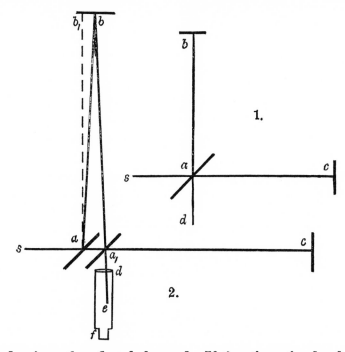

and *ca* is partly reflected along *ad*. If then the paths *ab* and *ac* are equal, the two rays interfere along *ad*. Suppose now, the ether being at rest, that the whole apparatus moves in the direction *sc*, with the velocity of the earth in its orbit, the direc-

* De l'Influence du Mouvement de la Terre sur les Phen. Lum. Archives Néerlandaises, xxi, 2ᵐᵉ livr., 1886.

tions and distances traversed by the rays will be altered thus :—
The ray sa is reflected along ab, fig. 2; the angle $bab_{,}$ being
equal to the aberration $=a$, is returned along $ba_{,,}$ $(aba_{,} =2a)$, and
goes to the focus of the telescope, whose direction is unaltered.
The transmitted ray goes along ac, is returned along $ca_{,,}$ and is
reflected at $a_{,,}$ making $ca_{,}e$ equal $90-a$, and therefore still coinciding with the first ray. It may be remarked that the rays $ba_{,}$
and $ca_{,,}$ do not now meet exactly in the same point $a_{,,}$ though
the difference is of the second order; this does not affect the
validity of the reasoning. Let it now be required to find the
difference in the two paths $aba_{,}$, and $aca_{,}$.

Let $V=$ velocity of light.

$v=$ velocity of the earth in its orbit.

$D=$ distance ab or ac, fig. 1.

$T=$ time light occupies to pass from a to c.

$T_{,}=$ time light occupies to return from c to $a_{,,}$ (fig. 2.)

Then $T=\dfrac{D}{V-v}$, $T_{,}=\dfrac{D}{V+v}$. The whole time of going and com-

ing is $T+T_{,}=2D\dfrac{V}{V^{2}-v^{2}}$, and the distance traveled in this time

is $2D\dfrac{V^{2}}{V^{2}-v^{2}}= 2D\left(1+\dfrac{v^{2}}{V^{2}}\right)$, neglecting terms of the fourth order.

The length of the other path is evidently $2D\sqrt{1+\dfrac{v^{2}}{V^{2}}}$, or to the

same degree of accuracy, $2D\left(1+\dfrac{v^{2}}{2V^{2}}\right)$. The difference is there-

fore $D\dfrac{v^{2}}{V^{2}}$. If now the whole apparatus be turned through $90°$,
the difference will be in the opposite direction, hence the displacement of the interference fringes should be $2D\dfrac{v^{2}}{V^{2}}$. Considering only the velocity of the earth in its orbit, this would
be $2D\times10^{-8}$. If, as was the case in the first experiment,
$D=2\times10^{6}$ waves of yellow light, the displacement to be
expected would be 0.04 of the distance between the interference
fringes.

In the first experiment one of the principal difficulties encountered was that of revolving the apparatus without producing distortion; and another was its extreme sensitiveness to
vibration. This was so great that it was impossible to see the
interference fringes except at brief intervals when working in
the city, even at two o'clock in the morning. Finally, as before remarked, the quantity to be observed, namely, a displacement of something less than a twentieth of the distance between the interference fringes may have been too small to be
detected when masked by experimental errors.

The first named difficulties were entirely overcome by mounting the apparatus on a massive stone floating on mercury; and the second by increasing, by repeated reflection, the path of the light to about ten times its former value.

The apparatus is represented in perspective in fig. 3, in plan in fig. 4, and in vertical section in fig. 5. The stone a (fig. 5) is about 1·5 meter square and 0·3 meter thick. It rests on an annular wooden float bb, 1·5 meter outside diameter, 0·7 meter inside diameter, and 0·25 meter thick. The float rests on mercury contained in the cast-iron trough cc, 1·5 centimeter thick, and of such dimensions as to leave a clearance of about one centimeter around the float. A pin d, guided by arms $gggg$, fits into a socket e attached to the float. The pin may be pushed into the socket or be withdrawn, by a lever pivoted at f. This pin keeps the float concentric with the trough, but does not bear any part of the weight of the stone. The annular iron trough rests on a bed of cement on a low brick pier built in the form of a hollow octagon.

3.

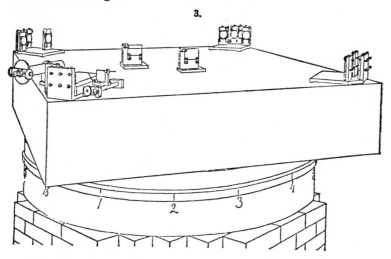

At each corner of the stone were placed four mirrors dd ee fig. 4. Near the center of the stone was a plane-parallel glass b. These were so disposed that light from an argand burner a, passing through a lens, fell on b so as to be in part reflected to $d_{,}$; the two pencils followed the paths indicated in the figure, $bdedbf$ and $bd_{,}e_{,}d_{,}bf$ respectively, and were observed by the telescope f. Both f and a revolved with the stone. The mirrors were of speculum metal carefully worked to optically plane surfaces five centimeters in diameter, and the glasses b and c were plane-parallel and of the same thickness, 1·25 centimeter;

their surfaces measured 5·0 by 7·5 centimeters. The second of
these was placed in the path of one of the pencils to compensate for the passage of the other through the same thickness of
glass. The whole of the optical portion of the apparatus was
kept covered with a wooden cover to prevent air currents and
rapid changes of temperature.

The adjustment was effected as follows: The mirrors having been adjusted by screws in the castings which held the

4.

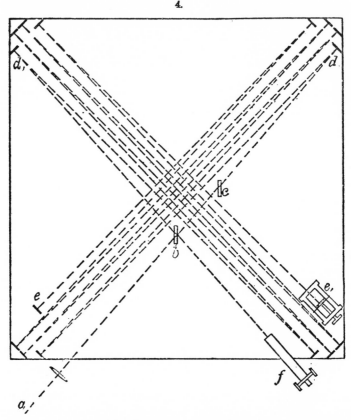

mirrors, against which they were pressed by springs, till light
from both pencils could be seen in the telescope, the lengths of
the two paths were measured by a light wooden rod reaching
diagonally from mirror to mirror, the distance being read from
a small steel scale to tenths of millimeters. The difference in
the lengths of the two paths was then annulled by moving the
mirror $e_{,}$. This mirror had three adjustments; it had an adjustment in altitude and one in azimuth, like all the other mirrors,

but finer; it also had an adjustment in the direction of the incident ray, sliding forward or backward, but keeping very accurately parallel to its former plane. The three adjustments of this mirror could be made with the wooden cover in position.

The paths being now approximately equal, the two images of the source of light or of some well-defined object placed in front of the condensing lens, were made to coincide, the telescope was now adjusted for distinct vision of the expected interference bands, and sodium light was substituted for white light, when the interference bands appeared. These were now made as clear as possible by adjusting the mirror e_i; then white light was restored, the screw altering the length of path was very slowly moved (one turn of a screw of one hundred threads to the inch altering the path nearly 1000 wave-lengths) till the colored interference fringes reappeared in white light. These were now given a convenient width and position, and the apparatus was ready for observation.

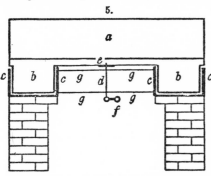

The observations were conducted as follows: Around the cast-iron trough were sixteen equidistant marks. The apparatus was revolved very slowly (one turn in six minutes) and after a few minutes the cross wire of the micrometer was set on the clearest of the interference fringes at the instant of passing one of the marks. The motion was so slow that this could be done readily and accurately. The reading of the screw-head on the micrometer was noted, and a very slight and gradual impulse was given to keep up the motion of the stone; on passing the second mark, the same process was repeated, and this was continued till the apparatus had completed six revolutions. It was found that by keeping the apparatus in slow uniform motion, the results were much more uniform and consistent than when the stone was brought to rest for every observation; for the effects of strains could be noted for at least half a minute after the stone came to rest, and during this time effects of change of temperature came into action.

The following tables give the means of the six readings; the first, for observations made near noon, the second, those near six o'clock in the evening. The readings are divisions of the screw-heads. The width of the fringes varied from 40 to 60 divisions, the mean value being near 50, so that one division

means 0·02 wave-length. The rotation in the observations at·
noon was contrary to, and in the evening observations, with,
that of the hands of a watch.

NOON OBSERVATIONS.

	16.	1.	2.	3.	4.	5.	6.	7.	8.	9.	10.	11.	12.	13.	14.	15.	16·
July 8.......	44·7	44·0	43·5	39·7	35·2	34·7	34·3	32·5	28·2	26·2	23·8	23·2	20·3	18·7	17·5	16·8	13·7
July 9.......	57·4	57·3	58·2	59·2	58·7	60·2	60·8	62·0	61·5	63·3	65·8	67·3	69·7	70·7	73·0	70·2	72·2
July 11..	27·3	23·5	22·0	19·3	19·2	19·3	18·7	18·8	16·2	14·3	13·3	12·8	13·3	12·3	10·2	7·3	6·5
Mean........	43·1	41·6	41·2	39·4	37·7	38·1	37·9	37·8	35·3	34·6	34·3	34·4	34·4	33·9	33·6	31·4	30·8
Mean in w. l.	·862	·832	·824	·788	·754	·762	·758	·756	·706	·692	·686	·688	·688	·678	·672	·628	·616
	·706	·692	·686	·688	·688	·678	·672	·628	·616								
Final mean..	·784	·762	·755	·738	·721	·720	·715	·692	·661								

P. M. OBSERVATIONS.

	16.	1.	2.	3.	4.	5.	6.	7.	8.	9.	10.	11.	12.	13.	14.	15.	16·
July 8.......	61·2	63·3	63·3	68·2	67·7	69·3	70·3	69·8	69·0	71·3	71·3	70·5	71·2	71·2	70·5	72·5	75·7
July 9.......	26·0	26·0	28·2	29·2	31·5	32·0	31·3	31·7	33·0	35·8	36·5	37·3	38·8	41·0	42·7	43·7	44·0
July 12......	66·8	66·5	66·0	64·3	62·2	61·0	61·3	59·7	58·2	55·7	53·7	54·7	55·0	58·2	58·5	57·0	56·0
Mean	51·3	51·9	52·5	53·9	53·8	54·1	54·3	53·7	53·4	54·3	53·8	54·2	55·0	56·8	57·2	57·7	58·6
Mean in w. l.	1·026	1·038	1·050	1·078	1·076	1·082	1·086	1·074	1·068	1·086	1·076	1·084	1·100	1·136	1·144	1·154	1·172
	1·068	1·086	1·076	1·084	1·100	1·136	1·144	1·154	1·172								
Final mean.	1·047	1·062	1·063	1·081	1·088	1·109	1·115	1·114	1·120								

The results of the observations are expressed graphically in
fig. 6. The upper is the curve for the observations at noon,
and the lower that for the evening observations. The dotted
curves represent *one-eighth* of the theoretical displacements. It
seems fair to conclude from the figure that if there is any dis·

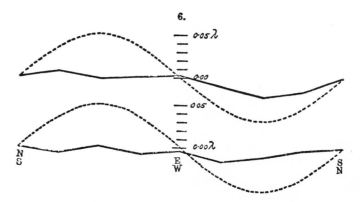

6.

placement due to the relative motion of the earth and the
luminiferous ether, this cannot be much greater than 0·01 of
the distance between the fringes.

Considering the motion of the earth in its orbit only, this

displacement should be $2D\frac{v^2}{V^2}=2D\times10^{-8}$. The distance D was about eleven meters, or 2×10^7 wave-lengths of yellow light; hence the displacement to be expected was 0·4 fringe. The actual displacement was certainly less than the twentieth part of this, and probably less than the fortieth part. But since the displacement is proportional to the square of the velocity, the relative velocity of the earth and the ether is probably less than one sixth the earth's orbital velocity, and certainly less than one-fourth.

In what precedes, only the orbital motion of the earth is considered. If this is combined with the motion of the solar system, concerning which but little is known with certainty, the result would have to be modified; and it is just possible that the resultant velocity at the time of the observations was small though the chances are much against it. The experiment will therefore be repeated at intervals of three months, and thus all uncertainty will be avoided.

It appears, from all that precedes, reasonably certain that if there be any relative motion between the earth and the luminiferous ether, it must be small; quite small enough entirely to refute Fresnel's explanation of aberration. Stokes has given a theory of aberration which assumes the ether at the earth's surface to be at rest with regard to the latter, and only requires in addition that the relative velocity have a potential; but Lorentz shows that these conditions are incompatible. Lorentz then proposes a modification which combines some ideas of Stokes and Fresnel, and assumes the existence of a potential, together with Fresnel's coefficient. If now it were legitimate to conclude from the present work that the ether is at rest with regard to the earth's surface, according to Lorentz there could not be a velocity potential, and his own theory also fails.

Supplement.

It is obvious from what has gone before that it would be hopeless to attempt to solve the question of the motion of the solar system by observations of optical phenomena *at the surface of the earth*. But it is not impossible that at even moderate distances above the level of the sea, at the top of an isolated mountain peak, for instance, the relative motion might be perceptible in an apparatus like that used in these experiments. Perhaps if the experiment should ever be tried in these circumstances, the cover should be of glass, or should be removed.

It may be worth while to notice another method for multiplying the square of the aberration sufficiently to bring it within the range of observation, which has presented itself during the preparation of this paper. This is founded on the fact that re-flection from surfaces in motion varies from the ordinary laws of reflection.

Let ab (fig. 1) be a plane wave falling on the mirror mn at an incidence of 45°. If the mirror is at rest, the wave front after reflection will be ac.

Now suppose the mirror to move in a direction which makes an angle a with its normal, with a velocity ω. Let V be the velocity of light in the ether supposed stationary, and let cd be the increase in the distance the light has to travel to reach d. In this time the mirror will have moved a distance $\dfrac{cd}{\sqrt{2}\cos a}$.

We have $\dfrac{cd}{ad} = \dfrac{\omega\sqrt{2}\cos a}{V}$ which put $= r$, and $\dfrac{ac}{ad} = 1 - r$.

In order to find the new wave front, draw the arc fg with b as a center and ad as radius; the tangent to this arc from d will be the new wave front, and the normal to the tangent from b will be the new direction. This will differ from the direction ba by the angle θ which it is required to find. From the equality of the triangles adb and edb it follows that $\theta = 2\varphi$, $ab = ac$,

$$\tan adb = \tan\left(45° - \frac{\theta}{2}\right) = \frac{1 - \tan\dfrac{\theta}{2}}{1 + \tan\dfrac{\theta}{2}} = \frac{ac}{ad} = 1 - r,$$

or neglecting terms of the order r^3,

$$\theta = r + \frac{r^2}{2} = \frac{\sqrt{2}\omega\cos a}{V} + \frac{\omega^2}{V^2}\cos^2 a.$$

Now let the light fall on a parallel mirror facing the first, we should then have $\theta_{,} = \dfrac{-\sqrt{2}\omega\cos a}{V} + \dfrac{\omega^2}{V^2}\cos^2 a$, and the total de-viation would be $\theta + \theta_{,} = 2\rho^2\cdot\cos^2 a$ where ρ is the angle of aberration, if only the orbital motion of the earth is considered. The maximum displacement obtained by revolving the whole apparatus through 90° would be $\varDelta = 2\rho^2 = 0.004''$. With fifty such couples the displacement would be $0.2''$. But astronomical observations in circumstances far less favorable than those in which these may be taken have been made to hundredths of a second; so that this new method bids fair to be at least as sensitive as the former.

The arrangement of apparatus might be as in fig. 2; s in the focus of the lens a, is a slit; bb cc are two glass mirrors optically plane and so silvered as to allow say one-twentieth of the light to pass through, and reflecting say ninety per cent. The intensity of the light falling on the observing telescope df

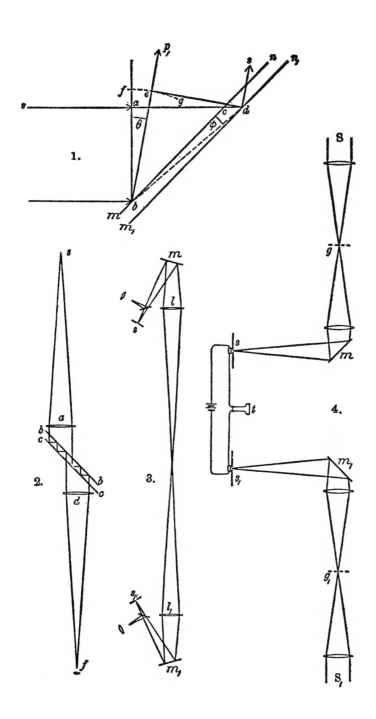

would be about one-millionth of the original intensity, so that if sunlight or the electric arc were used it could still be readily seen. The mirrors $bb_{,}$ and $cc_{,}$ would differ from parallelism sufficiently to separate the successive images. Finally, the apparatus need not be mounted so as to revolve, as the earth's rotation would be sufficient.

If it were possible to measure with sufficient accuracy the velocity of light without returning the ray to its starting point, the problem of measuring the first power of the relative velocity of the earth with respect to the ether would be solved. This may not be as hopeless as might appear at first sight, since the difficulties are entirely mechanical and may possibly be surmounted in the course of time.

For example, suppose (fig. 3) m and $m_{,}$ two mirrors revolving with equal velocity in opposite directions. It is evident that light from s will form a stationary image at $s_{,}$ and similarly light from $s_{,}$ will form a stationary image at s. If now the velocity of the mirrors be increased sufficiently, their phases still being exactly the same, both images will be deflected from s and $s_{,}$ in inverse proportion to the velocities of light in the two directions; or, if the two deflections are made equal, and the difference of phase of the mirrors be simultaneously measured, this will evidently be proportional to the difference of velocity in the two directions. The only real difficulty lies in this measurement. The following is perhaps a possible solution: $gg_{,}$ (fig. 4) are two gratings on which sunlight is concentrated. These are placed so that after falling on the revolving mirrors m and $m_{,,}$ the light forms images of the gratings at s and $s_{,}$ two very sensitive selenium cells in circuit with a battery and a telephone. If everything be symmetrical, the sound in the telephone will be a maximum. If now one of the slits s be displaced through half the distance between the image of the grating bars, there will be silence. Suppose now that the two deflections having been made exactly equal, the slit is adjusted for silence. Then if the experiment be repeated when the earth's rotation has turned the whole apparatus through 180°, and the deflections are again made equal, there will no longer be silence, and the angular distance through which s must be moved to restore silence will measure the required difference in phase.

There remain three other methods, all astronomical, for attacking the problem of the motion of the solar system through space.

1. The telescopic observation of the proper motions of the stars. This has given us a highly probably determination of the direction of this motion, but only a guess as to its amount.

2. The spectroscopic observation of the motion of stars in the line of sight. This could furnish data for the relative

motions only, though it seems likely that by the immense improvements in the photography of stellar spectra, the information thus obtained will be far more accurate than any other.

3. Finally there remains the determination of the velocity of light by observations of the eclipses of Jupiter's satellites. If the improved photometric methods practiced at the Harvard observatory make it possible to observe these with sufficient accuracy, the difference in the results found for the velocity of light when Jupiter is nearest to and farthest from the line of motion will give, not merely the motion of the solar system with reference to the stars, but with reference to the luminiferous ether itself.

NOTE ON SOURCES

This history of the Michelson-Morley aether-drift experiments may seem disturbing to those who have so long assumed so much from the reputation that Michelson's experiment acquired after 1887. On the other hand, a few scholars and specialists concerned with the origins of relativity theory and twentieth-century physics will recognize immediately many interpretive problems in this study that have been oversimplified, partly because of the narrative form and focus. Synthesis has forced the reduction of analysis in this work, and the author must leave to others the broader task of reconstruction of the historical record where need be.

The history of science, as critical narrative, has barely started to examine the era of our grandfathers, 1880–1930, covered roughly by the period of this essay. Although the literature relating to the Einsteinian revolution is immense, little of it can be considered serious history, very little is devoted to experimental physics, and still less considers expressly the experimental background to relativity theory. "Internalist" historiography, written by scientists turned historians, rightly dominates the recent past, but "external-ist" historiography, written by historians who have turned their attention toward science, should soon complement that dominance as the need for cultural integration increases. There are countless pedagogical, theoretical, and philosophical descriptions and discussions of Michelson-Morley, but most attempts to recount the aether-drift tests have shown such eagerness to get to Einstein and relativity that they have seriously mismanaged Michelson and the concept of an aethereal substratum. My purpose in this bibliographic essay is to chart a course through the sources that others may follow and so contribute to the social history of science that we may see science itself as a dynamic human endeavor rather than as a static mountain of accumulated knowledge.

Books and articles by persons listed in the preface are too numerous to list exhaustively here, but the most inspirational studies for me would have to include George Sarton, *The History of Science and the New Humanism*; James B. Conant, ed., *Harvard Case Histories in Experimental Science*; Gerald Holton and Duane H. D. Roller, *Foundations of Modern Physical Science*; Herbert Butterfield, *The Origins of Modern Science*; I. Bernard Cohen, *Franklin and Newton*; Charles C. Gillispie, *The Edge of Objectivity*; Thomas S. Kuhn, *The Structure of Scientific Revolutions*; Bernard Jaffe, *Men of Science in America*; Michael Polanyi, *Personal Knowledge*; Karl R. Popper, *The Logic of Scientific Discovery*; Mary B. Hesse, *Forces and Fields*; Adolf Grünbaum, *Philosophical Problems of Space and Time*; and, most recently, Stanley L. Jaki, *The Relevance of Physics*.

Other, secondary works, less inspirational but indispensable in leading back toward the primary sources, would have to include the following dozen: Peter G. Bergmann, *Introduction to the Theory of Relativity*; Max Born and Emil Wolf, *Principles of Optics*; René Dugas, *A History of Mechanics*; Philipp Frank, *Philosophy of Science*; Max Jammer, *Concepts of Space*; W. K. H. Panofsky and M. Phillips, *Classical Electricity and Magnetism*; Wolfgang Pauli, *Theory of Relativity*; Hans Reichenbach, *Philosophy of Space and Time*; Paul A. Schilpp, ed., *Albert Einstein: Philosopher-Scientist*; J. L. Synge, *Relativity: The Special Theory*; Sir Edmund Whittaker, *A History of the Theories of Aether and Electricity*; and Hermann Weyl, *Space, Time, Matter*. In addition to these books, two articles by Robert S. Shankland, both entitled "The Michelson-Morley Experiment," have recently reinforced the success-story interpretation: see *American Journal of Physics* 32: 16–35 (January, 1964), and *Scientific American* 211: 107–114 (November, 1964).

Perhaps the best starting point for bibliographic purposes as well as for the history of Michelson's experiment is still Dayton C. Miller's own major paper on this aspect of his career: "The Ether-Drift Experiment and the Determination of the Absolute Motion of the Earth," *Reviews of Modern Physics* 5:203–241 (July, 1933). The warning need hardly be given that this singular paper is to be read critically in the light of something like Peter G. Bergmann's "Fifty Years of Relativity," *Science* 123:487–494 March 23, 1956), or André Mercier's article with the same title in *Nature* 175:919–921 (May 28, 1955). Certainly Whittaker's aforementioned classic *History . . . of Aether and Electricity* slights Einstein and Michelson-Morley alike, but its opulent citations make it a basic guide. Also valuable is the 1911 bibliography compiled by Max Iklé and appended to his German translation of Michelson's *Light Waves and Their Uses*.

Research into manuscript and primary documentary materials on Michelson and the aether-drift experiments leads to the discovery of diffuse resources, none of which is adequate alone. Mrs. Dorothy Michelson Livingston of New York, A. A. Michelson's youngest daughter, has been collecting materials for a biography of her father for over a decade now and has accumulated originals or facsimile copies of much of Michelson's professional and personal correspondence. Gathering of family memorabilia had to begin anew upon Michelson's death in 1931 after the dispersion of what scant papers existed. I have been privileged to see Mrs. Livingston's collection, and I commend her efforts to publish her "Portrait" of her father.

In the Mojave Desert, at China Lake, California, the primary research building at the U.S. Naval Weapons Center has been named the "Michelson Laboratory" and a portion of its space has been dedicated to a museum honoring America's first Nobel Prize winner in science. William B. Plum, who began this collection (with the help of T. J. O'Donnell, long-time superintendent of the physical sciences development shops at the University of Chicago), has described the China Lake collection in "The Michelson Museum," *American Journal of Physics* 22:177–181 (April, 1954). D. T. McAllister, the present curator, has been a generous guide to me on several visits.

Also in southern California, the administrative offices, library, and optical shops of the Mount Wilson Observatory, operated under the aegis of the Carnegie Institute of Washington and located at 813 Santa Barbara Street, Pasadena, California, hold the bulk of personal papers of the directors and many associates of this, the world's leading astronomical center in the first half of the twentieth century. Mount Wilson was the site for much of the story told here in the last three chapters. Conversations with several members of the staff, past and present, as well as access to the directors' correspondence with Michelson and Miller and to the fine collection of journal literature were most helpful in this study.

In and near the nation's capital are several other caches of Michelson materials. The Morley Papers, in the Manuscript Division of the Library of Congress, were collected and deposited there by H. R. Williams after his biographical research. Together with the Simon Newcomb Papers, also in the Library of Congress, and the Records of the Naval Observatory, especially Record Group 78, in the National Archives, these manuscript records have been particularly valuable. The Naval Academy at Annapolis, along with the Office of Naval History at the Department of the Navy, under Admiral Ernest M. Eller, have also shown a renewed interest in Michelson materials.

The Smithsonian Institution and the Museum of History and Technology likewise have acquired, outright or on loan, materials and artifacts for Michelson exhibits from time to time. I am indebted to Nathan Reingold, formerly science bibliographer at the Library of Congress, for my first introduction to the scholarly resources around Washington, D. C.

In New England there are several other sources of unpublished materials. The Harvard University Archives contain a few letters from Michelson in the Pickering Papers related to the Harvard Observatory; the Boston Athenaeum contains the older records of the American Academy of Arts and Sciences; and the Air Force Cambridge Research Laboratories, at Hanscom Field, Bedford, Massachusetts, contain the Rayleigh Archives in care of John Howard. At Worcester, Massachusetts, the library of Clark University has several items relating to Michelson's short residence there in the 1890's. At Williams College in Williamstown are a few materials relating Morley to his alma mater. Yale University's Sterling Library holds the J. Willard Gibbs Papers, including six letters from Michelson between 1884 and 1886. Of Michelson's correspondence during this period about a dozen items have been preserved in the Henry A. Rowland Papers at Johns Hopkins University Library.

Case Western Reserve University in Cleveland still houses a few of the remains from, as well as the primary site for, the works of Michelson, Morley, and Miller. Before the merger of Case Institute of Technology and Western Reserve University in 1967, a Casiana Collection, consisting in large part of reprints of faculty publications, was in the care of H. R. Young. Miller's office files and laboratory records are in the personal custody of Robert S. Shankland. I have not had access to these Miller papers, but the materials at Mount Wilson and at the Niels Bohr Library of the American Institute of Physics in New York seem adequate to cover the thrust of Miller's work in the 1920's. To Dr. Charles Weiner, director of the Center for the History and Philosophy of Physics, and to Joan Warnow, librarian at the Bohr Library, all students of modern science can be grateful.

The last major repository of Michelson materials and unpublished documents is the manuscript division of the University of Chicago Library where there are some thirty-three items, consisting mostly of letters from Michelson to the president of the university. Most of these relate merely to routine academic protocol but about one-third of them cast some light on Michelson's work in progress from 1895 through 1930.

The most important source for any research into the history of modern science must be the original journal articles by which the work of individual scientists is made available to their fellow specialists. The principal journals consulted in the course of this study have been those in which the major contributions of the principal actors appeared. Beginning in the 1870's with the *American Journal of Science* (long known colloquially as *Silliman's Journal*) and the *Philosophical Magazine* of Britain, and continuing through the specialization process that produced the *Physical Review* and the *Astrophysical Journal* toward the turn of the century, the gist of the early aether-drift tests may be found in these pages. Prior to Michelson's researches, the main line of nineteenth-century optical advances can be traced through the journal of the French Academy of Sciences, *Comptes Rendus*; and well before Einstein the prime focus of continental physics shifted to the pages of *Annalen der Physik* (once known as *Poggendorf's Annalen* and as *Drude's Annalen*). In the twentieth century, as professional societies and their journals proliferate, relevant papers become more widely scattered, but this study has concentrated on the *American Journal of Physics* (formerly the *American Physics Teacher*) and on *Reviews of Modern Physics*, in addition to the eclectic journals, *Science* and *Nature*.

History of science has accepted too seriously and perhaps too often the dictum that history is biography writ large, but the biographical approach to the aether-drift experiments is virtually unavoidable. Michelson, Morley, and Miller carried the burden of their enigma for almost a half century before it was truly completed, or repeated by others. As experimentalists, their work was manifest more in action than in writings; theoreticians, who naturally write more books, therefore gained the audiences, and the few books and articles by the aether-drift experimenters themselves have been generally neglected as primary sources.

Michelson produced two small books, *Light Waves and Their Uses* and *Studies in Optics*, and seventy-eight papers, during his lifetime. All are listed in Harvey B. Lemon's bio-bibliography, "Albert Abraham Michelson: The Man and the Man of Science," *American Journal of Physics* 4:1–11 (February, 1936).

Morley wrote no books, but produced forty-eight papers as sole author, and a total of sixty-four papers, including those with collaborators. H. R. Williams's biography, *Edward Williams Morley: His Influence on Science in America*, contains an appendix that gives full particulars.

Miller was the most prolific member of the trio in publishing, having produced six books and ninety-eight papers before his death in 1941. Two of his books were done in the thirties, well after the aether-drift contest, and they are still highly regarded for their pedagogic value: *Anecdotal History of the Science of Sound* and *Sparks, Lightning, and Cosmic Rays: An Anecdotal History of Electricity.* Shankland's bio-bibliography carries a complete checklist: "Dayton Clarence Miller: Physics across Fifty Years," *American Journal of Physics* 9:281–288 (October, 1941).

Besides obituaries and a few memorial sketches, Michelson has had two book-length biographies published about him, Morley one, and Miller none. Bernard Jaffe has expanded an earlier sketch into a short biography entitled *Michelson and the Speed of Light.* Although fairly reliable factually, Jaffe's interpretation of the aether-drift tests accepts uncritically the linear progression to relativity thesis. The other biography of Michelson, by John H. Wilson, Jr., belongs to the genre of juvenilia and not the best of that. Williams's biography of Morley, already noted, is shallow but comprehensive.

With biography we slip back into the category of secondary source materials. The traditional taxonomy for the historian's craft, dividing all sources into primary or secondary according to their proximity to firsthand testimony, suffers much strain in a study of this sort where experimental science is the focus of interest and theoretical science is the arena of impact. Therefore, in the bibliographical listing following acknowledgments to personal informants, I have sacrificed evaluation of testimony in order to facilitate information retrieval.

Personal interviews—conversations and correspondence with a number of people who were more or less closely associated with the principals of this drama—have been of inestimable value. Conversations with Mrs. Dorothy Michelson Livingston, Mrs. Julius Pearson, and Miss Alma Pearson have helped me gain some warm human tints for my images of Michelson and the two brothers who were his technical assistants, Julius and Fred Pearson. Talks with Drs. Max Mason, John A. Anderson, Edison Pettit, Gustaf Strömberg, Walter T. Whitney, Robert S. Shankland, Alfred B. Focke, Edwin C. Kemble, Noel Little, C. J. Humphreys, and Mr. Joseph O. Hickox have furnished a number of personal reminiscences and personal estimates from men who had firsthand contact with Michelson, Morley, or Miller in their professional capacities, either as teachers or as experimenters. Likewise, through talks and correspondence with Horace W. Babcock, William V. Houston, Philip M. Morse, William B. Plum, Thomas J. O'Donnell, Harold A. Wilson, and Joseph H. Purdy, I have enjoyed some further indirect contact with the subjects of this study.

The list of references that follows is neither a complete catalogue of sources used in this study nor a set of suggested readings. Rather it represents selections from the contextual literature that should prove helpful for further studies in the social and intellectual history of physical science since 1850. Section I contains the most important reports of aether-drift experiments, arranged chronologically from 1880 to 1930, with a few more recent ones for good measure. Section II is a selected list of primary and secondary books relevant in some way to the aether, to Michelson's experiment, to Einstein's theory, or to the history of modern science. Section III is an alphabetical listing by author of published articles, and IV lists a few scholarly theses that seem most significant to a historical understanding of the experimental origins of relativity theory.

BIBLIOGRAPHY

I. Aether-Drift Experiments, Primary Reports: *Chronological List*

1881. Michelson, Albert A. "The Relative Motion of the Earth and the Luminiferous Ether." *American Journal of Science,* 3d ser. 22:120–129 (August).

1886. Michelson, Albert A., and Edward W. Morley. "Influence of Motion of the Medium on the Velocity of Light." *American Journal of Science,* 3d ser. 31:377–386 (May).

1887. Michelson, Albert A., and Edward W. Morley. "On the Relative Motion of the Earth and the Luminiferous Ether." *American Journal of Science,* 3d ser. 34:333–345 (November). Also published almost in parallel as "On the Relative Motion of the Earth and the Luminiferous Aether." *Philosophical Magazine,* 5th ser. 24:449–463 (December).

1893. Lodge, Oliver J. "Aberration Problems. A Discussion concerning the Motion of the Ether near the Earth and concerning the Connexion between Ether and Gross Matter; with Some New Experiments." *Philosophical Transactions of the Royal Society* 184: 727–804.

1897. Michelson, Albert A. "The Relative Motion of the Earth and the Ether." *American Journal of Science,* 4th ser. 3:475–478 (June).

1898. Morley, Edward W., and Dayton C. Miller. "On the Velocity of Light in a Magnetic Field." *Physical Review* 7:283–295 (December).

1902. Rayleigh, Lord. "Does Motion through the Aether Cause Double Refraction?" *Philosophical Magazine,* 6th ser. 4: 678–683 (December).

1903. Trouton, F. T., and H. R. Noble. "The Mechanical Forces Acting on a Charged Electric Condenser Moving through Space." *Philosophical Transactions of the Royal Society* 202:165–181.

1904. Brace, D. B. "On Double Refraction in Matter Moving through the Aether." *Philosophical Magazine,* 6th ser. 7:317–329 (April).

1905. Morley, Edward W., and Dayton C. Miller. "Report of an Experiment to Detect the FitzGerald-Lorentz Effect." *Proceedings of the American Academy of Arts and Sciences* 41:321–328 (August).

1907. Morley, Edward W., and Dayton C. Miller. "Final Report on Ether-Drift Experiments." *Science* 25:525 (April 5).

1913. Sagnac, M.G. "L'Éther lumineux démontré par l'effet du vent relatif d'éther dans interféromètre en rotation uniforme." *Comptes Rendus* 157:708–710 (October 27).

1913. Sagnac, M. G. "Sur la preuve de la réalité de l'éther lumineux par l'expérience de l'interférographe tournant." *Comptes Rendus* 157: 1410–1413 (December 22).

1915. Zeeman, Pieter. "Optical Investigation of Ether-Drift." *Nature* 96: 430–431 (December 16).

1918. Majorana, Quirino. "On the Second Postulate of the Theory of Relativity: Experimental Demonstration of the Constancy of Velocity of Light Reflected from a Moving Mirror." *Philosophical Magazine,* 6th ser. 35:163–174 (February).

1919. Majorana, Quirino. "Experimental Demonstration of the Constancy of Velocity of the Light emitted by a Moving Source." *Philosophical Magazine,* 6th ser. 37:145–149 (February).

1919. Righi, Augusto. "L'Expérience de Michelson et son interprétation." *Comptes Rendus* 168:837–842 (April 28).

1920. Righi, Augusto. "Sur les bases expérimentales de la Théorie de la Relativité." *Comptes Rendus* 170:497–501 (March 1); 170:1550–1554 (June 28).

1922. Miller, Dayton C. "Ether-Drift Experiments at Mount Wilson Solar Observatory." *Physical Review* 19:407–408 (April).

1924. Tomaschek, Rudolf. "Über das Verhalten des Lichtes ausserirdischer Lichtquellen." *Annalen der Physik* 73:105–126.

1925. Michelson, Albert A. (Part II with Henry G. Gale and Fred Pearson.) "The Effect of the Earth's Rotation on The Velocity of Light." *Astrophysical Journal* 61:137–145 (April).

1925. Miller, Dayton C. "Ether-Drift Experiments at Mount Wilson." *Proceedings of National Academy of Sciences* 11:306–314 (June).

1926. Miller, Dayton C. "Significance of the Ether-Drift Experiments of 1925 at Mt. Wilson." *Science* 63:433–443 (April 30).

1926. Kennedy, Roy J. "A Refinement of the Michelson-Morley Experiment." *Proceedings of National Academy of Sciences* 12:621–629.

1926. Piccard, A., and E. Stahel. "L'Expérience de Michelson, realisée en ballon libre." *Comptes Rendus* 183:420–421 (August 17).

1927. Piccard, A., and E. Stahel. "Sur le vent d'éther." *Comptes Rendus* 184:152 (January 17).

1927. Piccard, A., and E. Stahel. "L'Absence du vent d'éther au Rigi." *Comptes Rendus* 185:1198–1200 (November 28).

1927. Illingworth, K. K. "A Repetition of the Michelson-Morley Experiment Using Kennedy's Refinement." *Physical Review* 30:692–696 (November).

1928. Michelson, Albert A., and H. A. Lorentz et al. "Conference on the Michelson-Morley Experiment." *Astrophysical Journal* 68:341–402 (December).

1929. Michelson, Albert A., F. G. Pease, and F. Pearson. "Repetition of the Michelson-Morley Experiment." *Journal of the Optical Society of America* 18:181–182 (March).

1930. Pease, F. G. "Ether Drift Data." *Publication of the Astronomical Society of the Pacific* 42:197–202 (August).

1930. Joos, Georg. "Die Jenaer Wiederholung des Michelsonversuchs." *Annalen der Physik*, 5th ser. 7:385–407.

1932. Kennedy, Roy J., and Edward M. Thorndyke. "Experimental Establishment of the Relativity of Time." *Physical Review* 42:400–418 (November).

1933. Miller, Dayton C. "The Ether-Drift Experiment and the Determination of the Absolute Motion of the Earth." *Reviews of Modern Physics* 5:203–234 (July).

1938. Ives, Herbert E., and G. R. Stilwell. "An Experimental Study of the Rate of a Moving Atomic Clock." *Journal of the Optical Society of America* 28:215–226 (July).

1952. Ditchburn, R. W., and O. S. Heavens. "Relativistic Theory of a Rigid Body." Letter to the Editor, *Nature* 170:705 (October 25).

1955. Shankland, R. S., S. W. McCuskey, F. C. Leone, and G. Kuerti. "A New Analysis of the Interferometer Observations of Dayton C. Miller." *Reviews of Modern Physics* 27:167–178 (April).

1955. Essen, L. "A New Aether-Drift Experiment." *Nature* 175:793–794 (May 7).

1958. Cedarholm, J. P., G. F. Bland, B. L. Havens, and C. H. Townes. "New Experimental Test of Special Relativity." *Physical Review Letters* 1:342–343 (November 1).

1959. Cedarholm, J. P., and C. H. Townes. "A New Experimental Test of Special Relativity." *Nature* 184:1350–1351 (October 31).

1964. Turner, K. C., and H. A. Hill. "New Experimental Limit on Velocity-Dependent Interactions of Clocks and Distant Matter." *Physical Review* 134-B:252–256 (April 13).

II. BOOKS

Abraham, Max. *Theorie der Elektrizität.* 5th ed. Leipzig: Teubner, 1912. And see A. Föppl.

Aharoni, J. *The Special Theory of Relativity.* Oxford: Clarendon, 1959.

Airy, George Biddell. *On The Undulatory Theory of Optics.* London: Macmillan, 1866. (First published 1831.)

Anderson, David L. *The Discovery of the Electron: The Development of the Atomic Concept of Electricity.* Princeton: Van Nostrand, 1964.

Anderson, James L. *Principles of Relativity Physics.* New York: Academic Press, 1967.

Andrade, E. D. da C. *An Approach to Modern Physics.* Garden City: Doubleday, Anchor Books, 1957.

Andrews, C. F. *Optics of the Electromagnetic Spectrum.* Englewood Cliffs: Prentice-Hall, 1960.

Asher, Harry. *Experiments in Seeing.* New York: Basic Books, 1961.

Asimov, Isaac. *Biographical Encyclopedia of Science and Technology.* Garden City: Doubleday, 1964.

Atkinson, E., ed. and trans. See Ganot, Adolphe.

Badash, Lawrence, ed. *Rutherford and Boltwood: Letters on Radioactivity.* New Haven: Yale University Press, 1969

Baker, Robert H. *Astronomy: A Textbook for University and College Students.* 7th ed. Princeton: Van Nostrand, 1959. (First published 1930.)

Ballif, Jae R., and William E. Dibble. *Conceptual Physics: Matter in Motion.* New York: Wiley, 1969.

Barber, Bernard, and Walter Hirsch, eds. *The Sociology of Science.* New York: Free Press of Glencoe, 1962.

Barlow, William. *New Theories of Matter and Force.* London: Sampson Low, 1885.

Barnett, Lincoln. *The Universe and Dr. Einstein.* New York: Mentor Books, 1952. (First published New York: Harper, 1948.)

Barter, E. G. *Relativity and Reality: A Re-Interpretation of Anomalies Ap-*

pearing in the Theories of Relativity. New York: Philosophical Library, 1953.

Barus, Carl. *Interferometer Experiments in Acoustics and Gravitation.* 3 parts. Washington, D.C.: Carnegie Institution Publication No. 310, 1921–1925.

Basalla, George, William Coleman, and Robert H. Kargon, eds. *Victorian Science: A Self-Portrait from the Presidential Addresses of the British Association for the Advancement of Science.* Garden City: Doubleday, 1970.

Bates, D. R., ed. *Space Research and Exploration.* New York: William Sloane, 1958.

Bauer, Edmond. *L'Électromagnétisme: Hier et aujourd'hui.* Paris: Albin Michel, 1949.

Bavink, Bernhard. *The Anatomy of Modern Science: An Introduction to the Scientific Philosophy of Today.* Translated by H. S. Hatfield. London: G. Bell, 1932.

Beauregard, O. Costa de. *Précis of Special Relativity.* Translated by Banesh Hoffman. New York: Academic Press, 1966.

Bedini, Silvio. *Early Scientific Instruments and Their Makers.* Washington, D.C.: Smithsonian Institution, 1964.

Bell, Arthur Ernest. *Christian Huygens and the Development of Science in the Seventeenth Century.* London: Arnold, 1947.

Bentley, Arthur F. *Relativity in Man and Society.* New York: Putnam, 1926.

Bergmann, Peter Gabriel. *Introduction to the Theory of Relativity.* Englewood Cliffs: Prentice-Hall, 1942.

Bergson, Henri. *Duration and Simultaneity.* Translated by Leon Jacobson. New York: Bobbs-Merrill, 1965.

Bernal, J. D. *Science and Industry in the Nineteenth Century.* London: Routledge & Kegan Paul, 1953.

———. *Science in History.* 2d ed. London: Watts, 1957. (First published 1954.)

Bernstein, Aaron. *Naturwissenschaftliche Volksbücher* [People's Books on Natural Science]. Berlin: Gustav Hempel, 1880.

Berry, Arthur. *A Short History of Astronomy: From Earliest Times through the Nineteenth Century.* London: John Murray, 1898. New York: Dover, 1961.

Bickley, W. G., and R. E. Gibson. *Via Vector to Tensor: An Introduction to the Concepts and Techniques of the Vector and Tensor Calculus.* London: English Universities, 1962. New York: Wiley, 1962.

Bird, James Malcom, ed. *Einstein's Theories of Relativity and Gravitation.* New York: Scientific American, 1921.

Birkhoff, George David. *The Origin, Nature, and Influence of Relativity.* New York: Macmillan, 1925.

Blake, Ralph M., Curt J. Ducasse, and Edward H. Madden. *Theories of Scientific Method: The Renaissance through the Nineteenth Century.* Seattle: University of Washington Press, 1960.

Blanco, V. M., and S. W. McCuskey. *Basic Physics of the Solar System.* Reading, Mass.: Addison-Wesley, 1961.

Blin-Stoyle, R. J., et al. *Turning Points in Physics: A Series of Lectures Given at Oxford University in Trinity Term 1958.* New York: Harper & Row, Torchbooks, 1961. (First published Amsterdam: North Holland, 1959.)

Blondlot, R. *"N" Rays: A Collection of Papers Communicated to the Academy of Sciences.* Translated by J. Garcin with added notes. London: Longmans, Green, 1905.

Bohm, David. *The Special Theory of Relativity.* New York: Benjamin, 1965.

Bohr, Niels. *Atomic Physics and Human Knowledge.* New York: Wiley, 1958.

————. *On the Constitution of Atoms and Molecules: Papers of 1913 Reprinted from the* Philosophical Magazine *with an Introduction by L. Rosenfeld.* Copenhagen: Munksgaard, 1963.

Bondi, Hermann, *Cosmology.* Cambridge: University Press, 1952.

————. *Relativity and Common Sense: A New Approach to Einstein.* Garden City: Doubleday, Anchor Books, 1962.

Boni, Nell, Monique Russ, and Dan H. Laurence, comps. *A Bibliographical Checklist and Index to the Published Writings of Albert Einstein.* Paterson, N.J.: Pageant Books, 1960.

Bonner, William. *The Mystery of the Expanding Universe.* New York: Macmillan, 1964.

Boorse, Henry A., and Lloyd Motz, eds. *The World of the Atom.* 2 vols. New York: Basic Books, 1966.

Borel, Émile. *L'Espace et le temps.* Paris: Presses Universitaires de France, 1923. Édition Définitive, 1949.

Born, Max. *Physics in My Generation.* New York: Springer Verlag, 1969.

————. *Einstein's Theory of Relativity.* Revised edition prepared in collaboration with G. Leibfried and W. Biem. New York: Dover, 1962. (First published London: Methuen, 1924.)

————. *Experiment and Theory in Physics.* New York: Dover, 1956.

————. *My Life and My Views.* New York: Scribner, 1968.

————, and Emil Wolf. *Principles of Optics: Electromagnetic Theory of Propagation, Interference and Diffraction of Light.* London: Pergamon, 1959.

Bothezat, George de. *Back to Newton: A Challenge to Einstein's Theory of Relativity.* New York: G. E. Stechert, 1936.

Bourbon, B. *L'Éther: Essai de synthèse des théories de la physique moderne.* Paris: Dumond, 1948.

Boyer, Carl B. *A History of Mathematics.* New York: Wiley, 1968.

————. *The Rainbow: From Myth to Mathematics.* New York: Yoseloff, 1959.

Bradbury, Savile, and G. l'E. Turner, eds. *Historical Aspects of Microscopy. Papers Read at a Conference Held by the Royal Microscopical Society, Oxford, March 18, 1966.* Cambridge: Heffner, 1967.

Bragg, Sir William H. *Electrons & Ether Waves.* London: Oxford University Press, 1921.

————. *The Universe of Light.* London: G. Bell, 1947.

Braithwaite, Richard Bevan. *Scientific Explanation: A Study of the Function of Theory, Probability, and Law in Science.* New York: Harper & Row, Torchbooks, 1960. (First published 1953.)

Brandt, John C., and Paul W. Hodge. *Solar System Astrophysics.* New York: McGraw-Hill, 1964.

Brashear, J. A. *The Autobiography of a Man Who Loved the Stars.* Edited by W. L. Scaife. New York: American Society of Mechanical Engineers, 1924.

Brewster, David. *A Treatise on Optics.* London: Longmans, Green, 1831.

Bridgman, P. W. *A Sophisticate's Primer of Relativity.* Middletown, Conn.: Wesleyan University Press, 1962.

Bright, Laurence. *Whitehead's Philosophy of Physics.* New York: Sheed & Ward, 1958.

Broad, C. D. *Scientific Thought.* New York: Harcourt, Brace, 1923.

————. *The Mind and Its Place in Nature.* New York: Harcourt, Brace, 1925.

Brock, W. H., ed. *The Atomic Debates: Brodie and the Rejection of the Atomic Theory: Three Studies.* Leicester: University Press, 1967.

Broglie, Louis V. de. *Physics and Microphysics.* Translated by Martin Davidson. New York: Harper & Row, Torchbooks, 1960. (First published 1955, New York: Pantheon.)

————. *The Current Interpretation of Wave Mechanics: A Critical Study.* Amsterdam: Elsevier, 1964.

————. *The Revolution in Physics: A Non-Mathematical Survey of Quanta.* Translated by R. W. Niemeyer. New York: Noonday, 1953.

Brush, Stephen G. *Kinetic Theory.* 2 vols. I: *The Nature of Gases and Heat.* II: *Irreversible Processes.* Oxford: Pergamon, 1965–1966.

Buckley, Arabella B. *A Short History of Natural Science and of the Progress of Discovery from the Time of the Greeks to the Present Day, for the Use of Schools and Young Persons.* New York: Appleton, 1876.

Buckley, H. *A Short History of Physics.* New York: Van Nostrand, 1928.

Burlingame, Roger. *Engines of Democracy: Inventions and Society in Mature America.* New York: Scribner, 1940.

Burtt, Edwin Arthur. *The Metaphysical Foundations of Modern Physical Science.* Garden City: Doubleday, Anchor Books, 1955. (First published London, 1924; revised ed., 1932.)

Butler, S. T., and H. Messel, eds. *The Universe of Time and Space: A Course of Selected Lectures in Astronomy, Cosmology, and Physics.* Oxford: Pergamon, 1963; New York: Macmillan, 1963.

Butterfield, Herbert. *The Origins of Modern Science.* London: G. Bell, 1949; New York: Macmillan, 1960.

————, et al. *A Short History of Science: Origins and Results of the Scientific Revolution: A Symposium.* Garden City: Doubleday, Anchor Books, 1959.

Byrn, Edward W. *The Progress of Invention in the Nineteenth Century.* New York: Munn, 1900.

Cahn, William. *Einstein: A Pictorial Biography.* New York: Citadel, 1955.

Cajori, Florian. *A History of Physics: In its Elementary Branches Including the Evolution of Physical Laboratories.* Rev. ed. New York: Macmillan, 1929. (First published 1899.)

Campbell, Norman R. *Foundations of Science: The Philosophy of Theory and Experiment* [formerly titled: *Physics: The Elements*]. New York: Dover, 1957. (First published Cambridge: University Press, 1919[?].)

————."Relativity." Supplementary chapter 16 to his *Modern Electrical Theory.* Cambridge: University Press, 1923.

————. *What Is Science?* New York: Dover, 1952. (First published 1921.)

Campbell, William Wallace. *Stellar Motions: With Special Reference to Motions Determined by Means of the Spectrograph.* New Haven: Yale University Press, 1913.

Candler, C. *Modern Interferometers.* Glasgow: Hilger & Watts, 1951.

Čapek, Milič. *The Philosophical Impact of Contemporary Physics.* Princeton: Van Nostrand, 1961.

Carmichael, R. D. *The Theory of Relativity: Mathematical Monograph 12.* New York: Wiley, 1913.

————, et al. *A Debate on the Theory of Relativity.* Chicago: Open Court, 1927.

Carr, H. Wildon. *The General Principle of Relativity in Its Philosophical and Historical Aspect.* 2d ed. London: Macmillan, 1922. (First published 1920.)

Carus, Paul. *The Principle of Relativity in the Light of the Philosophy of Science.* Chicago: Open Court, 1913.

Cassirer, Ernst. *Determinism and Indeterminism in Modern Physics: Historical and Systematic Studies of the Problem of Causality.* New Haven: Yale University Press, 1956.

————. *Substance and Function* and *Einstein's Theory of Relativity.* Both books bound as one. Translated by W. C. and M. C. Swabey. New York: Dover, 1953. [Authorized translation; Einstein critical reader.] (First published Chicago: Open Court, 1923.)

————. *The Problem of Knowledge: Philosophy, Science, and History since Hegel.* Translated by W. H. Woglon and C. W. Hendel. New Haven: Yale University Press, 1950.

Chalmers, Thomas W. *Historic Researches: Chapters in the History of Physical and Chemical Discovery.* London: Morgan, 1949.

Chase, Carl Trueblood. *A History of Experimental Physics.* New York: Van Nostrand, 1932.

————. *The Evolution of Modern Physics.* New York: Van Nostrand, 1947.

Chiu, Hong-Yee, and William F. Hoffmann, eds. *Gravitation and Relativity.* New York: Benjamin, 1964.

Clagett, Marshall, ed. *Critical Problems in the History of Science: Proceedings of the Institute for the History of Science at the University of Wisconsin, September 1–11, 1957.* Madison: University of Wisconsin Press, 1959.

Clark, Ronald W. *Einstein: The Life and Times.* New York: World, 1971.

Clerke, Agnes M. *A Popular History of Astronomy During the Nineteenth Century.* 2d ed. Edinburgh: A. & C. Black, 1887. (First published 1885.)

Cloud, John Wills. *Castles in the Ether.* Rev. ed. New York: Charles Francis, 1928.

Cochrane, Rexmond G. *Measures for Progress: A History of the National*

Bureau of Standards. Washington: Government Printing Office, 1966.

Cohen, I. Bernard. *Franklin and Newton: An Inquiry into Speculative Newtonian Experimental Science and Franklin's Work in Electricity as an Example Thereof*. Philadelphia: American Philosophical Society, 1956.

————. *Roemer and the First Determination of the Velocity of Light*. New York: Burndy Library, 1944.

Cohen, Morris R. *Reason and Nature: An Essay on the Meaning of Scientific Method*. Glencoe, Ill.: Free Press, 1959. (First published 1931.)

Coleman, James A. *Relativity for the Layman: A Simplified Account of the History, Theory, and Proofs of Relativity*. New York: Signet Science Books, 1954.

Collier, Katherine B. *Cosmogonies of Our Fathers: Some Theories of the 17th and 18 Centuries*. New York: Columbia University Press, 1934.

Combridge, J. T., comp., *Bibliography of Relativity and Gravitation Theory: 1921–1937*. London: King's College [offprint], 1965.

Conant, James Bryant. *On Understanding Science: An Historical Approach*. New York: Mentor Books, 1951.

————, ed. *Harvard Case Histories in Experimental Science*. 2 vols. Cambridge, Mass.: Harvard University Press, 1957.

Conn, George K. T., and H. D. Turner. *The Evolution of the Nuclear Atom*. London: Iliffe, 1965; vol. I, New York: American Elsevier, 1966.

Cooper, Herbert J. *Scientific Instruments*. Brooklyn: Chemical Publishing Co., 1946.

Cooper, Lane. *Aristotle, Galileo and the Tower of Pisa*. Ithaca: Cornell University Press, 1935.

Couderc, Paul. *The Expansion of the Universe*. Translated by J. B. Sidgwick. New York: Macmillan, 1952.

Crew, Henry. *The Rise of Modern Physics: A Popular Sketch*. Baltimore: Williams and Wilkins, 1928.

————, ed. *The Wave Theory of Light: Memoirs by Huygens, Young, and Fresnel*. New York: American Book, 1900.

Crombie, A. C. *Robert Grosseteste and the Origins of Experimental Science*. Oxford: Clarendon, 1953.

————, ed. *Scientific Change*. New York: Basic Books, 1963.

Crowe, Michael J. *A History of Vector Analysis: The Evolution of the Idea of a Vectorial System*. Notre Dame, Ind.: University of Notre Dame Press, 1967.

Cullwick, E. G. *Electromagnetism and Relativity: With Particular Reference to Moving Media and Electromagnetic Induction.* 2d ed. London: Longmans, Green, 1959. (First published 1957.)

Cunningham, E. *Relativity and the Electron Theory.* Monographs on Physics Series, edited by J. J. Thomson and F. Horton. London: Longmans, Green, 1915.

————. *The Principle of Relativity.* Cambridge: University Press, 1914.

d'Abro, A. *The Decline of Mechanism (In Modern Physics).* New York: Van Nostrand, 1939.

————. *The Evolution of Scientific Thought: From Newton to Einstein.* 2d ed. New York: Dover, 1950. (First published 1927.)

Dampier, William Cecil. *A History of Science And its Relations with Philosophy and Religion.* 4th ed. Cambridge: University Press, 1966. (First published 1929.)

————, and Margaret Dampier, eds. *Readings in the Literature of Science: Being Extracts from the Writings of Men of Science to Illustrate the Development of Scientific Thought.* New York: Harper & Row, Torchbooks, 1959.

Dampier-Whetham, William Cecil. *The Recent Development of Physical Science.* 3d ed. Philadelphia: Blakiston, 1906.

Dana, Edward S., et al. *A Century of Science in America*: With Special Reference to the American Journal of Science, 1818–1918. New Haven: Yale University Press, 1918.

Danto, Arthur, and Sidney Morgenbesser, eds. *Philosophy of Science.* New York: Meridian Books, 1960.

Dantzig, Tobias. *Henri Poincaré: Critic of Crisis.* New York: Scribner, 1954.

Darrow, Floyd L. *The New World of Physical Discovery.* New York: Blue Ribbon Books, 1930.

Darrow, Karl K. *The Renaissance of Physics.* New York: Macmillan, 1936.

Dewey, John. *The Quest for Certainty: A Study of the Relation of Knowledge and Action.* (Gifford Lectures, 1929.) New York: Putnam, Capricorn Books, 1960. (First published 1929, Putnam.)

Dibner, Bern. *Ten Founding Fathers of Electrical Science.* Norwalk, Conn.: Burndy Library, 1954.

————. *The New Rays of Professor Roentgen.* Norwalk, Conn.: Burndy Library, 1963.

Dickie, Robert H. *The Theoretical Significance of Experimental Relativity.* New York: Gordon & Breach, 1964.

Dicks, D. R. *Early Greek Astronomy: To Aristotle.* Ithaca, New York: Cornell University Press, 1970.

Dingle, Herbert. *The Scientific Adventure: Essays in the History and Philosophy of Science.* London: Pitman, 1952.

————. *The Special Theory of Relativity.* 3rd ed. London: Methuen, 1950. (First published 1940.)

————, ed. *A Century of Science 1851–1951.* London: Hutchinson, 1951.

Dirac, P. A. M. *Lectures on Quantum Field Theory.* New York: Academic Press, 1966.

Dircks, Henry. *Perpetuum Mobile: Or Search for Self-Motive Power, during the 17th, 18th, and 19th Centuries.* London: Spon, 1861.

Ditchburn, R. W. *Light.* New York: Interscience, 1953.

Dolbear, A. E. *Matter, Ether, and Motion.* 2d ed. Boston: Lee & Shepard, 1894.

————. *Modes of Motion, or Mechanical Conceptions of Physical Phenomena.* Boston: Lee & Shepard, 1897.

Doppler, Christian. *Abhandlungen* (1842–1847). In *Ostwald's Klassiker Der Exakten Wissenschaften.* No. 161. Leipzig: Engelman, 1907.

Drake, Stillman, ed. *The Controversy on the Comets of 1618: Galileo Galilei, Horatio Grassi, Mario Guiducci, and Johann Kepler.* Philadelphia: University of Pennsylvania Press, 1960. Translated with C. D. O'Malley.

Driesch, Hans. *Relativitätstheorie und Weltanschauung.* Leipzig: Quelle & Meyer, 1930. (First published 1924.)

Drude, Paul. *Physik des Aethers auf electromagnetischer Grundlage.* Stuttgart: F. Enke, 1894.

————. *Theory of Optics.* Translated by C. R. Mann and R. A. Millikan. New York: Longmans, Green, 1901.

Dugas, René. *A History of Mechanics.* Translated by J. R. Maddox. Neuchâtel: Griffen edition, 1955.

Duhem, Pierre. *The Aim and Structure of Physical Theory.* Translated by P. P. Weiner. New York: Atheneum, 1962. (First published 1906.)

Dupree, A. Hunter. *Science in the Federal Government: A History of Policies and Activities to 1940.* Cambridge, Mass.: Harvard University Press, Belknap Press, 1957.

Durell, Clement V. *Readable Relativity.* New York: Harper & Row, Torchbooks, 1960. (First published 1926: London.)

Eddington, Arthur S. *The Combination of Relativity Theory and Quantum Theory.* Dublin: Institute for Advanced Studies, 1943.

————. *The Mathematical Theory of Relativity.* Cambridge: University Press, 1923.

————. *Space, Time and Gravitation: An Outline of the General Relativity Theory.* New York: Harper & Row, Torchbooks, 1959. (First published 1920.)

————. *Stellar Movements and the Structure of the Universe.* London: Macmillan, 1914.

Ehrenfest, Paul. *Collected Scientific Papers.* Edited by Martin J. Klein. Amsterdam: North Holland, 1959.

Einstein, Albert. *Äther und Relativitätstheorie.* Berlin: Julius Springer Verlag, 1920.

————. *Ideas and Opinions, by Albert Einstein.* Translated by Sonja Bargmann. New York: Crown, 1954.

————. *Investigations on the Theory of Brownian Movement.* Edited by R. Furth. Translated by A. D. Cowper. New York: Dover, 1956 (from Methuen edition of 1926).

————. *Lettres à Maurice Solovine.* Facsimile and translated into French. Paris: Gauthier-Villars, 1956.

————. *Out of My Later Years.* New York: Philosophical Library, 1950.

————. *Relativity: The Special and General Theory.* Translated by Robert H. Lawson. New York: Henry Holt, 1920. (First published 1916.)

————. *Sidelights on Relativity.* London: Methuen, 1922.

————, and Leopold Infeld. *The Evolution of Physics: The Growth of Ideas from Early Concepts to Relativity and Quantum.* New York: Simon and Schuster, 1938.

Enriques, Federigo. *Problems of Science.* Translated by Katherine Royce. Chicago: Open Court, 1914. (First published 1906.)

Evans, Herbert M., ed. *Men and Moments in the History of Science.* Seattle: University of Washington Press, 1959.

Fabry, Charles. *Les Applications des interférences lumineuses.* Paris: Éditions de la revue d'optique théorique et instrumentale, 1923.

Faraday, Michael. *Faraday's Diary: Being the Various Philosophical Notes of Experimental Investigation . . . during the Years 1820–1862.* Edited by Thomas Martin. 7 vols. and Index. London: G. Bell, 1932–1936.

Favaro, Antonio, ed. *Le Opere di Galileo Galilei.* Edizione Nazionale. 20 vols. Florence: G. Barbera, 1890–1909.

Fedynskii, V. V., ed. *The Earth in the Universe [Zemlya vo vselennoi].* Translated by Israel Program for Scientific Translations. Jerusalem: I.P.S.T., 1968.

Feigl, Herbert, and May Brodbeck, eds. *Readings in the Philosophy of Science.* New York: Appleton-Century-Crofts, 1953.

Feigl, Herbert, and Grover Maxwell, eds., *Current Issues in the Philosophy of Science: Symposia of Scientists and Philosophers.* New York: Holt, Rinehart & Winston, 1961.

Feigl, Herbert, and W. Sellers. *Readings in Philosophical Analysis.* New York: Appleton-Century-Crofts, 1949.

Feynman, Richard P., et al. *The Feynman Lectures on Physics,* mainly mechanics, radiation, and heat. Edited by Robert B. Leighton and Matthew Sands. 3 vols. Reading, Mass.: Addison-Wesley, 1965–1966.

Fierz, Markus, ed. *Theoretical Physics in the Twentieth Century: A Memorial Volume to Wolfgang Pauli.* New York: Interscience, 1960.

Fiske, Bradley A. *From Midshipman to Rear Admiral.* New York: Century, 1919.

FitzGerald, George Francis. *The Scientific Writings of the Late George Francis FitzGerald.* Collected and edited with a Historical Introduction by Joseph Larmor. Dublin: Hodges, Figgis, 1902.

Flammarion, G. C., and André Danjon, eds. *The Flammarion Book of Astronomy.* Translated by A. & B. Pagel. New York: Simon & Schuster, 1964. First published as Camille Flammarion's *Astronomie Populaire,* Paris, 1880.

Fleming, Donald. *John William Draper and the Religion of Science.* Philadelphia: University of Pennsylvania Press, 1950.

Fleming, J. A. *Waves and Ripples in Water, Air and Aether: A Course of Christmas Lectures Delivered at the Royal Institution of Great Britain.* London: Society for Promoting Christian Knowledge, 1912.

Fock, V. *The Theory of Space, Time and Gravitation.* Translated by N. Kemmer. New York: Pergamon, 1959.

Föppl, August. *Einführung in die Maxwellsche Theorie der Elektrizität.* Leipzig: Teubner, 1894. See also Abraham, Max.

Frank, Philipp. *Einstein: His Life and Times.* Translated by George Rosen. Edited and revised by Shuichi Kusaka. New York: Knopf, 1947.

———. *Philosophy of Science: The Link between Science and Philosophy.* Englewood Cliffs: Prentice-Hall, 1957.

———. *Relativity—A Richer Truth.* Boston: Beacon, 1950.

Fraser, Charles G. *Half-Hours with Great Scientists: The Story of Physics.* New York: Reinhold, 1948.

Fraser, J. T., ed. *The Voices of Time: A Cooperative Survey of Man's Views of Time as Expressed by the Sciences and by the Humanities.* New York: Braziller, 1966.

Fresnel, Augustin. *Oeuvres complètes d'Augustin Fresnel,* publiées par Henri de Senarmont, et al. 3 vols. Paris: Imprimérie Impériale: 1866–1870.

Freundlich, Erwin. *The Foundation of Einstein's Theory of Gravitation.* Translated by H. L. Brose. Cambridge: University Press, 1920.

Froome, K. D., and L. Essen. *The Velocity of Light and Radio Waves.* New York: Academic Press, 1969.

Galilei, Galileo. *Dialogue Concerning the Two Chief World Systems— Ptolemaic and Copernican.* Translated by Stillman Drake. 2nd ed. Berkeley: University of California Press, 1967.

———. *Dialogues Concerning Two New Sciences.* Translated by Henry Crew and Alfonso de Salvio. New York: Dover, n.d. (First published in English in 1914; classic edition, 1638.)

Gamow, George. *Biography of Physics.* New York: Harper & Row, Torchbooks, 1964. (First published 1961.)

———. *My World Line: An Informal Autobiography.* New York: Viking Press, 1970.

———. *Thirty Years That Shook Physics: The Story of Quantum Theory.* Garden City: Doubleday, 1966.

Gandillot, Maurice. *L'Éthérique: essai de physique expérimentale.* Paris: Librairie Vuibert, 1923.

[Ganot, Adolphe]. *Elementary Treatise on Physics, Experimental, and Applied.* Translated and edited from *Ganot's Éléments de Physique* by E. Atkinson. 14th ed. Revised. New York: William Wood, 1893.

Gardner, Martin. *Fads and Fallacies in the Name of Science.* New York: Dover, 1957. (First published 1952.)

———. *Relativity for the Million.* New York: Macmillan, 1962.

Gaul, Harriet A., and Ruby Eiseman. *John Alfred Brashear: Scientist and Humanitarian, 1840–1920.* Philadelphia: University of Pennsylvania Press, 1940.

Gauss, Karl F. *Theory of the Motion of Heavenly Bodies: Moving about the Sun in Conic Sections.* Translated by Charles H. Davis. New York: Dover, 1963. (First published in Latin, 1809; in German, 1857.)

Gerholm, Tor Ragnar. *Physics and Man: An Invitation to Modern Physics.* Totowa, N.J.: Bedminster, 1967.

Gershenson, Daniel E., and Daniel A. Greenberg, eds. *The Natural Philos-*

opher: A Series of Volumes Containing Papers Devoted to the History of Physics and to the Influence of Physics on Human Thought and Affairs through the Ages. 3 vols. New York: Blaisdell, 1963–1964.

Geymonat, Ludovico. *Galileo Galilei: A Biography and Inquiry into His Philosophy of Science.* 2d ed. New York: McGraw-Hill, 1965.

Gibbs, J. Willard. *Elementary Principles in Statistical Mechanics.* New York: Dover, 1960. (First published 1902; New Haven: Yale University Press.)

————. *The Collected Works of J. Willard Gibbs.* 2 vols. Edited by Arthur Haas. New York: Longmans, Green, 1928.

Gill, T. P. *The Doppler Effect: An Introduction to the Theory of the Effect.* London: Academic Press, 1965.

Gillispie, Charles Coulston. *The Edge of Objectivity: An Essay in the History of Scientific Ideas.* Princeton: Princeton University Press, 1960.

Gluck, Irving D. *Optics: The Nature and Applications of Light.* New York: Holt, Rinehart & Winston, 1964.

Goethe, J. W. von. *Theory of Colours.* Translated by Charles L. Eastlake. Cambridge, Mass.: M.I.T. Press, 1970. (First published in German in 1840.)

Goldstein, Herbert. *Classical Mechanics.* Cambridge, Mass.: Addison-Wesley, 1950.

Goode, G. B. *Origin of the National Scientific and Educational Institutions of the United States.* New York: Putnam, 1890.

[Goode, George Brown]. *A Memorial of George Brown Goode: Together with a Selection of His Papers on Muesums and on the History of Science in America.* Washington: U.S. Government Printing Office, 1901.

Goodspeed, Thomas Wakefield. *A History of the University of Chicago.* Chicago: University of Chicago Press, 1916.

Goran, Morris. *The Story of Fritz Haber.* Norman: University of Oklahoma Press, 1967.

Grant, Robert. *History of Physical Astronomy: From the Earliest Ages to the Middle of the Nineteenth Century.* Reprinted from the London Edition [1852]. New York: Johnson Reprint, 1966.

Greene, John C. *The Death of Adam: Evolution and Its Impact on Western Thought.* New York: Mentor Books, 1961.

Grosseteste, Robert. *On Light [De Luce].* Translated by C. C. Riedl. Milwaukee: Marquette University Press, 1942.

Gruber, L. Franklin. *The Einstein Theory: Relativity and Gravitation with*

Some of the More Significant Implications for the General-Reader. Burlington, Iowa: The Lutheran Literary Board, 1923.

Grünbaum, Adolf. *Philosophical Problems of Space and Time.* New York: Knopf, 1963.

Guerlac, Henry, ed. *Ithaca: Proceedings of the 10th International Congress of the History of Science, 1962.* 2 vols. Paris: Hermann, 1964.

Guillemin, Amédée. *The Heavens: An Illustrated Handbook of Popular Astronomy.* Edited by J. N. Lockyer, R. A. Proctor. 9th ed. London: Bentley & Son, 1883.

Guillemin, Victor. *The Story of Quantum Mechanics.* New York: Scribner, 1968.

Haar, D. ter. *The Old Quantum Theory.* Oxford: Pergamon, 1967.

Haas, Arthur, ed. *A Commentary on the Scientific Writings of J. Willard Gibbs . . . in Two Volumes.* New Haven: Yale University Press, 1936.

Hadamard, Jacques. *An Essay on the Psychology of Invention in the Mathematical Field.* Princeton: Princeton University Press, 1945.

Haldane [of Cloan, Richard Burdon], Viscount. *The Reign of Relativity.* New Haven: Yale University Press, 1921.

Hale, George E. *National Academies and the Progress of Research.* Lancaster, Pa.: Am. Ass'n. Adv. Sci., 1915.

Hall, A. R. *The Scientific Revolution, 1500–1800: The Formation of the Modern Scientific Attitude.* Boston: Beacon, 1957. (First published 1954.)

Handel, S. *The Electronic Revolution.* Baltimore: Penguin, 1967.

Hanson, N. R. *The Concept of the Positron: A Philosophical Analysis.* Cambridge: University Press, 1963.

Hardy, Arthur C., and Fred H. Perrin. *The Principles of Optics.* New York: McGraw-Hill, 1932.

Hare, Michael M. *Microcosm and Macrocosm: An Approach to the Synthesis of the Real.* New York: Julian Press, 1966.

Harris, Errol E. *The Foundations of Metaphysics in Science.* New York: Humanities Press, 1965.

Harrow, Benjamin. *From Newton to Einstein: Changing Conceptions of the Universe.* New York: Van Nostrand, 1920.

Hartmann, Hans. *Schöpfer Des Neuen Weltbildes: Grosse Physiker Unserer Zeit.* Bonn: Athenäum Verlag, 1952.

Harvey, E. Newton. *A History of Luminescence: From the Earliest Times until 1900.* Philadelphia: American Philosophical Society, 1957.

Hawkins, Hugh. *Pioneer: A History of The Johns Hopkins University, 1874–1889.* Ithaca: Cornell University Press, 1960.

Heathcote, Niels H. de V. *Nobel Prize Winners in Physics: 1901–50.* New York: Schuman, 1953.

Heaviside, Oliver. *Electromagnetic Theory.* 2 vols. London: "The Electrician" Publishing Co., 1893 (I), 1899 (II).

Heisenberg, Werner. *Physics and Philosophy: The Revolution in Modern Science.* New York: Harper & Row, Torchbooks, 1962. (First published 1958.)

———. *The Physical Principles of the Quantum Theory.* Chicago: University of Chicago Press, 1930.

Helliwell, T. M. *Introduction to Special Relativity.* Boston: Allyn and Bacon, 1966.

Helmholtz, Hermann Ludwig Ferdinand von. *Helmholtz's Treatise on Physiological Optics.* Translated by J. P. C. Southall. 3 vols. 3d ed. Menasha, Wisc.: Optical Society of America, 1924.

———. *Popular Lectures on Scientific Subjects.* Translated by E. Atkinson. New York: Appleton, 1873.

Hempel, Carl G. *Philosophy of Natural Science.* Englewood Cliffs: Prentice-Hall, 1966.

Herneck, Friedrich. *Albert Einstein: Ein Leben fur Wahrheit Menschlichkeit und Frieden.* Berlin: Buchverlag der Morgen, 1967.

Hertz, Heinrich. *Electric Waves: Being Researches on the Propagation of Electric Action with Finite Velocity through Space.* Translated by D. E. Jones. London: Macmillan, 1900. (First published 1892.)

———. *Gesammelte Werke von Heinrich Hertz.* Edited by Ph. Lenard. 3 vols. Leipzig: Barth, 1895–1910.

———. *The Principles of Mechanics Presented in a New Form.* Translated by D. E. Jones and J. T. Wally. New York: Dover, 1956. (First published 1899.)

Herzberg, Gerhard. *Atomic Spectra and Atomic Structure.* Translated by J. W. T. Spinks. 2d ed. New York: Dover, 1944.

Hesse, Mary B. *Forces and Fields: The Concept of Action at a Distance in the History of Physics.* New York: Philosophical Library, 1962.

———. *Models and Analogies in Science.* Notre Dame, Ind.: University of Notre Dame Press, 1966.

Heyl, Paul R. *New Frontiers of Physics.* New York: Appleton, 1930.

———. *The Fundamental Concepts of Physics in the Light of Modern Dis-*

covery: Three Lectures at Carnegie Institute, Pittsburgh, January, 1925. Baltimore: Williams and Wilkins, 1926.

Hiebert, Erwin N. *Historical Roots of the Principle of the Conservation of Energy*. Madison: Historical Society of Wisconsin, 1962.

Hoffman, Banesh. *The Strange Story of the Quantum: An Account for the General Reader of the Growth of the Ideas Underlying Our Present Atomic Knowledge*. 2d ed. New York: Dover, 1959.

Holton, Gerald. *Introduction to Concepts and Theories in Physical Science*. Cambridge, Mass.: Addison-Wesley, 1953, 1962.

————, and Duane H. D. Roller. *Foundations of Modern Physical Science*. Reading, Mass.: Addison-Wesley, 1958.

Hooke, Robert. *Micrographia: or Some Physiological Descriptions of Minute Bodies Made by Magnifying Glasses with Observations and Inquiries Thereupon*. London: Jo Martyn and Ja. Allestry, 1665. Facsimile reprint, Brussels, 1966.

Hooper, William G. *Aether and Gravitation*. London: Chapman & Hall, 1903.

Hoppe, Edmund. *Geschichte der Optik*. Leipzig: J. J. Weber, 1926.

Hoskin, Michael A. *William Herschel: And the Construction of the Heavens*. London: Oldbourne, 1963; New York: Norton, 1964.

Houstoun, R. A. *A Treatise on Light*. London: Longmans, Green, 1930.

————. *Physical Optics*. New York: Interscience, 1957.

Howard, J. N. *The Rayleigh Archives Dedication. Special Report No. 63* (April, 1967). Bedford, Mass.: Air Force Cambridge Research Laboratories, 1967.

Howorth, Muriel. *Pioneer Research on the Atom: The Life Story of Frederick Soddy*. London: New World, 1958.

Hoyle, Fred. *The Nature of the Universe*. New York: Harper & Row, 1960. (First published 1950.)

Hubble, Edwin. *The Observational Approach to Cosmology*. Oxford: Clarendon, 1937.

————. *The Realm of the Nebulae*. New York: Dover, 1958. (First published New Haven: Yale University Press, 1936.)

Hutten, Ernest H. *The Language of Modern Physics: An Introduction to the Philosophy of Science*. London: George Allen & Unwin, 1956.

Huygens, Christiaan. *Treatise on Light: In Which Are Explained the Causes of That Which Occurs in Reflexion, & in Refraction and Particularly in the Strange Refraction of Iceland Crystal*. Translated by S. P. Thompson. London: Macmillan, 1912.

Ihde, Aaron J. *The Development of Modern Chemistry*. New York: Harper & Row, 1964.

Infeld, Leopold. *Albert Einstein: His Work and Its Influences on Our World*. New York: Scribner, 1950.

————. *Quest: The Evolution of a Scientist*. New York: Doubleday, Doran, 1941.

Israël, Hans, Erich Ruckhaber, and Rudolf Weinmann, eds. *Hundert Autoren gegen Einstein*. Leipzig: Voigtlander, 1931.

Jaffe, Bernard. *Men of Science in America: The Role of Science in the Growth of Our Country*. New York: Simon and Schuster, 1944.

————. *Michelson and The Speed of Light*. Garden City: Doubleday, Anchor Books, 1960.

Jaki, Stanley L. *The Relevance of Physics*. Chicago: University of Chicago Press, 1966.

Jammer, Max. *Concepts of Force: A Study in the Foundations of Dynamics*. Cambridge, Mass.: Harvard University Press, 1957.

————. *Concepts of Mass in Classical and Modern Physics*. Cambridge, Mass.: Harvard University Press, 1961.

————. *Concepts of Space: The History of Theories of Space in Physics*. New York: Harper & Row, Torchbooks, 1960. (First published Cambridge, Mass.: Harvard University Press, 1954.)

————. *The Conceptual Development of Quantum Mechanics*. New York: McGraw-Hill, 1966.

Janossy, Lajos. *Cosmic Rays*. Dublin: Institute for Advanced Studies, 1947.

Jeans, Sir James. *Physics and Philosophy*. New York: Macmillan, 1943.

————. *The Mathematical Theory of Electricity and Magnetism*. 4th ed. Cambridge: University Press, 1920. (First published 1908.)

Jeffreys, Sir Harold. *Scientific Inference*. 2d ed. Cambridge: University Press, 1957. (First published 1931.)

Jellinek, Karl. *Weltsystem, Weltäther und die Relativitätstheorie: eine Einführung für experimentelle Naturwissenschaftlichen*. Basel: Wepf, 1949.

Jenkins, Francis A., and Harvey E. White. *Fundamentals of Optics*. 2d ed. New York: McGraw-Hill, 1950. (First published 1937.)

Jevons, W. Stanley. *The Principles of Science: A Treatise on Logic and Scientific Method*. New York: Dover, 1958. (First published 1873.)

Jewkes, John, David Sawers, and Richard Stillerman. *The Sources of Invention*. London: Macmillan, 1958.

Johnson, Martin. *Time, Knowledge and the Nebulae: An Introduction to the Meanings of Time in Physics, Astronomy and Philosophy, and the Relativities of Einstein and of Milne.* London: Faber & Faber, 1945.

Johnston, Majorie, ed. *The Cosmos of Arthur H. Compton.* New York: Knopf, 1967.

Joos, Georg. *Theoretical Physics.* Translated by I. M. Freeman. New York: Hafner, 1934. (First published 1932.)

Jordan, David Starr, ed. *Leading American Men of Science.* New York: Henry Holt, 1910.

Kaplan, Flora. *Nobel Prize Winners: Charts; Indexes; Sketches.* 2d ed. Chicago: Nobelle, 1941. (First published 1939.)

Kaplan, Norman, ed. *Science and Society.* Chicago: Rand McNally, 1965.

Kaplan, S. A., and S. B. Pikelner. *The Interstellar Medium.* Cambridge, Mass.: Harvard University Press, 1970.

Katz, Robert. *An Introduction to the Special Theory of Relativity.* Princeton: Van Nostrand, 1964.

Kelvin, Lord (Sir William Thomson). *Baltimore Lectures on Molecular Dynamics and the Wave Theory of Light.* London: C. J. Clay, 1904. (First published Baltimore, 1884.)

————. *Notes of Lectures on Molecular Dynamics and The Wave Theory of Light: . . . Stenographically Reported by A. S. Hathaway, Lately Fellow in Mathematics of The Johns Hopkins University.* Baltimore: "papyrograph" reproduction, 1884.

————. *Popular Lectures and Addresses.* 3 vols. Nature Series. London: Macmillan, 1891–1894.

————, and Peter Guthrie Tait. *Treatise on Natural Philosophy.* 2 vols. Cambridge: University Press, 1923. (First published 1879.)

Kilmister, C. W., ed. *Men of Physics: Sir Arthur Eddington.* Oxford: Pergamon Press, 1966.

Kopff, August. *Grundzüge der Einsteinschen Relativitätstheorie.* Leipzig: S. Hirzel, 1923.

Korzybski, Alfred. *Science and Sanity: An Introduction to Non-Aristotelian Systems and General Semantics.* 2d ed. Lancaster, Pa.: Science Press, 1941. (First published 1933.)

Koslow, Arnold, ed. *The Changeless Order: The Physics of Space, Time, and Motion.* New York: Braziller, 1967.

Koyré, Alexandre. *From the Closed World to the Infinite Universe.* Baltimore: Johns Hopkins Press, 1957.

————. *Mélanges Alexandre Koyré.* 2 vols. Paris: Hermann, 1964. See also Woolf, Harry (article by).

Krafft, Carl Frederick. *Ether and Matter.* Richmond, Va.: Dietz Printing Co., 1945.

Kranzberg, Melvin, and Carroll W. Pursell, Jr., eds. *Technology in Western Civilization.* 2 vols. New York: Oxford University Press, 1967.

Krogdahl, Wasley S. *The Astronomical Universe: An Introductory Text in College Astronomy.* New York: Macmillan, 1952.

Kuhn, Thomas S. *The Copernican Revolution: Planetary Astronomy in the Development of Western Thought.* New York: Modern Library, 1957.

————. *The Structure of Scientific Revolutions.* Chicago: University of Chicago Press, Phoenix Books, 1962.

————, John L. Heilbron, Paul Forman, and Lini Allen, compilers. *Sources for History of Quantum Physics: An Inventory and Report.* Philadelphia: American Philosophical Society, 1967.

Kuznetsov, B. *Einstein.* Translated by V. Talmy. Moscow: Progress Publishers, 1965.

Langevin, Paul. *Oeuvres Scientifiques de Paul Langevin.* Paris: Centre national de la recherche scientifique, 1950.

Lanczos, Cornelius. *Albert Einstein and the Cosmic World Order.* New York: Wiley, 1965.

LaPlace, Pierre S. (Marquis de). *Celestial Mechanics.* Translated by Nathaniel Bowditch. 4 vols. New York: Chelsea, 1966. (First published in French 1799–1825; in English, 1829–1839: Boston.)

Larmor, [Sir] Joseph. *Aether and Matter: A Development of the Dynamical Relations of the Aether to Material Systems on the Basis of the Atomic Constitution of Matter, Including a Discussion of the Influence of the Earth's Motion on Optical Phenomena, Being an Adams Prize Essay in the University of Cambridge.* Cambridge: University Press, 1900.

LaRosa, Michele. *Der Äther, geschichte einer hypothese.* Translated (from Italian) by K. Muth. Leipzig: Barth, 1912.

Larson, Dewey B. *New Light on Space and Time.* Portland, Ore.: North Pacific Publishers, 1965.

Laue, Max von. *Die Relativitätstheorie.* 2 vols. Braunschweig: Fredrich Vieweg, 1955.

————. *History of Physics.* Translated by Ralph Oesper. New York: Academic Press, 1950.

Laurence, Lionel, and H. Oscar Wood. *General and Practical Optics.* 4th ed. London: School of Optics, 1932.

LeBon, Gustave. *The Evolution of Forces*. New York: Appleton, 1908.

Lecky, S. T. S. *Wrinkles in Practical Navigation*. 2d ed. London: G. Philip & Son, Ltd., 1884. (First published 1881.)

Lemon, Harvey B. *From Galileo to the Nuclear Age: An Introduction to Physics*. 2d ed. Chicago: University of Chicago Press, 1946. (First Published 1934.)

Lenard, Philipp. *Great Men of Science: A History of Scientific Progress*. Translated by H. S. Hatfield. New York: Macmillan, 1933.

Lewis, Edwin Herbert. *University of Chicago Poems*. Chicago: University of Chicago Press, 1923.

Lieber, Lillian R. *The Einstein Theory of Relativity*. New York: Rinehart, 1936.

Lindsay, Robert Bruce. *Lord Rayleigh—The Man and His Work*. Oxford: Pergamon, 1970.

———, and Henry Margenau. *Foundations of Physics*. 2d ed. New York: Dover, 1959. (First published New York: Wiley, 1936.)

Lipetz, Ben-Ami. *A Guide to Case Studies of Scientific Activity*. Carlisle, Mass.: Intermedia, 1965.

Lives in Science: A Scientific American Book. New York: Simon and Schuster, 1957.

Lloyd, Humphrey. *Elementary Treatise on the Wave Theory of Light*. 2d ed. London: Longmans, 1857.

Lodge, Sir Oliver. *Advancing Science: Being Personal Reminiscences of the British Association in the Nineteenth Century*. London: E. Benn, 1931.

———. *Beyond Physics, or The Idealisation of Mechanism*. London: Allen & Unwin, 1930.

———. *Continuity*. London: J. M. Dent, 1913.

———. *Electrons, or The Nature and Properties of Negative Electricity*. London: G. Bell, 1907.

———. *Ether & Reality: A Series of Discourses on the Many Functions of the Ether of Space*. London: Hodder & Stoughton, 1925.

———. *The Ether of Space*. London: Harper, 1909.

———. *Modern Views of Electricity*. 2d ed. London: Macmillan, 1892. (First published 1889.)

———. *My Philosophy: Representing My Views on the Many Functions of the Ether of Space*. London: E. Benn, 1933.

———. *Relativity: A Very Elementary Exposition*. New York: Doran, 1926.

————. *Der Weltäther von Sir Oliver Lodge.* Translated by Hilde Bark-hausen. Braunschweig: F. Vieweg & Sohn, 1911.

Lorentz, H. A. *Collected Papers.* 9 vols. The Hague: Martinus Nijhoff, 1935–1939. Edited by P. Zeeman and A. D. Fokker.

————. *The Einstein Theory of Relativity: A Concise Statement.* New York: Brentano, 1920.

————. *Lectures on Theoretical Physics: Delivered at the University of Leiden.* 3 vols. Translated by L. Silberstein and A. P. H. Trivelli. London: Macmillan, 1927–1931.

————. *Problems of Modern Physics: A Course of Lectures Delivered in the California Institute of Technology.* Boston: Ginn, 1927.

————. *The Theory of Electrons and Its Application to the Phenomena of Light and Radiant Heat.* 2d ed. New York: Dover, 1952. (First published in 1909 from lectures delivered in 1906.)

————. *Versuch einer Theorie der electrischen und optischen Erscheinungen in bewegten Körpern.* Leiden: E. J. Brill, 1895.

————, A. Einstein, H. Minkowski, and H. Weyl. *The Principle of Relativity: A Collection of Original Memoirs on the Special and General Theory of Relativity.* Translated by W. Perrett and G. B. Jeffrey. New York: Dover, 1923.

Lorentz, G. L. de Haas, ed. *H. A. Lorentz: Impressions of His Life and Work.* Amsterdam: North Holland, 1957.

Ludwig, Gunter. *Wave Mechanics.* Oxford: Pergamon, 1968.

Lynch, Arthur. *The Case Against Einstein.* London: Philip Allan, 1932.

[McAllister, D. T.]. *Albert Abraham Michelson: The Man Who Taught A World to Measure.* Publication of the Michelson Museum No. 3, China Lake, Calif.: Naval Weapons Center, 1970.

McAllister, D. T., ed. *The Albert A. Michelson Nobel Prize and Lecture.* Publication of the Michelson Museum, No. 2, China Lake, Calif.: U.S. Naval Ordnance Test Station, 1966.

McCabe, Joseph. *The Religion of Sir Oliver Lodge.* London: Watts, 1914.

McCormmach, Russell, ed. *Historical Studies in the Physical Sciences.* Philadelphia: University of Pennsylvania Press, 1969.

McCrea, W. L. *Relativity Physics.* London: Methuen, 1949. (First published 1935.)

MacCullagh, James. *The Collected Works of James MacCullagh.* Edited by John H. Jellett and Samuel Haughton. Dublin: Hodges, Figgis, 1880.

Macfarlane, A. *Lectures on Ten British Physicists of the Nineteenth Century.* New York: Wiley, 1919.

McGucken, William. *Nineteenth-Century Spectroscopy: Development of the Understanding of Spectra, 1802–1897.* Baltimore: Johns Hopkins Press, 1969.

Mach, Ernst. *Popular Scientific Lectures.* Translated by Thomas J. McCormack. 5th ed., LaSalle: Open Court, 1943. (First published 1894–1898.)

———. *The Principles of Physical Optics: An Historical and Philosophical Treatment.* Translated by J. S. Anderson and A. F. A. Young (in 1926). New York: Dover, n.d. (First published 1916.)

———. *The Science of Mechanics: A Critical and Historical Account of Its Development.* Translated by Thomas J. McCormack. 4th ed. Chicago: Open Court, 1907. (First published 1883.)

———. *Supplement to the Third English Edition of The Science of Mechanics: A Critical and Historical Account of its Development.* Translated by P. E. B. Jourdain. Chicago: Open Court, 1915.

Maclaurin, Richard C. *The Theory of Light: A Treatise on Physical Optics.* Cambridge: University Press, 1908.

McVittie, G. C. *Fact and Theory in Cosmology.* London: Eyre & Spottiswoode, 1961.

Magie, William F. *A Source Book in Physics.* New York: McGraw-Hill, 1935.

Mallik, D. N. *Optical Theories: Based on Lectures Delivered before the Calcutta University.* Cambridge: University Press. 1917.

Mandelker, Jakob. *Relativity and the New Energy Mechanics.* New York: Philosophical Library, 1966.

Mann, C. Riborg. *Manual of Advanced Optics.* Chicago: Scott, Foresman, 1902.

Margenau, Henry. *The Nature of Physical Reality: A Philosophy of Modern Physics.* New York: McGraw-Hill, 1950.

———. *Open Vistas: Philosophical Perspectives of Modern Science.* New Haven: Yale University Press, 1961.

Mascart, M. E. *Traité d'optique.* 3 vols. Paris: Gauthier-Villars, 1889–1893.

Mason, Max, and Warren Weaver. *The Electromagnetic Field.* New York: Dover, n.d. (First published in 1929; Chicago: University of Chicago Press.)

Mason, Stephen F. *A History of the Sciences: Main Currents of Scientific Thought.* London: Routledge & Kegan Paul, 1953.

Massey, Sir Harrie. *The New Age in Physics.* New York: Harper & Row, 1960.

Maxwell, James Clerk. *Matter and Motion*. Edited by Sir Joseph Larmor. New York: Dover, n.d. (First published 1877.)

———. *Origins of Clerk Maxwell's Electric Ideas, as Described in Familiar Letters to William Thomson*. Edited by Sir Joseph Larmor. Cambridge: University Press, 1937.

———. *The Scientific Papers of James Clerk Maxwell*. Edited by W. D. Niven. 2 vols. in one. New York: Dover, n.d.

———. *A Treatise on Electricity and Magnetism*. 2 vols. Oxford: Clarendon, 1873. 3d ed. unabridged. New York: Dover, 1954.

Mead, George Herbert. *Movements of Thought in the Nineteenth Century*. Chicago: University of Chicago Press, 1936.

Mélanges Alexandre Koyré. See Koyré, Alexandre.

Melsen, Andrew G. Van. (See Van Melsen.)

Mendeléeff, Professor D. [Dmitri]. *An Attempt Towards a Chemical Conception of the Ether*. Translated by George Kamensky. London: Longmans, Green, 1904.

Mermin, N. David. *Space and Time in Special Relativity*. New York: McGraw-Hill, 1968.

Merz, John Theodore. *A History of European Thought in the Nineteenth Century*. 4 vols. Edinburgh: William Blackwood, 1896–1914.

Metraux, Guy S., and François Crouzet, eds. *The Evolution of Science: Readings from the History of Mankind*. New York: Mentor Books, 1963.

Meyer, Wilhelm F., et al., eds. *Encyklopädia der Mathematischen Wissenschaften: mit Einschluss ihrer Anwendungen*. Leipzig: Teubner, 1898–1935.

Michelmore, Peter. *Einstein: Profile of the Man*. New York: Dodd, Mead, 1962.

Michelson, Albert A. *Light Waves and Their Uses*. Chicago: University of Chicago Press, 1903. See also German translation with added 54-page bibliography by Max Iklé, *Lichtwellen und ihre Anwendungen*. Leipzig: J. A. Barth, 1911.

———. *Studies in Optics*. Chicago: University of Chicago Press, 1927.

Michelson, Charles. *The Ghost Talks*. New York: Putnam, 1944.

Michelson, Miriam. *The Madigans*. New York: Century, 1904.

Mill, John Stuart. *A System of Logic: Ratiocinative and Inductive, Being a Connected View of the Principles of Evidence and the Methods of Scientific Investigation*. 2 vols. 10th ed. London: Longmans, Green, 1879.

Miller, Dayton C. *Anecdotal History of the Science of Sound*. New York: Macmillan, 1935.

———. *The Dayton C. Miller Collections Relating to the Flute*. Cleveland: Privately printed, 1935.

———. *The Science of Musical Sounds*. New York: Macmillan, 1916.

———. *Sparks, Lightning, and Cosmic Rays: An Anecdotal History of Electricity*. New York: Macmillan, 1939.

Miller, Perry, ed. *American Thought: Civil War to World War I*. New York: Rinehart, 1954.

Millikan, Robert Andrews. *Autobiography*. New York: Prentice-Hall, 1950.

———. *Electrons (+ and —), Protons, Photons, Neutrons, and Cosmic Rays*. 2d ed. revised. Chicago: University of Chicago Press, 1947. (First published 1935.)

———, et al. *Time and Its Mysteries*. New York: New York University Press, 1936–1949.

Milne, Edward Arthur. *Relativity, Gravity, and World Structure*. Oxford: Clarendon, 1935.

———. *Kinematic Relativity*. Oxford: Clarendon, 1948.

Møller, C. C. *The Theory of Relativity*. Oxford: Clarendon, 1960. (First published 1951.)

Montagu, M. F. A., ed. *Studies and Essays in the History of Science and Learning. Offered in Homage to George Sarton on the Occasion of His Sixtieth Birthday, August 31, 1944*. New York: Schuman, 1944.

Moore, Ruth. *Niels Bohr: The Man, His Science, & the World They Changed*. New York: Knopf, 1966.

More, Louis Trenchard. *The Limitations of Science*. New York: Holt, 1915.

Moszkowski, Alexander. *Einstein the Searcher: His Work Explained from Dialogues with Einstein*. Translated by Henry L. Brose. London: Methuen, 1921.

Mottelay, Paul Fleury. *Bibliographical History of Electricity and Magnetism*. London: Griffin, 1922.

Müller, Aloys. *Die Problem des absoluten Raumes und seine Beziehung zum allgemeinen Raumproblem*. Braunschweig: F. Vieweg & Sohn, 1911.

Müller-Markus, Siegfried. *Einstein und die Sowjetphilosophie Krisis Einer Lehre*. 2 vols. Dordrecht, Holland: D. Reidel, 1960–1966.

Muller, Herbert J. *Science and Criticism: The Humanistic Tradition in Contemporary Thought*. New Haven: Yale University Press, 1943.

Munitz, Milton K., ed. *Theories of the Universe: From Babylonian Myth to Modern Science*. Glencoe, Ill.: Free Press, 1957.

Muses, Charles A. *An Evaluation of Relativity Theory: After a Half Century.* New York: S. Weiser, 1953.

Nagel, Ernest. *The Structure of Science: Problems in the Logic of Scientific Explanation.* New York: Harcourt, Brace & World, 1961.

Nash, Leonard K. *The Nature of the Natural Sciences.* Boston: Little, Brown, 1963.

Nathan, Otto, and Heinz Norden, eds. *Einstein on Peace.* London: Methuen, 1963.

Nef, John U. *War and Human Progress: An Essay on the Rise of Industrial Civilization.* Cambridge, Mass.: Harvard University Press, 1950.

Nernst, H. Walther. *The New Heat Theorem: Its Foundations in Theory and Experiment.* Translated by Guy Barr. New York: Dutton, 1924. (First published 1917.)

Neugebauer, Otto. *The Exact Sciences in Antiquity.* Princeton: Princeton University Press, 1952.

Newcomb, Simon. "Measures of the Velocity of Light made under the direction of The Secretary of the Navy during the years 1880–82," *Astronomical Papers* prepared for the use of the *American Ephemeris and Nautical Almanac* 2, Parts III–IV. Washington: U.S. Government Printing Office, 1885.

———. *The Relation of Scientific Method to Social Progress: An Address Delivered before the Philosophical Society of Washington, December 4th, 1880.* Washington: Philosophical Society of Washington, 1880.

———. *The Reminiscences of An Astronomer.* Boston: Houghton Mifflin, 1903.

Newman, James R. *Science and Sensibility.* Abridged edition. Garden City: Doubleday, Anchor Books, 1963. (First published 1961.)

———, ed. *What is Science? Twelve Eminent Scientists and Philosophers Explain Their Various Fields to the Layman.* New York: Simon and Schuster, 1955.

———, ed. *The World of Mathematics: A Small Library of the Literature of Mathematics from A'h-mose the Scribe to Albert Einstein.* 4 vols. New York: Simon and Schuster, 1956.

Newton, Isaac. *Isaac Newton's Papers & Letters on Natural Philosophy, and related documents.* Edited by I. Bernard Cohen. Cambridge, Mass.: Harvard University Press, 1958.

———. *Opticks: or A Treatise of the Reflections, Refractions, Inflections & Colours of Light.* New York: Dover, 1952 (based on the 4th edition, London: 1730).

————. *Sir Isaac Newton's Account of the Aether with Some Additions by Way of Appendix*. Edited by B. R., M.D. [Bryan Robinson]. Dublin: S. Powell, 1745.

————. *Sir Isaac Newton's Mathematical Principles of Natural Philosophy and His System of the World*. Translated by Andrew Motte (in 1729). The translations revised, and supplied with an historical and explanatory appendix by Florian Cajori. Berkeley: University of California Press, 1946.

Ney, E. P. *Electromagnetism and Relativity*. New York: Harper & Row, 1962.

Nordmann, Charles. *Einstein and the Universe: A Popuar Exposition of the Famous Theory*. Translated by Joseph McCabe. New York: Holt, 1922. (First published Paris, 1921.)

North, John David. *The Measure of the Universe: A History of Modern Cosmology*. Oxford: Clarendon, 1965.

Northrop, F. S. C. *The Logic of the Sciences and the Humanities*. New York: Meridian Books, 1959. (First published New York: Macmillan, 1947.)

Oldham, Frank. *Thomas Young, F.R.S.: Philosopher and Physician*. London: Arnold, 1933.

O'Rahilly, Alfred. *Electromagnetics: A Discussion of Fundamentals*. London: Longmans, Green, 1938.

Ostwald, Wilhelm. *Die Überwindung des Wissenschaftlichen Materialismus*. Leipzig: Veit & Co., 1895.

Otis, Arthur S. *Light Velocity and Relativity: The Problem of Light Velocity*. Yonkers-on-Hudson: Burcket & Assoc., 1963.

Page, Thornton, and Lou Williams Page, eds. *Stars and Clouds of the Milky Way: The Structure and Motion of Our Galaxy*. New York: Macmillan, 1968.

Page, Thornton, and Lou Williams Page, eds. *Wanderers in the Sky: The Motions of Planets and Space Probes*. Vol. I: *Sky and Telescope*. Library of Astronomy. New York: Macmillan, 1965.

Palter, Robert M. *Whitehead's Philosophy of Science*. Chicago: University of Chicago Press, 1960.

Pannekoek, A. *A History of Astronomy*. New York: Interscience, 1961.

Panofsky, W. K. H., and M. Phillips. *Classical Electricity and Magnetism*. Cambridge: Addison-Wesley, 1955.

Pauli, Wolfgang. *Theory of Relativity*. Translated by G. Field. New York: Pergamon, 1958. (First published Leipzig: B. G. Teubner, 1921.)

Peacock, George. *Life of Thomas Young, M.D., F.R.S., &c.* London: John Murray, 1855.

Pearson, Karl. *The Grammar of Science.* 3d ed., reissue (1911). New York: Meridian Books, 1957. (First published 1892.)

Peierls, R .E. *The Laws of Nature.* New York: Scribner, 1956.

Petzoldt, Joseph. *Das Weltproblem vom Standpunkte Des Relativistischen Positivismus Aus.* Leipzig: Teubner, 1924.

Pla, Cortes. *El Enigma De La Luz.* Buenos Aires: Kraft, 1949.

Planck, Max. *The Origin and Development of the Quantum Theory: Being the Nobel Prize Address Delivered before The Royal Swedish Academy of Science at Stockholm, 2 June 1920.* Translated by H. T. Clarke and L. Silberstein. Oxford: Clarendon, 1922.

————. *The Philosophy of Physics.* Translated by W. H. Johnston. New York: Norton, 1936.

————. *Scientific Autobiography and Other Papers.* Translated by Frank Gaynor. New York: Philosophical Library, 1949.

————. *The Universe in the Light of Modern Physics.* Translated by W. H. Johnston. New York: Norton, 1931.

————. *Where is Science Going?* Translated by James Murphy. New York: Norton, 1932.

Pledge, H. T. *Science Since 1500: A Short History of Mathematics, Physics, Chemistry, Biology.* New York: Harper & Row, Torchbooks, 1959. (First published London, 1939.)

Podobed, V. V. *Fundamental Astrometry: Determination of Stellar Co-ordinates.* Edited by A. N. Vyssotsky. Chicago: Chicago University Press, 1965. (First published in Russian, 1962.)

Poincaré, Henri. *Mathematics and Science: Last Essays.* Translated by J. W. Bolduc. New York: Dover, 1963. (First published as *Dernières Pensées,* 1913.)

————. *Les Méthodes nouvelles de la mécanique céleste.* 3 vols. Paris: 1892, 1893, 1899. New York: Dover, 1957.

————. *Science and Hypothesis.* Translated by W. J. G. [*sic*] New York: Dover, 1952. (First published 1902.)

————. *Science and Method.* Translated by Francis Maitland. New York: Dover, n.d. (First published 1904[?].)

————. *The Value of Science.* Translated by G. B. Halsted. New York: Dover, 1958. (First published 1905.)

Polanyi, Michael. *Personal Knowledge: Towards a Post-Critical Philosophy.* Chicago: University of Chicago Press, 1958.

Poor, Charles Lane. *Gravitation Versus Relativity: A Non-Technical Explanation of the Fundamental Principles of Gravitational Astronomy and a Critical Examination of the Astronomical Evidence Cited as Proof of the Generalized Theory of Relativity.* New York: Putnam, 1922.

Popper, Karl R. *Conjectures and Refutations: The Growth of Scientific Knowledge.* New York: Harper & Row, Torchbooks, 1968. (First published 1962.)

———. *The Logic of Scientific Discovery.* New York: Basic Books, 1959.

Preston, S. Tolver. *Physics of the Ether.* London: Spon, 1875.

Preston, Thomas. *The Theory of Light.* Edited by Charles J. Joly. 3d ed. London: Macmillan, 1901. (First published 1890.)

Price, Derek J. de Solla. *Little Science, Big Science.* New York: Columbia University Press, 1963.

———. *Science since Babylon.* New Haven: Yale University Press, 1961.

Priestley, Joseph. *The History and Present State of Discoveries relating to Vision, Light, and Colours.* 2 vols. London, 1772.

Proctor, Richard A. *Other Worlds Than Ours: The Plurality of Worlds Studied under the Light of Recent Scientific Researches.* New York: J. A. Hill, 1904.

Prokhovnik, S. J. *The Logic of Special Relativity.* New York: Cambridge University Press, 1967.

Pupin, Michael. *From Immigrant to Inventor.* New York: Scribner, 1930.

Rainich, George Y. *Mathematics of Relativity: Lecture Notes.* New York: Wiley, 1950. (First published 1933, Ann Arbor: Edwards.)

Ramsauer, Carl. *Grundversuche der Physik in Historischer Darstellung.* Berlin: Springer-Verlag, 1953.

Ramsay, William. *The Gases of the Atmosphere: The History of Their Discovery.* 3rd ed. New York: Macmillan, 1905. (First published London, 1896.)

Randall, John H., Jr. *The Making of the Modern Mind: A Survey of the Intellectual Background of the Present Age.* Rev. ed. Boston: Houghton Mifflin, 1940. (First published 1926.)

Rankine, W. J. M. *Miscellaneous Scientific Papers.* London: Griffin, 1881.

Rasmussen, Eigil. *Matter and Gravity: Logic Applied to Physics in a Common-Sense Approach to the Classical Ether Theory.* New York: Exposition, 1958.

Rayleigh, Lord. *Scientific Papers, by John William Strutt, 3rd Baron Rayleigh.* 6 vols. Cambridge: University Press, 1899–1920.

Reichenbach, Hans. *Atom and Cosmos: The World of Modern Physics.* Translated by E. S. Allen. New York: Macmillan, 1933.

————. *From Copernicus to Einstein.* New York: The Wisdom Library, 1942.

————. *The Philosophy of Space and Time.* Translated by Maria Reichenbach and John Freund. New York: Dover, 1958.

————. *The Rise of Scientific Philosophy.* Berkeley: University of California Press, 1953.

————. *The Theory of Relativity and A Priori Knowledge.* Translated by Maria Reichenbach. Berkeley: University of California Press, 1965.

Reingold, Nathan, ed. *Science in Nineteenth Century America: A Documentary History.* New York: Hill & Wang, 1966.

Reiser, Anton [pseudonym for Rudolf Kayser]. *Albert Einstein: A Biographical Portrait.* New York: A. & C. Boni, 1930.

Resnick, Robert. *Introduction to Special Relativity.* New York: Wiley, 1968.

Reynolds, Osborne. *On an Inversion of Ideas as to the Structure of the Universe.* Cambridge: University Press, 1903.

————. *Papers on Mechanical and Physical Subjects.* 3 vols. Cambridge: University Press, 1900–1903. Vol. III: *The Sub-mechanics of the Universe.* 1903.

Rice, J. *Relativity: A Systematic Treatment of Einstein's Theory.* London: Longmans, Green, 1923.

Rindler, W. *Special Relativity.* Edinburgh: Oliver & Boyd, 1960.

Rodwell, G. F. *Dictionary of Science: Comprising Astronomy, Chemistry, Dynamics, Electricity, Heat, Hydrodynamics, Hydrostatics, Light, Magnetism, Mechanics, Meteorology, Pneumatics, Sound, and Statics; Preceded by an Essay on the History of the Physical Sciences.* Philadelphia: Franklin News Co., 1886.

Roe, Joseph W. *English and American Tool Builders.* New York: McGraw-Hill, 1916.

Rohr, Moritz von. *Geometrical Investigation of the Formation of Images in Optical Instruments: Embodying the Results of Scientific Researches Conducted in German Optical Workshops.* Translated by R. Kanthack. London: H.M. Stationery Office, 1920.

Romer, Alfred. *Radiochemistry and the Discovery of Isotopes.* New York: Dover, 1970.

————, ed. *The Discovery of Radioactivity and Transmutation.* New York: Dover, 1964.

Ronan, Colin A. *Edmond Halley: Genius in Eclipse*. Garden City: Doubleday, 1969.

Ronchi, Vasco. *Optics: The Science of Vision*. Translated by Edward Rosen. New York: New York University Press, 1957. (First published 1955, in Italian.)

————. *Storia della Luce*. Bologna: Zanichelli, 1939.

Rossi, Bruno. *Optics*. Reading, Mass.: Addison-Wesley, 1957.

Rosser, W. G. V. *An Introduction to the Theory of Relativity*. London: Butterworth, 1964.

Rouse, Hunter, and Simon Ince. *History of Hydraulics*. New York: Dover, 1957.

Routledge, Robert. *Discoveries and Inventions of the Nineteenth Century*. London: Geo. Routledge & Sons, 1879.

Rowland, H. A. *The Physical Papers of Henry Augustus Rowland*. Baltimore: Johns Hopkins, 1902.

Rukeyser, Muriel. *Willard Gibbs*. Garden City: Doubleday, Doran, 1942.

Russell, Alexander. *Lord Kelvin: His Life and Work*. London: T. C. & E. C. Jack, 1912.

Russell, Bertrand. *The A B C of Relativity*. New York: Harper, 1925.

Sabra, A. I. *Theories of Light: From Descartes to Newton*. London: Oldbourne, 1967.

Safford, Truman H. *The Development of Astronomy in the United States*. Williamstown, Mass.: Published by the college, 1888.

Salisbury, (Lord) Robert Arthur Talbot Gascoyne-Cecil 3rd, Marquess of. *Evolution: A Retrospect*. London: Roxburghe, 1894.

Samuel, Viscount. *Essay in Physics*. Oxford: Blackwell, 1951.

Sanders, J. H. *Velocity of Light*. Oxford: Pergamon, 1965.

Sarton, George. *The History of Science and the New Humanism*. 3d ed. New York: Braziller, 1956.

————. *Horus: A Guide to the History of Science: A First Guide for the Study of the History of Science, with Introductory Essays on Science and Tradition*. New York: Ronald, 1952.

————. *The Study of the History of Mathematics* and *The Study of the History of Science*. 2 vols. bound as one. New York: Dover, 1954. (Both first published 1936.)

Schilpp, Paul Arthur, ed. *Albert Einstein: Philosopher-Scientist*. 2 vols. New York: Harper & Row, Torchbooks, 1959. (First published 1949–1951, in *Library of Living Philosophers*.)

Schlick, Moritz. *Space and Time in Contemporary Physics: An Introduction to the Theory of Relativity and Gravitation.* Translated by H. L. Brose. New York: Oxford University Press, 1920.

Schmidt, Harry. *Relativity and the Universe: A Popular Introduction into Einstein's Theory of Space and Time.* Translated by K. Wichmann. 2d ed. London: Methuen, 1922. (First published 1921.)

Schneer, Cecil J. *Mind and Matter: Man's Changing Concepts of the Material World.* New York: Grove, 1969.

———. *The Search for Order: The Development of the Major Ideas in the Physical Sciences from the Earliest Times to the Present.* New York: Harper, 1960.

Schoeck, Helmut, and James W. Wiggins. *Relativism and the Study of Man.* Princeton: Van Nostrand, 1961.

Schofield, Robert E. *Mechanism and Materialism: British Natural Philosophy in the Age of Reason.* Princeton: Princeton University Press, 1970.

Schonland, Basil. *The Atomists (1805–1933).* New York: Oxford University Press, 1968.

Schrödinger, Erwin. *Science and the Human Temperament.* Translated by J. Murphy and W. J. Johnston. New York: Norton, 1935.

———. *Space-Time Structure.* Cambridge: University Press, 1950.

Schuster, Sir Arthur. *Biographical Fragments.* London: Macmillan, 1932.

———. *The Progress of Physics: During 33 Years 1875–1908.* Cambridge: University Press, 1911.

Schwartz, Herman M. *Introduction to Special Relativity.* New York: McGraw-Hill, 1968.

Schwartz, Jacob T. *Relativity in Illustrations.* New York: New York University Press, 1962.

Sciama, D. W. *The Unity of the Universe.* Garden City: Doubleday, Anchor Books, 1961. (First published 1959.)

Seabrook, William. *Doctor Wood: Modern Wizard of the Laboratory.* New York: Harcourt, Brace, 1941.

Seelig, Carl. *Albert Einstein: A Documentary Biography.* Translated by Mervyn Savill. London: Staples, 1956. (First Published 1954, Zurich.)

Sen, D. K. *Fields and/or Particles.* Toronto: Ryerson, 1968.

Shadowitz, A. *Special Relativity.* Philadelphia: Saunders, 1968.

Shamos, M. W., ed. *Great Experiments in Physics.* New York: Holt, 1959.

Shapere, Dudley, ed. *Philosophical Problems of Modern Science.* New York: Macmillan, 1965.

Shapley, Harlow. *Ad astra per aspera: Through Rugged Ways to the Stars.* New York: Scribner, 1969.

————. *Flights from Chaos: A Survey of Material Systems from Atoms to Galaxies.* New York: McGraw-Hill, 1930.

————, ed. *Source Book in Astronomy, 1900–1950.* Cambridge, Mass. Harvard University Press, 1960.

Shapley, Harlow, Helen Wright, and Samuel Rapport, eds. *Readings in the Physical Sciences.* New York: Appleton-Century-Crofts, 1948.

Sharlin, Harold I. *The Convergent Century: The Unification of Science in the Nineteenth Century.* New York: Abelard-Schuman, 1966.

————. *The Making of the Electrical Age: From the Telegraph to Automation.* London: Abelard-Schuman, 1963.

Silberstein, Ludwik. *Elements of the Electromagnetic Theory of Light.* London: Longmans, Green, 1918.

————. *The Theory of General Relativity and Gravitation.* Toronto: University of Toronto Press, 1922.

————. *The Theory of Relativity.* London: Macmillan, 1914.

Singer, Charles. *A Short History of Scientific Ideas to 1900.* Oxford: Clarendon, 1959.

————, et al. *A History of Technology.* 5 vols. New York: Oxford University Press, 1954–1958.

Sitter, Willem de. *Kosmos: A Course of Six Lectures on the Development of Our Insight into the Structure of the Universe, Delivered for the Lowell Institute, in Boston in November, 1931.* Cambridge, Mass: Harvard University Press, 1932.

Smart, J. J. C. *Problems of Space and Time.* New York: Macmillan, 1964.

Smart, W. M. *Stellar Kinematics.* New York: Wiley, 1968.

Soddy, Frederick. *Matter and Energy.* London: Williams and Norgate, 1912.

Sommerfeld, Arnold. *Optics: Lectures on Theoretical Physics.* Vol. IV. New York: Academic Press, 1954.

Speiro, H. L. *The Fluid Ether.* London: T. J. Winterson, 1958.

Stallo, J. B. *The Concepts and Theories of Modern Physics.* Edited by Percy W. Bridgman. Cambridge, Mass.: Harvard University Press, 1960. (First published New York: Appleton, 1881.)

Stebbing, L. Susan. *Philosophy and the Physicists.* London: Methuen, 1937.

Steel, W. H. *Interferometry.* London: Cambridge University Press, 1962.

Steinmetz, Charles Proteus. *Four Lectures on Relativity and Space.* New York: McGraw-Hill, 1923.

Stephenson, G., and C. W. Kilmister. *Special Relativity for Physicists*. London: Longmans, Green, 1958.

Stevens, Blamey. *The Psychology of Physics*. Manchester, Eng.: Sherratt & Hughes, 1939.

Stokes, George Gabriel. *Mathematical and Physical Papers*. 5 vols. Cambridge: University Press, 1880–1905.

————. *On Light: First Course On the Nature of Light, Delivered at Aberdeen in November, 1883*. [Burnett Lectures.] London: Macmillan, 1884.

————, G. Forbes et al. *Science Lectures at South Kensington*. 2 vols. London: Macmillan, 1878–1879.

Strong, John. *Concepts of Physical Optics*. San Francisco: Freeman, 1958.

————, et al. *Procedures in Experimental Physics*. New York: Prentice-Hall, 1938.

Strutt, Robert John. *John William Strutt, Third Baron Rayleigh*. London: Arnold, 1924.

Styzhkin, N. I. *History of Mathematical Logic from Leibniz to Peano*. Cambridge, Mass.: M.I.T. Press, 1969.

Sullivan, J. W. N. *The Limitations of Science*. New York: Viking, 1933.

Süsskind, Charles, ed. *The Encyclopedia of Electronics*. New York: Reinhold, 1962.

Swann, W. F. G. *The Architecture of the Universe*. New York: Macmillan, 1934.

Swenson, Loyd S., Jr., James M. Grimwood, and Charles C. Alexander. *This New Ocean: A History of Project Mercury*. Washington: Government Printing Office [NASA SP-4201], 1966.

Synge, John Lighton. *Relativity: the General Theory*. New York: Interscience, 1960.

————. *Relativity: The Special Theory*. Amsterdam: North Holland, 1956.

Tait, P. G. *Lectures on Some Recent Advances in Physical Science*. London: Macmillan, 1876.

————. *Light*. Edinburgh: Black, 1884.

Taton, René. *Reason and Chance in Scientific Discovery*. Translated by A. J. Pomerans. New York: Science Editions, Inc., 1962. (First published 1957.)

————, ed. *History of Science*. Translated by A. J. Pomerans. 4 vols.: I. *Ancient and Medieval Science*; II. *The Beginnings of Modern Science*; III. *Science in the Nineteenth Century*; IV. *Science in the Twentieth Century*. New York: Basic Books, 1963–1966. (First published 1957–1964.)

Taylor, Edwin F., and John A. Wheeler. *Spacetime Physics*. San Francisco: Freeman, 1963.

Taylor, F. Sherwood. *A Short History of Science and Scientific Thought: With Readings from the Great Scientists from the Babylonians to Einstein*. New York: Norton, 1963. (First published London, 1949 as *Science Past and Present*.)

Taylor, Lloyd W. *Physics: The Pioneer Science*. Boston: Houghton Mifflin, 1941.

Thackray, Arnold. *Atoms and Powers: An Essay on Newtonian Matter-Theory and the Development of Chemistry*. Cambridge: Harvard University Press, 1970.

Theobald, D. W. *The Concept of Energy*. London: Spon, 1966.

Thiel, Rudolf. *And There Was Light: The Discovery of the Universe*. Translated by R. and C. Winston. London: André Deutsch, 1958.

Thirring, J. H. *The Ideas of Einstein's Theory: The Theory of Relativity in Simple Language*. Translated by R. A. B. Russell. London: Methuen, 1921.

Thompson, Silvanus P. *The Life of William Thomson, Baron Kelvin of Largs*. 2 vols. London: Macmillan, 1910.

————. *Light: Visible and Invisible: A Series of Lectures Delivered at the Royal Institution of Great Britain, at Christmas, 1896*. New York: Macmillan, 1903.

Thomson, George P. *J. J. Thomson: And the Cavendish Laboratory in His Day*. London: Nelson, 1964.

Thomson, J. Arthur. *Introduction to Science*. (Home University Library of Modern Knowledge, No. 21.) New York: Holt, 1911.

————. *Progress of Science in the Century*. The Nineteenth Century Series. London: W. & R. Chambers, 1906.

Thomson, J. J. *Beyond the Electron*. Cambridge: University Press, 1928.

————. *The Corpuscular Theory of Matter*. London: Constable, 1907.

————. *Electricity and Matter*. New York: Scribner, 1904.

————. *Notes on Recent Researches in Electricity and Magnetism*. Oxford: Clarendon, 1893.

————. *Recollections and Reflections*. New York: Macmillan, 1937.

————, et al. *James Clerk Maxwell: A Commemoration Volume, 1831–1931*. Cambridge: University Press, 1931.

Thomson, Sir William. See Kelvin, Lord.

Thornton, Jesse E., comp. *Science and Social Change*. Washington: Brookings Institution, 1939.

Tolansky, S. *Curiosities of Light Rays and Light Waves.* New York: American Elsevier, 1964.

———. *An Introduction to Interferometry.* London: Longmans, Green, 1955.

Tolman, Richard C. *Relativity, Thermodynamics and Cosmology.* Oxford: Clarendon, 1934.

———. *The Theory of the Relativity of Motion.* Berkeley: University of California Press, 1917.

Tonnelat, M. A. *Einstein's Unified Field Theory.* Translated by Richard Akerib. New York: Gordon & Breach, 1966.

Toulmin, Stephen. *The Philosophy of Science: An Introduction.* New York: Harper & Row, Torchbooks, 1960. (First published 1953.)

———, ed. *Physical Reality: Philosophical Essays on Twentieth-Century Physics.* New York: Harper & Row, Torchbooks, 1970.

———, and June Goodfield. *The Architecture of Matter.* New York: Harper & Row, 1962.

———, and June Goodfield. *The Discovery of Time.* London: Hutchinson, 1965.

Tricker, R. A. R. *The Contributions of Faraday and Maxwell to Electrical Science.* Oxford: Pergamon, 1966.

———. *Early Electrodynamics: The First Law of Circulation.* Oxford: Pergamon, 1965.

Trowbridge, John. *What Is Electricity?* New York: Appleton, 1900.

True, Frederick W., ed. *A History of the First Half-Century of the National Academy of Sciences 1863–1913.* Washington: Lord Baltimore, 1913.

Truesdell, Clifford A. *Essays in the History of Mechanics.* New York: Springer-Verlag, 1968.

———. *Six Lectures in Modern Natural Philosophy.* New York: Springer-Verlag, 1966.

Trumpler, Robert J., and Harold F. Weaver. *Statistical Astronomy.* Berkeley: University of California Press, 1953.

Turner, Herbert Hall. *Astronomical Discovery.* Berkeley: University of California Press, 1963. (First published 1904.)

Tyndall, John. *Fragments of Science: A Series of Detached Essays, Addresses and Reviews.* 2 vols. 6th ed. New York: Appleton, 1897.

———. *Lectures on Light: Delivered in the United States in 1872–73.* New York: Appleton, 1873.

———. *Light and Electricity: Notes of Two Courses of Lectures before the Royal Institution of Great Britain.* New York: Appleton, 1881.

Tyson, Howell N. *Kinematics.* New York: Wiley, 1966.

Unsöld, Albrecht. *The New Cosmos.* Translated by William H. McCrea. New York: Springer-Verlag, 1969.

Vallentin, Antonina. *The Drama of Albert Einstein.* New York: Doubleday, 1954.

Van Melsen, Andrew G. *From Atomos to Atom: The History of the Concept* Atom. New York: Harper & Row, Torchbooks, 1960. (First published 1952.)

Vasiliev, A. V. *Space Time Motion: An Historical Introduction to the General Theory of Relativity.* Translated by H. M. Lucas and C. P. Sanger. New York: Knopf, 1924.

Very, Frank W. *The Luminiferous Ether. Part I: Its Relation to the Electron and to a Universal Interstellar Medium; Part II: Its Relation to the Atom.* Boston: The Four Seas Co., 1919.

Wallace, Alfred Russel, et al. *The Progress of the Century.* New York: Harper, 1901.

Warner, Deborah J. *Alvan Clark & Sons: Artists in Optics.* U.S. National Museum Bulletin No. 274. Washington: Smithsonian Institution, 1968.

Watson, David L. *Scientists Are Human.* London: Watts, 1938.

Watson, W. H. *On Understanding Physics.* New York: Harper & Row, Torchbooks, 1959.

Weil, Ernest. *Albert Einstein: A Bibliography of His Scientific Papers 1901–1930.* London: Goldschmidt, 1937.

Weinburg, Carlton B. *Mach's Empirio-Pragmatism in Physical Science.* New York: Albee, 1937.

Wertheimer, Max. *Productive Thinking.* Edited by Michael Wertheimer. Enlarged edition. New York: Harper, 1959. (First published 1945.)

Weyl, Hermann. *Philosophy of Mathematics and Natural Science.* Translated by Olaf Helmer. Princeton: Princeton University Press, 1949.

———. *Space, Time, Matter.* Translated by Henry L. Brose. 4th ed. New York: Dover, n.d. (First published 1918.)

Wheeler, John A. *Einstein's Vision: Wie steht es heute mit Einstein's Vision alles als Geometrie aufzufassen.* Berlin: Springer-Verlag, 1968.

Wheeler, Lynde Phelps. *Josiah Willard Gibbs: The History of a Great Mind.* Revised edition. New Haven: Yale University Press, 1952.

Whewell, William. *History of the Inductive Sciences from the Earliest to the Present Time*. 2 vols. 3d ed. rev. New York: Appleton, 1859.

Whitehead, Alfred North. *The Concept of Nature*. Ann Arbor: University of Michigan, 1957. (First published 1920.)

————. *An Enquiry Concerning The Principles of Natural Knowledge*. Cambridge: University Press, 1919.

————. *The Principle of Relativity with Application to Physical Science*. Cambridge: University Press, 1922.

Whiteman, Michael. *Philosophy of Space and Time and the Inner Constitution of Matter: A Phenomenological Study*. London: George Allen & Unwin, 1967.

Whitford, Robert H. *Physics Literature: A Reference Manual*. Washington, D.C.: Scarecrow, 1954.

Whittaker, [Sir] Edmund T. *From Euclid to Eddington: A Study of Conceptions of the External World*. London: Cambridge University Press, 1949.

————. *A. History of the Theories of Aether and Electricity*. London: Longmans, Green, 1910. Revised and reissued in 2 vols. New York: Harper & Row, Torchbooks, 1960. (Vol. I: *The Classical Theories*, first published London 1910, revised and enlarged, 1951; vol. II: *The Modern Theories*, first published London 1953.)

Whyte, Lancelot Law. *Critique of Physics*. New York: Norton, 1931.

————. *Essay on Atomism: From Democritus to 1960*. New York: Harper & Row, 1961.

Wien, Wilhelm. *Die Relativitätstheorie vom Standpunkt der Physik und Erkenntnislehre*. Leipzig: Barth, 1921.

Wiener, Philip P., ed. *Readings in Philosophy of Science: Introduction to the Foundations and Cultural Aspects of the Sciences*. New York: Scribner, 1953.

————, and Aaron Noland, eds. *Roots of Scientific Thought: A Cultural Perspective*. New York: Basic Books, 1957.

Wightman, William P. D. *The Growth of Scientific Ideas*. Edinburgh: Oliver & Boyd, 1950.

————. *The Story of Nineteenth-Century Science*. New York: Harper, 1901.

Wigner, Eugene P. *Symmetries and Reflections: Scientific Essays*. Edited by W. J. Moore and M. Scriven. Bloomington: Indiana University Press, 1967.

Williams, Howard R. *Edward Williams Morley: His Influence on Science in America*. Easton, Pa.: Chemical Education Publishing Co., 1957.

Williams, L. Pearce. *Michael Faraday: A Biography*. New York: Basic Books, 1965.

———. *The Origins of Field Theory*. New York: Random House, 1966.

———, ed. *Relativity Theory: Its Origins and Impact on Modern Thought*. New York: Wiley, 1968.

Williams, W. Ewart. *Applications of Interferometry*. London: Methuen, 1928.

Wilson, John H., Jr. *Albert A. Michelson: America's First Nobel Prize Physicist*. New York: Messner, 1958.

Wilson, Mitchell. *American Science and Invention: A Pictorial History. The Fabulous Story of How American Dreamers, Wizards and Inspired Tinkerers Converted a Wilderness into the Wonder of the World*. New York: Bonanza Books, 1960.

Wilson, William. *A Hundred Years of Physics*. London: Gerald Duckworth, 1950.

Witten, Louis, ed. *Gravitation: An Introduction to Current Research*. New York: Wiley, 1962.

Wolf, A. *A History of Science, Technology, and Philosophy in the 16th & 17th Centuries*. 2 vols. 2d ed. New York: Harper & Row, Torchbooks, 1950. (First published 1934.)

Wood, Alexander. *Thomas Young: Natural Philosopher, 1773–1829*. London: Cambridge University Press, 1954.

Wood, De Volson. *The Luminiferous Aether*. New York: Van Nostrand, 1886.

Wood, Robert W. *Physical Optics*. New York: Macmillan, 1905.

Woodbury, Robert S. *History of the Gear-Cutting Machine*. Cambridge, Mass.: M. I. T. Press, 1958.

Woodruff, L. L., ed. *The Development of the Sciences*. New Haven: Yale University Press, 1923.

Woods, Hugh. *Aether: A Theory of the Nature of Aether and of Its Place in the Universe*. London: The Electrician Printing and Publishing Co., 1906.

Woolf, Harry, ed. *Quantification: A History of the Meaning of Measurement in the Natural and Social Sciences*. Indianapolis: Bobbs-Merrill, 1961.

Wright, Helen. *Explorer of the Universe: A Biography of George Ellery Hale*. New York: Dutton, 1966.

Yilmaz, Hüseyin. *Introduction to the Theory of Relativity and the Principles of Modern Physics*. New York: Blaisdell, 1965.

Yolton, John W. *The Philosophy of Science of A. S. Eddington.* The Hague: Martinus Nijhoff, 1960.

Young, A. P. *Lord Kelvin.* (Pamphlet.) London: Longmans, Green, 1948.

Young, Thomas. *A Course of Lectures on Natural Philosophy and the Mechanical Arts.* 2 vols. Edited by P. Kelland. New edition. London: Taylor & Walton, 1845.

———. *Miscellaneous Works of the late Thomas Young, M.D., F.R.S., &c.* Edited by George Peacock. 2 vols. London: John Murray, 1855.

III. PUBLISHED ARTICLES

Airy, George Biddell. "On a Supposed Alteration in the Amount of Astronomical Aberration of Light, Produced by the Passage of the Light through a Considerable Thickness of Refracting Medium." *Proceedings of the Royal Society,* London 20:35–39 (November 23, 1871).

Aitken, Robert G. *Our Journey through Space.* Leaflet #43. *Astronomical Society of the Pacific* (1932), I, 175–178.

Ames, Joseph S. "Henry Augustus Rowland." *Astrophysical Journal* 13: 241–248 (May, 1901).

[Appleyard, Rollo]. "Clerk Maxwell and the Michelson Experiment." *Nature* 125: 566–567 (April 12, 1930).

Articolo, George A. "The Michelson-Morley Experiment and the Phase Shift upon a 90° Rotation." *American Journal of Physics* 37: 215–216 (February, 1969).

Bartlett, Albert A. "Compton Effect: Historical Background." *American Journal of Physics* 32:120–127 (February, 1964).

Barus, C. "The Compressibility of Colloids, with Applications to the Jelly Theory of the Ether." *American Journal of Science,* 4th ser. 6:285–298 (October, 1898).

Beach, F. E. "Albert Abraham Michelson." *American Journal of Science,* 5th ser. 22:97–99 (August, 1931).

Bennett, James O'Donnell. "Superlative Americans. Second Article. Albert Abraham Michelson at 70." the *Chicago Tribune,* rotogravure section (1923), pp. 22–23. (Fotofax copy, n.d.)

Bergmann, Peter G. "Fifty Years of Relativity." *Science* 123:487–494 (March 23, 1956).

Blitzer, Leon. "On the Meaning of the Fresnel Coefficient of Ether Drag in Relativity." *American Journal of Physics* 15:446–448 (November, 1947).

Boas, Marie. "The Establishment of the Mechanical Philosophy." *Osiris* 10:412–541 (1952).

Bork, Alfred M. "Maxwell and the Electromagnetic Wave Equations." *American Journal of Physics* 35:844–848 (September, 1967).

———. "The 'FitzGerald Contraction.'" *Isis* 57:199–207 (Summer, 1966).

———. "The Fourth Dimension in 19th Century Physics." *Isis* 55:326–338 (September, 1964).

———. "Maxwell and the Vector Potential." *Isis* 58:210–222 (Summer, 1967).

———. "Physics Just Before Einstein." *Science* 152:597–603 (April 29, 1966).

———. " 'Vectors Versus Quaternions'—The Letters in *Nature*." *American Journal of Physics* 34: 202–211 (March, 1966).

Born, Max. "Physics and Relativity." *Proceedings* [of the Conference of the] *Jubilee of Relativity Theory.* Helvetica Physica Acta, Supplement 4. Basel: Birkhauser Verlag, 1956.

Boyajian, A. "A. A. Michelson Visits Immanuel Kant." *Scientific Monthly* 59: 438–450 (December, 1944).

Boyer, Carl B. "Early Estimates of the Velocity of Light." *Isis* 33: 24–40 (March, 1941).

Brill, Dieter R., and Robert C. Perisho. "Resource Letter GR-1 on General Relativity." *American Journal of Physics* 36:1–8 (February, 1968).

Bromberg, Joan. "Maxwell's Displacement Current and His Theory of Light." *Archive for History of Exact Sciences* 4:218–234 (1967).

Bronowski, J. "The Clock Paradox." *Scientific American* 208:134–144 (February, 1963).

Brush, Stephen G. "Foundations of Statistical Mechanics, 1845–1915." *Archive for History of Exact Sciences* 4:145–183 (1967).

———. "Mach and Atomism." *Synthese* 18:192–215 (1968).

———. "Note on the History of the FitzGerald-Lorentz Contraction." *Isis* 58:230–232 (Summer, 1967).

———. "Science and Culture in the Nineteenth Century: Thermodynamics and History." *The Graduate Journal* [The University of Texas] 7:477–565 (Spring, 1967).

———. "The Wave Theory of Heat: A Forgotten Stage in the Transition from the Caloric Theory of Thermodynamics." *Br. J. Hist. Sci.* 5(18): 145–167 (1970).

Bucherer, A. H. "Die experimentelle Bestätigung des Relativitätsprinzips." *Annalen der Physik* 28:513–536 (1909).

Buchwald, E. "Hundert Jahre Fizeauscher Mitführungsversuch." *Die Naturwissenschaften* 38:519–536 (November, 1951).

Campbell, Norman. "The Common Sense of Relativity." *Philosophical Magazine*, 6th ser. 21:502–517 (April, 1911).

Carmichael, R. D. "On the Theory of Relativity: Mass, Force, and Energy." *Physical Review*, 2d ser. 1:161–178 [part I] (February, 1913).

————. "On the Theory of Relativity: Philosophical Aspects." *Physical Review*, 2d ser. 1:179–196 [part II] (March, 1913).

Caws, Peter. "The Structure of Discovery." *Science* 166:1375–1380 (December 12, 1969).

Chandrasekhar, S. "The Richtmyer Memorial Lecture: Some Historical Notes." *Am. J. Phys.* 37: 577–584 (June, 1969).

Chappell, John E., Jr. "Georges Sagnac and the Discovery of the Ether." *Archives internationales d'histoire des sciences* 18:175–190 (July–December, 1965).

Clarke, Frank W. "Edward Williams Morley, 1838–1923." *Biographical Memoirs*, National Academy of Sciences 21(10) (1927), 6 pp., with 2 pp. bibliography by Olin F. Tower.

Cohen, I. Bernard. "American Physicists at War." Part I: "From the Revolution to the World Wars." *American Journal of Physics* 13:223–235 (August 1945); part II: "From the First World War to 1942." 13:333–346 (October, 1945).

————. "The First Explanation of Interference." *American Journal of Physics* 8:99–106 (April, 1940).

————. "An Interview with Einstein." *Scientific American* 193:68–73 (July, 1955).

————. "Newton in the Light of Recent Scholarship." *Isis* 51:489–514 (December, 1960).

————. "A Sense of History in Science." *American Journal of Physics* 18: 343–359 (September, 1950).

Compton, Arthur H. "Nobel Prize Winners in Physics." *Current History* 34:699–705 (August, 1931).

Comstock, Daniel F. "The Principle of Relativity." *Science* 31:767–772 (May 20, 1910).

————. "Reasons for Believing in the Ether." *Science* (new series) 25:432–433 (March 15, 1907).

Crew, William H. "The Researches of Professor A. A. Michelson." *U.S. Naval Institute Proceedings* 56: 38–41 (January, 1930).

Daniels, George H. "The Process of Professionalization in American Science: The Emergent Period, 1820–1860." *Isis* 58: 151–166 (Summer, 1967).

Darrow, Karl D. "Elementary Notions of Quantum Mechanics." *Reviews of Modern Physics* 6:23–68 (January, 1934).

De Benedetti, Sergio. "The Mossbauer Effect." *Scientific American* 202:72–80 (April, 1960).

de Bray, M. E. J. Gheury. "The Velocity of Light: History of its determination from 1849 to 1933." *Isis* 25:437–448 (September, 1936).

Dicke, R. H. "The Eötvös Experiment." *Scientific American* 205:84–94 (December, 1961).

Dingle, Herbert. "The Case against Special Relativity." *Nature* 216:113–117 (October 14, 1967).

——. "The Interpretation of the Michelson-Morley and Kennedy-Thorndyke Experiments." *Philosophical Magazine*, 7th ser. 27:693–702 (June, 1939).

——. "Reason and Experiment in Relation to the Special Relativity Theory." *British Journal of Philosophy of Science* 15:41–61 (May, 1964).

Dirac, P. A. M. "The Evolution of the Physicist's Picture of Nature." *Scientific American* 208:45–53 (May, 1963).

——. "Is There an Aether?" Letter to the Editor. *Nature* 168:906 (November 24, 1951).

Dyson, Freeman J. "Innovation in Physics." *Scientific American* 199:74–82 (September, 1958).

Eddington, A. S. "Gravitation—I. And the Principle of Relativity." *Scientific American Supplement* no. 2218:2–3 (July 6, 1918).

——. "Gravitation—II. And the Principle of Relativity." *Scientific American Supplement* no. 2219:22–23 (July 13, 1918).

——. "Professor Albert Abraham Michelson, Foreign Member of the Royal Society." *Nature* 127:825 (May 30, 1931).

Einstein, Albert. "Dr. Einstein's Address." *Proceedings of the American Philosophical Society* 93: 544–545 (1949).

——. "Field Theories, Old and New." The *New York Times*, February 3, 1929. First Reprint in pamphlet form. New York: Readex Microprint Corp., 1960.

——. Über die von der molekularkinetischen Theorie der Warme geforderte Bewegung von in ruhenden Flussigkeiten suspendierten Teilchen." *Annalen der Physik* 17:549–560 (1905).

——. "Über einen die Erzeugung und Verwandlung des Lichtes betreffenden heuristischen Gesichtspunkt." *Annalen der Physik* 17:132–148 (1905).

———. "Zur Elektrodynamik bewegter Körper." *Annalen der Physik* 17: 891–921 (1905).

Eisele, Carolyn. "The Scientist-Philosopher C. S. Peirce at the Smithsonian." *Journal of the History of Ideas* 18:537–547 (October, 1957).

Everitt, C. W. F. "Maxwell's Scientific Papers." *Applied Optics* 6:639–646 (April, 1967).

Feinberg, Gerald. "Light." *Scientific American* 219:50–59 (September, 1968).

Finlay-Freundlich, E. "Cosmology." In Otto Neurath et al., eds., *International Encyclopedia of Unified Science*, I, part II, pp. 506–565. Chicago: The University of Chicago Press, 1955.

FitzGerald, George Francis. "The Ether and the Earth's Atmosphere." *Science* 13:390 (May 17, 1889).

———. "On the Structure of Mechanical Models Illustrating Some Properties of the Aether." *Philosophical Magazine* (5) 19:438–443 (June, 1885).

Fizeau, H. "Sur les hypothèses relatives à l'éther lumineux." *Annales de Chimie et de Physique*, 3d ser. 57:385–404 (1859).

———. "Sur une expérience relative à la vitesse de propagation de la lumière." *Comptes Rendus* 29:90–92 (July, 1849).

———, and L. Breguet. "Note sur l'expérience relative à la vitesse comparative de la lumière dans l'air et dans l'eau." *Comptes Rendus* 30:562–563 (May 6, 1850).

Fletcher, Harvey. "Dayton Clarence Miller, 1866–1941." *Biographical Memoirs*, National Academy of Sciences 23:61–74 (1945).

Foucault, Léon. "Détermination expérimentale de la vitesse de la lumière; parallaxe du soleil." *Comptes Rendus* 55:501–503 (September 22, 1862).

———. "Méthode générale pour mesurer la vitesse de la lumière dans l'air et les milieux transparents. Vitesses relatives de la lumière dans l'air et dans l'eau. Projet d'expérience sur la vitesse de propagation du calorique rayonnant." *Comptes Rendus* 30:551–560 (May 6, 1850).

Fox, J. G. "Evidence against Emission Theories." *American Journal of Physics* 33:1–17 (January, 1965).

Gale, Henry G. "Albert Abraham Michelson." *Astrophysical Journal* 74: 1–9 (July, 1931).

Gardner, G. H. F. "Rigid-Body Motions in Special Relativity." Letter to the Editor. *Nature* 170:243–244 (August 9, 1952).

Gardner, James P. "Reality, Relativity, and Common Sense." *Space Journal of the Astro-Sciences* 1:14–23 (March–May, 1959).

Glazebrook, R. T. ["Review of Optical Theories."] In *British Association for the Advancement of Science, Annual Report* (1893), 671–681.

Goldberg, Stanley. "Henri Poincaré and Einstein's Theory of Relativity." *American Journal of Physics* 35:934–944 (October, 1967).

———. "The Lorentz Theory of Electrons and Einstein's Theory of Relativity." *American Journal of Physics* 37:982–994 (October, 1969).

Gordon, James B. "The Maser." *Scientific American* 199:42–50 (December, 1958).

Grünbaum, Adolf. "The Bearing of Philosophy on the History of Science." *Science* 143:1406–1412 (March 27, 1964).

Haar, Charles M. "E. L. Youmans: A Chapter in the Diffusion of Science in America." *Journal of the History of Ideas* 9:193–213 (April, 1948).

Hale, George E. "Alvan G. Clark." *Astrophysical Journal* 6:136–138 (August, 1897).

———. "Some of Michelson's Researches." *Astronomical Society of the Pacific* 43:175–185 (June, 1931).

Hanson, Norwood R. "Waves, Particles, and Newton's 'Fits'." *Journal of the History of Ideas* 21:370–391 (July–September, 1960).

Hasenöhrl, Fritz. "Zur Theorie der Strahlung in bewegten Körpern." *Annalen der Physik*, 4th ser. 15:344–370 (October 25, 1904).

Hay, William H. "Paul Carus: A Case-Study of Philosophy on the Frontier." *Journal of the History of Ideas* 17:498–519 (October, 1956).

Heisenberg, Werner. "From Plato to Max Planck: The Philosophical Problems of Atomic Physics." *Atlantic* 204:109–113 (November, 1959).

Hesse, Mary B. "Models in Physics." *British Journal for the Philosophy of Science* 4:198–214 (November, 1953).

Heyl, Paul R. "The History and Present Status of the Physicists' Concept of Light." *Journal of the Optical Society of America* 18:183–192 (March, 1929).

Hicks, William M. "On the Michelson-Morley Experiment Relating to the Drift of Ether." *Philosophical Magazine*, 6th ser. 3:9–36 (January, 1902). [See also Hicks letters on pp. 256, 555.]

———. ["Theories of the Aether."] *British Association for the Advancement of Science Annual Report for 1895*, pp. 595–606.

Higgins, Thomas J. "Book-Length Biographies of Physicists and Astronomers." *American Journal of Physics* 12:234–236 (August, 1944).

Hirosige, Tetu. "Lorentz's Theory of Electrons and the Development of the

Concept of Electromagnetic Field." *Japanese Studies in the History of Science* 1:101–110 (1963).

Holton, Gerald. "Einstein and the 'Crucial' Experiment." *American Journal of Physics* 37:968–982 (October, 1969).

———. "Einstein, Michelson, and the 'Crucial Experiment'." *Isis* 60:133–197 (Summer, 1969).

———. "Influences on Einstein's Early Work in Relativity Theory." *American Scholar* 37:59–79 (Winter, 1967–1968).

———. "Mach, Einstein, and the Search for Reality." *Daedalus (Proceedings of the American Academy of Arts and Sciences)* 97:636–673 (Spring, 1968).

———. "On the Origins of the Special Theory of Relativity." *American Journal of Physics* 28:627–636 (October, 1960).

———. "Resource Letter SRT-1 on Special Relativity Theory." *American Journal of Physics* 30:462–469 (June, 1962).

———. "Stil und Verwirklichung in der Physik." *Eranos Jahrbuch* 33:319–363 (1964).

———. "The Metaphor of Space-Time Events in Science."*Eranos Jahrbuch* 34:33–78 (1965).

———. "Über die Hypothesen, welche der Naturwissenschaft zugrunde liegen." *Eranos Jahrbuch* 31:351–425 (1962).

Howard, John N. "The Michelson-Rayleigh Correspondence of the AFCRL Rayleigh Archives." *Isis* 58:88–89 (Spring, 1967).

Ives, Herbert E. "Aberration of Clocks and the Velocity of Light Paradox." *Journal of the Optical Society of America* 27:305 (September, 1937).

———. "Graphical Exposition of the Michelson-Morley Experiment." *Journal of the Optical Society of America* 27:177 (May, 1937).

———. "Light Signals Sent around a Closed Path." *Journal of the Optical Society of America* 28:296 (August, 1938).

Jamin, J. C. " Mémoire sur la mesure des indices de réfraction des gaz." *Annales de chimie et de physique*, 3d ser. 49:282–303 (1857).

———. "Mémoire sur les variations de l'indice de réfraction de l'eau à diverses pressions." *Annales de chimie et de physique*, 3d ser. 52:163–171 (1858).

———. "Mémoire sur les variations de l'indice de réfraction de la vapeur d'eau." *Annales de chimie et de physique*, 3d ser. 52:171–188 (1858).

Jauncey, G. E. M. "The Birth and Early Infancy of X-Rays." *American Journal of Physics* 13:362–379 (December, 1945).

Joncich, Geraldine. "Scientists and the Schools of the 19th Century: The

Case of American Physicists." *American Quarterly* 18:667–685 (Winter, 1966).

Joos, Georg. "Wiederholungen des Michelson-Versuchs." *Die Naturwissenschaften* 38:784–789 (September 18, 1931).

Jourdain, Philip E. B. "Newton's Hypotheses of Ether and Gravitation: From 1672 to 1679." *The Monist* 25:79–106 (January, 1915).

―――. "Newton's Hypotheses of Ether and Gravitation: From 1679 to 1693." *The Monist* 25:234–254 (April, 1915).

―――. "Newton's Hypotheses of Ether and Gravitation: From 1693 to 1726." *The Monist* 25:418–440 (July, 1915).

Kaufmann, W. "Über die Konstitution des Elektrons." *Annalen der Physik*, 4th ser 19:487–553 (1906).

[Kelvin, Lord]. "Nineteenth Century Cloud over the Dynamical Theory of Heat and Light." *American Journal of Science*, 4th ser. 12:391–392 (November, 1901).

Kelvin, Lord. "The Wave Theory of Light." *The Harvard Classics*. Edited by Charles W. Eliot. Vol. 30: *Scientific Papers*. New York: Collier and Son, 1910. (First published New York: Harper, 1906.)

Kendall, James. "The Adventures of an Hypothesis." *Proceedings of the Royal Society of Edinburgh* 63: 182–198 (July 3, 1950).

Kennard, E. H. "The Trouton-Noble Experiment." *Bulletin of the National Research Council* 4:162–172 (December, 1922).

Kennedy, Roy J. "Another Ether-Drift Experiment." *Physical Review*, 2d ser. 20:26–33 (July, 1922).

―――. "Simplified Theory of the Michelson-Morley Experiment." *Physical Review* 47:965–968 (June, 1935).

Keswani, G. H. "Origin and Concept of Relativity." *British Journal for the Philosophy of Science* (part I) 15:286–306 (February, 1965); and (part II) 16:19–32 (May, 1965).

Kevles, Daniel J. "George Ellery Hale, the First World War, and the Advancement of Science in America." *Isis* 59:427–437 (Winter, 1968).

Klein, Martin J. "Thermodynamics in Einstein's Thought." *Science* 157: 509–516 (August 4, 1967).

Kondo, Herbert. "History and the Physicist." *American Journal of Physics* 23:430–436 (October, 1955).

Kuhn, Thomas S. "The Function of Measurement in Modern Physical Science." *Isis* 52:161–193 (June, 1961).

―――. "The Historical Structure of Scientific Discovery." *Science* 136: 760–764 (June, 1962).

————. "The Turn to Recent Science." *Isis* 58:409–419 (Fall, 1967).

Langevin, P. "Sur la théorie de relativité et l'expérience de M. Sagnac." *Comptes Rendus* 173:831–834 (November 7, 1921).

Langley, S. P. "The History of a Doctrine." *American Journal of Science,* 3d ser. 37:1–23 (January, 1889).

Larmor, [Sir] Joseph. "Obituary Notice of Fellow Deceased, Lord Kelvin." *Proceedings of the Royal Society* (A) 81:Appendix (1908).

Laue, Max von. "Einstein und die Relativitätstheorie." *Die Naturwissenschaften* 43:1–8 (January, 1956).

Lemon, Harvey B. "Albert Abraham Michelson: The Man and the Man of Science." *The American Physics Teacher* (now *American Journal of Physics*) 4:1–11 (February, 1936).

Lenard, P. "Über Äther und Uräther." *Jahrbuch der Radioaktivität und Elektronick* 17:307–356 (July 19, 1921).

Lewis, Gilbert N., and Richard C. Tolman. "The Principle of Relativity and Non-Newtonian Mechanics." *Philosophical Magazine*, 6th ser. 18: 510–523 (October, 1909).

Livingston, Dorothy Michelson. "Michelson in the Navy; The Navy in Michelson." *U.S. Naval Institute Proceedings* 95(6):72–79 (June, 1969).

Lodge, [Sir] Oliver J. "A. A. Michelson." No. 64 of *Scientific Worthies. Nature* 117:1–6 (January 2, 1926).

————. "Aberration Problems. A Discussion concerning the Motion of the Ether near the Earth and concerning the Connexion between Ether and Gross Matter; with Some New Experiments." *Transactions of the Royal Society* 184:727–804 (July, 1893).

————. "Aether and Matter: Being Remarks on Inertia, and on Radiation, and on the Possible Structure of Atoms." *Nature* 104:15–19 (September 4, 1919); and 104:82–87 (September 25, 1919).

————. "The Geometrisation of Physics, and Its Supposed Basis on the Michelson-Morley Experiment." *Nature* 106:795–800 (February, 1921).

Lorentz, H. A. "De l'influence du mouvement de la terre sur les phénomènes lumineux." *Archives néerlandaises des sciences exactes et naturelles* 21: 103–176 ([Haarlem] 1887).

Lovering, Joseph. "Michelson's Recent Researches on Light." *Annual Report, . . . Smithsonian Institution, . . . 1889* (1890), pp. 449–468.

McCormmach, Russell, "Henri Poincaré and the Quantum Theory." *Isis* 58:37–55 (Spring, 1967).

McCrea, W. H. "Why the Special Theory of Relativity Is Correct." *Nature* 216:117–125 (October 14, 1967).

Magie, William F. "The Primary Concepts of Physics." *Science* (new series) 35:281–293 (February 23, 1912).

Mercier, André. "Fifty Years of the Theory of Relativity." *Nature* 175: 919–921 (May 28, 1955).

Michelson, Albert A. "Effect of Reflection from a Moving Mirror on the Velocity of Light." *Astrophysical Journal* 37:190–193 (April, 1913).

———. "Light Waves as Measuring Rods for Sounding the Infinite and the Infinitesimal." *The University* [of Chicago] *Record* 11:136–153 (April, 1925).

———. "A Plea for Light Waves." In *Proceedings of American Association for the Advancement of Science* 37:67–78 (May, 1889).

———. "Relative Motion of the Earth and Aether." *Philosophical Magazine*, 6th ser. 8:716–719 (December, 1904).

———. "A Theory of the 'X-Rays.' " *American Journal of Science*, 4th ser. 1:312–314 (April, 1896).

———. "The Velocity of Light." *The Decennial Publications of The University of Chicago*, IX. Chicago: University of Chicago Press, 1904.

———, and Edward W. Morley. "On a Method of Making the Wave Length of Sodium Light the Actual and Practical Standard of Length." *American Journal of Science*, 3d ser. 34:427–430 (December, 1887).

Michelson, Albert A., F. G. Pease, and F. Pearson. "Repetition of the Michelson-Morley Experiment." *Nature* 123:88 (January 19, 1929).

Miller, Dayton C. "Comments on Dr. Georg Joos's Criticism of the Ether-Drift Experiment." *Physical Review*, 2d ser. 45:114 (January 15, 1934).

———. "The Ether-Drift Experiment and the Determination of the Absolute Motion of the Earth." *Nature* 133:162–164 (February 3, 1934).

Millikan, Robert A. "Albert Abraham Michelson: The First American Nobel Laureate." *Scientific Monthly* 48:16–27 (January, 1939).

———. "Albert A. Michelson." *Biographical Memoirs*, National Academy of Sciences 19(4) (1938).

———. "Albert Einstein on His Seventieth Birthday." *Reviews of Modern Physics* 21:343–345 (July, 1949).

———. "The Evolution of Twentieth Century Physics." *Annual Report, . . . Smithsonian Institution, . . . 1927*, pp. 191–199 (1928).

———. "Michelson Memorial Address." *Commemorating Michelson Laboratory Dedication: May Seventh and Eighth, Nineteen Forty-Eight*. Pp. 17–26. China Lake, Calif. U.S. Naval Ordnance Test Station (1949).

————. "Michelson's Economic Value." *Science* 69:481–485 (May 10, 1929).

Mineur, H. "The Experiment of Miller and the Hypothesis of the Dragging Along of the Ether." Translated by A. F. Miller. *Journal of The Royal Astronomical Society of Canada* 21:206–214 (June, 1927).

Morley, Edward W. "A Completed Chapter in the History of the Atomic Theory." *Proceedings of American Association for the Advancement of Science* 45:1–22 (January, 1897).

————, and D. C. Miller. Extract from a Letter Dated August 4, 1904, Cleveland, Ohio, to Lord Kelvin. *Philosophical Magazine*, 6th ser. 8: 753–754 (December, 1904).

————, and D. C. Miller. "On the Theory of Experiments to Detect Aberrations of the Second Degree." *Philosophical Magazine*, 6th ser. 9:669–680 (May, 1905).

Moulton, Forest R. "Albert Abraham Michelson." *Popular Astronomy* 39: 307–310 (June–July, 1931).

Mountcastle, H. W. "Dayton Clarence Miller." *Science* 93:270–272 (March 21, 1941).

H. F. N. [*sic*]. "Albert Abraham Michelson." *Obituary Notices of Fellows of the Royal Society* 1:18–25 (December, 1932).

Nassau, J. J., and P. M. Morse. "A Study of Solar Motion by Harmonic Analysis." *Astrophysical Journal* 65:73–85 (March, 1927).

Newman, James R. "William Kingdon Clifford." *Scientific American* 188: 78–84 (February, 1953).

Nicholson, J. W. "On Uniform Rotation, the Principle of Relativity, and the Michelson-Morley Experiment." *Philosophical Magazine*, 6th ser. 24:820–827 (1912).

Ockert, Carl E. "Speed of Light." *American Journal of Physics* 36:158–161 (February, 1968).

O'Donnell, Thomas J. "An Introduction to Interferometry." *American Machinist* 104:123–146 (June 13, 1960).

Oppolzer, Egon R. v. "Erdbewegung und Aether." *Annalen der Physik*, 4th ser. 8:898–907 (July 10, 1902).

Page, Leigh. "Relativity and the Ether." *American Journal of Science*, 4th ser. 38:169–187 (August, 1914).

————, and C. M. Sparrow. "Relativity and Miller's Repetition of the Michelson-Morley Experiment." *Physical Review* 28:384 (1926).

Palter, Robert, ed. "The *Annus Mirabilis* of Sir Isaac Newton: Tricentennial Celebration." *The Texas Quarterly* 10 (3) (Autumn, 1967).

Planck, Max. "Ueber die Verteilung der Energie Zwischen Aether und Materie." *Annalen der Physik*, 4th ser. 9:629–641 (October 21, 1902).

Platt, John R. "Commentary on Theological Resources from the Physical Sciences." *Zygon* 1:33–42 (March, 1966).

Poincaré, Henri. "The Connection between the Ether and Matter." *Annual Report, . . . Smithsonian Institution, . . . 1912.* Pp. 119–210 (1913).

———. "The Principles of Mathematical Physics." Translated by G. B. Halsted. *The Monist* 15:1–24 (January, 1905).

"Proceedings of the Michelson Meeting of the Optical Society of America." *Journal of the Optical Society of America* 18:143–286 (March, 1929).

Rayleigh, Lord. "Aberration." *Nature* 45:499–502 (March 24, 1892).

Rezneck, Samuel. "The Education of an American Scientist: H. A. Rowland, 1848–1901." *American Journal of Physics* 28:155–162 (March, 1960).

Ritz, Walther. "Recherches critiques sur l'électrodynamique générale." *Annales de chimie et de physique*, 8th ser. 13:145–275 (February, 1908).

Robertson, H. P. "Postulate versus Observation in the Special Theory of Relativity." *Reviews of Modern Physics* 21:378–382 (July, 1949).

———. "Relativistic Cosmology." *American Journal of Physics* 5:62–90 (January, 1933).

Ronchi, Vasco. "The Development of Optics and Its Impact on Society." *Impact of Science on Society* (UNESCO publication) 9:123–136 (1959).

Rothman, Milton A. "Things That Go Faster Than Light." *Scientific American* 203:142–152 (July, 1960).

Rowland, Henry A. "The Highest Aim of the Physicist." *American Journal of Science*, 4th ser. 8:401–411 (December, 1899).

"The Royal Society: Anniversary Meeting." *The Times* (London), December 2, 1907, p. 7.

Rubinowicz, A. "Thomas Young and the Theory of Diffraction." *Nature* 180:160–162 (July 27, 1957).

Rush, J. H. "The Speed of Light." *Scientific American* 193:62–67 (August, 1955).

St. John, Charles E. "Evidence for the Gravitational Displacement of Lines in the Solar Spectrum Predicted by Einstein's Theory." *Astrophysical Journal* 67:195–239 (April, 1928).

Sarton, George. "Discovery of the Aberration of Light." *Isis* 16:233–239 (November, 1931).

Scribner, Charles, Jr. "Henri Poincaré and the Principle of Relativity." *American Journal of Physics* 32:672–678 (September, 1964).

Shankland, Robert S. "A. A. Michelson, 1852–1931." *Nature* 171:101–102 (January 17, 1953).

——. "Albert A. Michelson at Case." *American Journal of Physics* 17: 487–490 (November, 1949).

——. "Conversations with Albert Einstein." *American Journal of Physics* 31:47–57 (January, 1963).

——. "Dayton Clarence Miller: Physics Across Fifty Years." *American Journal of Physics* 9:273–283 (October, 1941).

——. "Final Velocity-of-Light Measurements of Michelson." *American Journal of Physics* 35:1095–1096 (November, 1967).

——. "Michelson-Morley Experiment." *American Journal of Physics* 32: 16–35 (January, 1964).

——. "The Michelson-Morley Experiment." *Scientific American* 211: 107–114 (November, 1964).

——. "Rayleigh and Michelson." *Isis* 58:86–88 (Spring, 1967).

——, et al. "New Analysis of the Interferometer Observations of Dayton C. Miller." *Reviews of Modern Physics* 27:167–178 (April, 1955).

Shibayev, V. A. "Professor A. A. Michelson and his Contribution to Science." *Journal of Urusvati* 2:1–4. New York: Himalayan Research Institute, Roerich Museum (January, 1932).

Shirley, John W. "The Harvard Case Histories in Experimental Science: The Evolution of an Idea." *American Journal of Physics* 19:419–423 (October, 1951).

Silberstein, L. "The Recent Eclipse Results and Stokes-Planck's Aether." *Philosophical Magazine*, 6th ser. 39:161–170 (February, 1920).

Simpson, Thomas K. "Maxwell and the Direct Experimental Test of His Electromagnetic Theory." *Isis* 57:411–432 (Winter, 1966).

Smithson, J. R. "Michelson at Annapolis." *American Journal of Physics* 18: 425–428 (October, 1950).

Spencer, J. Brookes. "On the Varieties of Nineteenth-Century Magneto-Optical Discovery." *Isis* 61:34–51 (Spring, 1970).

Stewart, Albert B. "The Discovery of Stellar Aberration." *Scientific American* 210:100–108 (March, 1964).

Stewart, G. Walter. "Dayton Clarence Miller." *Acoustical Society of America Journal* 12:477–380 (April, 1941).

Stokes, George G. "On the Constitution of the Luminiferous Aether Viewed with Reference to the Phaenomenon of the Aberration of Light." *Philosophical Magazine* (3) 29:6–10 (July, 1846).

————. "On the Constitution of the Luminiferous Ether." *Philosophical Magazine* (3) 32:343–349 (May, 1848).

————. "The Luminiferous Aether." *Annual Report, . . . Smithsonian Institution, . . . 1893.* Pp. 113–119 (1894).

Strömberg, Gustaf. "Space Structure and Motion." *Science* 76:477–481 (November 25, 1932), and 504–508 (December 2, 1932).

Stuewer, Roger H. "A Critical Analysis of Newton's Work on Diffraction." *Isis* 61:188–205 (Summer, 1970).

Süsskind, Charles. "Observations of Electromagnetic Wave Radiation before Hertz." *Isis* 55:32–42 (March, 1964).

Swann, W. F. G. "The Relation of the Restricted to the General Theory of Relativity and the Significance of the Michelson-Morley Experiment." *Science* 62:145–148 (August 14, 1925).

Swenson, Loyd S., Jr. "The Michelson-Morley-Miller Experiments Before and After 1905." *Journal for the History of Astronomy* 1, part I:56–78 (February, 1970).

Synge, J. L. "Gardner's Hypothesis and the Michelson-Morley Experiment." Letter to the Editor. *Nature* 170:244 (August 9, 1952).

Taylor, Barry N., et al. "The Fundamental Physical Constants." *Sci. Am.* 223:62–78 (October, 1970).

Taylor, William. "The Manufacture of Optical Elements." *Proceedings of the Optical Convention 1926.* Part 2. London: Optical Convention, 1926.

"The Third American Association Prize." [Miller's AAAS Prize]. *Science* 63:105 (January 29, 1926).

Thomson, J. J. "On the Light Thrown by Recent Investigations on Electricity on the Relation between Matter and Ether." (The Adamson Lecture, November 4, 1907.) Manchester: University [Press] Lectures, no. 8, 1908.

Toulmin, Stephen E. "Criticism in the History of Science: Newton on Absolute Space, Time, and Motion." *The Philosophical Review* 68:1–29; 203–227 (1959).

————. "The Evolutionary Development of Natural Science." *American Scientist* 55:456–471 (December, 1967).

Tourin, R. H. "Optical History of a Machine Tool Builder" [Warner and Swasey]. *Applied Optics* 3:1385–1386 (December, 1964).

Tower, Olin F. "Edward Williams Morley." *Science* 57:431–434 (April 13, 1923).

————. "Edward Williams Morley, 1838–1923." *Proceedings of the Amer-*

ican Chemical Society issued with *Journal of the American Chemical Society* 45:93–98 (January, 1923).

Townes, Charles H. "Quantum Electronics, and Surprise in Development of Technology." *Science* 159:699–703 (February 16, 1968).

Turner, Joseph. "Maxwell and the Method of Physical Analogy." *British Journal for the Philosophy of Science* 6:226–238 (November, 1955).

Twyman, Frank. "Professor A. A. Michelson, Sc.D., LL.D., Ph.D." In *Physical Society of London: Proceedings* 43:625–632 (September 1, 1931).

Watson, E. C. "The Discovery of X-Rays." *American Journal of Physics* 12:281–291 (October, 1945).

Wheeler, John A. "Our Universe: The Known and the Unknown." *American Scholar* 37:248–274 (Spring, 1968).

Whittaker, E. T. "The Aether: Past and Present." *Endeavor* 2: 117–120 (July, 1943).

Whyte, Lancelot L. "A Forerunner of Twentieth Century Physics: A Re-View of Larmor's 'Aether and Matter.' " *Nature* 186:1010–1014 (June 25, 1960).

Wien, W. "Über die Differentialgleichungen der Elektrodynamik für bewegte Körper." *Annalen der Physik*, 4th ser. (part I) 13:641–662; (part II) 13:663–668 (March 8, 1904).

———. "Über positive Elektronen und die Existenz hoher Atomgewichte." *Annalen der Physik*, 4th ser 13:669–677 (March 8, 1904).

Wilson, Harold A. "Gravitation and the Ether." *Rice Institute Pamphlet* 8:23–33 (1921).

Woodruff, A. E. "Action at a Distance in Nineteenth Century Electrodynamics." *Isis* 53: 439–459 (December, 1962).

———. "The Contributions of Hermann von Helmholtz to Electrodynamics." *Isis* 59:300–311 (Fall, 1968).

Woolf, Harry. "The Beginnings of Astronomical Spectroscopy," in *L'Aventure de la science: Mélanges Alexandre Koyré*, I, 618–634. Paris: Hermann, 1964.

Zajac, A., et al. "Real Fringes in the Sagnac and Michelson Interferometers." *American Journal of Physics* 29:669–673 (October, 1961).

Zuckerman, Harriet. "The Sociology of Nobel Prizes." *Scientific American* 217:25–33 (November, 1967).

IV. THESES AND DISSERTATIONS

Badash, Lawrence. "The Early Developments in Radioactivity, with Emphasis on Contributions from the United States." Ph.D. diss., Yale University, 1964. Microfilmed, 1965.

Collins, Bert Kirkman. "The Michelson-Morley Experiment and the Theories of the Ether." Senior thesis, B.A., Harvard University, March, 1967.

Goldberg, Stanley. "The Early Response to Einstein's Special Theory of Relativity, 1905–1911: A Case Study in National Differences." Ph.D. diss., Harvard University, 1969.

Kevles, Daniel J. "The Study of Physics in America 1865–1916." Ph.D. diss., Princeton University, 1964.

Miller, Dennis Francis. "The Early Influence of Einstein and the Special Theory of Relativity in America." M.A. thesis, Ohio State University, 1965.

Offner, Abe. "A Critical Survey of Ether Drift Experiments." M.A. thesis, Western Reserve University, Cleveland, 1931.

Palter, Robert Monroe. "Philosophy and Theories of Relativity." Ph.D. diss., University of Chicago, 1952.

Ruddick, William M. "The Special Theory of Relativity and Conceptual Change." Ph.D. diss., Department of Philosophy, Harvard University, 1964.

Strong, Gordon Bartley. "The Metaphysical Residua from Relativity Theory." M.A. thesis, University of Chicago, 1930.

INDEX

aberration: Bradley on, 13–15; Stokes on, 23–24, 77; Airy on, 26; Maxwell and, 57; Lodge on, 104; Rayleigh on, 110–111; mentioned, 89–90, 94

Abraham, Max: and Einstein, 163

acoustics: analogies with light, 13, 16, 20, 23, 52, 59 n. 10; Miller's interest in, 52, 174

Adams, Walter S.: encourages Miller, 210; mentioned, 204 n. 35, 214, 225 n. 29

Adelbert College. SEE Western Reserve University.

aether: as imponderable fluid, 5; as stage for action, 6, 9, 12; luminiferous, 10–12, 17–19, 181 *passim*; electrical, 15, 85; "thermiferous," 19; as elastic-solid, 20–23, 76–81, 85, 127; electromagnetic, 26–28, 85, 102, 127; described in Ganot's text, 37: Michelson's assumption on, 54–55; Maxwell on, 57; Kelvin on, 76–81, 111; mechanical properties of, 80 n. 14, 81 n. 16, 107, 142, 169 n. 38; Gibbs on, 85; Rodwell's definition of, 89; ontology of, 101, 201; Lodge on, 103, 186, 201; colloidal conception of, 132; chemical conception of, 132–133; Hooper's description of, 135; pragmatists and realists on, 137–139; Salisbury on, 138; sketch of evolution of meaning of, 139–140; Poincaré on, 150; Einstein on, 153, 187–188; Larmor's "Aether" article on, 165; status of, in 1910, 175; semantics of, 186–189; Heyl on, 221–222; as synonym for curvature of space, 228

Aether and Matter (Larmor): 123–124

aether drag: Michelson and, 24; Stokes's hypothesis of, 24, 104, 111, 146; Michelson-Morley and, 74, 76, 81–87;

and Encke's Comet, 111; Silberstein's suggestion for test of, 196–197. SEE ALSO Fizeau; Michelson-Morley; "water-drag" experiments

aether-drift experiments: Michelson's initial hypothesis of, 54–55; Maxwell's suggestion for, 57–58, 66–68; Michelson's report to Bell on, 69–70; Lorentz's criticism of, 73; Potier's criticism of, 73; "classic" tests, 87–95; historical development of, 119–120; by Morley-Miller, 141–146, 151–153; by Rayleigh, 143; by Brace, 146–147; by Trouton and Noble, 147–148; by Sagnac, 181–182; by Miller, 194–206; by Kennedy, 214; by Piccard and Stahel, 214–215; by Illingworth, 215–216; by Michelson, Pease, and Pearson, 217–226; "summit conference" on, 217–219; Joos/Zeiss and, 225–226

Airy, George Biddell: on aberration, 26; mentioned, 21 n. 26, 104, 218

American Association for the Advancement of Science: 50, 97, 99, 116, 144, 212

Ampère, André M.: 20

Anderson, John A.: 217, 219–220

Arago, François: on wave theory, 20; mentioned, 24, 25, 64

astronomy: Newtonian, 7, 10; Ptolemaic, 7, 40; Copernican, 7, 21, 40, 106; and Bessel on determination of parallax of fixed star, 21; and celestial navigation, 40–41; Kapteyn and, 209; Shapley and, 209; mentioned, 18, 44, 49. SEE ALSO aberration

Balazs, N.: on Einstein and Michelson-Morley, 158

"Ballad of Ryerson, The": stanzas on Michelson in, 42–43